T0178445

Communications
in Computer and Information Science　　1721

Rationale

The CCIS series is devoted to the publication of proceedings of computer science conferences. Its aim is to efficiently disseminate original research results in informatics in printed and electronic form. While the focus is on publication of peer-reviewed full papers presenting mature work, inclusion of reviewed short papers reporting on work in progress is welcome, too. Besides globally relevant meetings with internationally representative program committees guaranteeing a strict peer-reviewing and paper selection process, conferences run by societies or of high regional or national relevance are also considered for publication.

Topics

The topical scope of CCIS spans the entire spectrum of informatics ranging from foundational topics in the theory of computing to information and communications science and technology and a broad variety of interdisciplinary application fields.

Information for Volume Editors and Authors

Publication in CCIS is free of charge. No royalties are paid, however, we offer registered conference participants temporary free access to the online version of the conference proceedings on SpringerLink (http://link.springer.com) by means of an http referrer from the conference website and/or a number of complimentary printed copies, as specified in the official acceptance email of the event.

CCIS proceedings can be published in time for distribution at conferences or as post-proceedings, and delivered in the form of printed books and/or electronically as USBs and/or e-content licenses for accessing proceedings at SpringerLink. Furthermore, CCIS proceedings are included in the CCIS electronic book series hosted in the SpringerLink digital library at http://link.springer.com/bookseries/7899. Conferences publishing in CCIS are allowed to use Online Conference Service (OCS) for managing the whole proceedings lifecycle (from submission and reviewing to preparing for publication) free of charge.

Publication process

The language of publication is exclusively English. Authors publishing in CCIS have to sign the Springer CCIS copyright transfer form, however, they are free to use their material published in CCIS for substantially changed, more elaborate subsequent publications elsewhere. For the preparation of the camera-ready papers/files, authors have to strictly adhere to the Springer CCIS Authors' Instructions and are strongly encouraged to use the CCIS LaTeX style files or templates.

Abstracting/Indexing

CCIS is abstracted/indexed in DBLP, Google Scholar, EI-Compendex, Mathematical Reviews, SCImago, Scopus. CCIS volumes are also submitted for the inclusion in ISI Proceedings.

How to start

To start the evaluation of your proposal for inclusion in the CCIS series, please send an e-mail to ccis@springer.com.

Christopher L. Buckley · Daniela Cialfi ·
Pablo Lanillos · Maxwell Ramstead · Noor Sajid ·
Hideaki Shimazaki · Tim Verbelen
Editors

Active Inference

Third International Workshop, IWAI 2022
Grenoble, France, September 19, 2022
Revised Selected Papers

 Springer

Editors
Christopher L. Buckley ⓘ
University of Sussex
Brighton, UK

Pablo Lanillos ⓘ
Donders Institute for Brain, Cognition
and Behaviour
Nijmegen, The Netherlands

Noor Sajid
Wellcome Centre for Human Neuroimaging
London, UK

Tim Verbelen ⓘ
Ghent University
Ghent, Belgium

Daniela Cialfi ⓘ
University of Chieti-Pescara
Pescara, Italy

Maxwell Ramstead
Wellcome Centre for Human Neuroimaging
London, UK

Hideaki Shimazaki ⓘ
Kyoto University
Kyoto, Japan

ISSN 1865-0929   ISSN 1865-0937 (electronic)
Communications in Computer and Information Science
ISBN 978-3-031-28718-3   ISBN 978-3-031-28719-0 (eBook)
https://doi.org/10.1007/978-3-031-28719-0

This Springer imprint is published by the registered company Springer Nature Switzerland AG
The registered company address is: Gewerbestrasse 11, 6330 Cham, Switzerland

Preface

The active inference framework, which first originated in neuroscience, can be assimilated to a theory of choice behavior and learning. The basic assumption of this new theoretical and methodological structure is the distinction between goal-directed and habitual behavior of an intelligent agent and how they contextualize each other. In this architecture, the intelligent agent's main goal is to minimize surprise or, more formally, its free energy. In particular, in contrast to other approaches, the resulting behavior has both explorative (epistemic) and exploitative (pragmatic) aspects that are sensitive to ambiguity and risk respectively. Thus, active inference offers an interesting framework not only for understanding behavior and the brain, but also for developing artificial intelligent agents and for investigating novel machine learning algorithms.

In this scope, the present volume presents some recent developments in active inference and its applications. These papers were presented and discussed at the 3rd International Workshop on Active Inference (IWAI 2022), which was held in conjunction with the European Conference on Machine Learning and Principles and Practice of Knowledge Discovery in Databases (ECML-PKDD). The workshop took place on September 19, 2022 in Grenoble, France. We received 31 submissions, out of which 25 papers were accepted for publication after a peer review process. The review process was double-blind, and three reviewers could score between -3 (strong reject) and 3 (strong accept). Papers were accepted when a net positive score was obtained. We also awarded Conor Heins, Ruben van Bergen, and Samuel Wauthier for their contributions, with an honorable mention for Justus Huebotter and Alex Kiefer.

The IWAI 2022 organizers would like to thank the Program Committee for all reviews, the authors for their outstanding contributions, VERSES for sponsoring the awards, Anjali Bhat for her awesome keynote, and all attendees for making it such a memorable event.

September 2022

Christopher L. Buckley
Daniela Cialfi
Pablo Lanillos
Maxwell Ramstead
Noor Sajid
Hideaki Shimazaki
Tim Verbelen

Organization

Chair

Tim Verbelen   Ghent University - imec, Belgium

Organizing Committee - Program Chairs

Christopher L. Buckley	University of Sussex, UK
Daniela Cialfi	University of Chieti-Pescara, Italy
Pablo Lanillos	Donders Institute, Netherlands
Maxwell Ramstead	VERSES Inc, USA; University College London, UK
Noor Sajid	University College London, UK
Hideaki Shimazaki	Kyoto University, Japan
Tim Verbelen	Ghent University - imec, Belgium

Program Committee

Christopher L. Buckley	University of Sussex, UK
Daniela Cialfi	University of Chieti-Pescara, Italy
Lancelot Da Costa	Imperial College London, UK
Cedric De Boom	Ghent University - imec, Belgium
Karl Friston	University College London, UK
Zafeirios Fountas	Huawei Technologies, UK
Conor Heins	Max Planck Institute of Animal Behavior, Germany
Natalie Kastel	Univeristy of Amsterdam, Netherlands
Brennan Klein	Northeastern University, USA
Pablo Lanillos	Donders Institute, Netherlands
Christoph Mathys	Aarhus University, Denmark
Mark Miller	Hokkaido University, Japan
Ayca Ozcelikkale	Uppsala University, Sweden
Thomas Parr	University College London, UK
Maxwell Ramstead	VERSES Inc, USA; University College London, UK
Noor Sajid	University College London, UK

Panos Tigas	Oxford University, UK
Hideaki Shimazaki	Kyoto University, Japan
Kai Ueltzhöffer	Heidelberg University, Germany
Ruben van Bergen	Radboud University, Netherlands
Tim Verbelen	Ghent University - imec, Belgium
Martijn Wisse	Delft University of Technology, Netherlands

Contents

Preventing Deterioration of Classification Accuracy in Predictive Coding Networks

Paul F. Kinghorn[1(✉)], Beren Millidge[2], and Christopher L. Buckley[1]

[1] School of Engineering and Informatics, University of Sussex, Brighton, UK
{p.kinghorn,c.l.buckley}@sussex.ac.uk
[2] MRC Brain Networks Dynamics Unit, University of Oxford, Oxford, UK
beren@millidge.name

Abstract. Predictive Coding Networks (PCNs) aim to learn a generative model of the world. Given observations, this generative model can then be inverted to infer the causes of those observations. However, when training PCNs, a noticeable pathology is often observed where inference accuracy peaks and then declines with further training. This cannot be explained by overfitting since both training and test accuracy decrease simultaneously. Here we provide a thorough investigation of this phenomenon and show that it is caused by an imbalance between the speeds at which the various layers of the PCN converge. We demonstrate that this can be prevented by regularising the weight matrices at each layer: by restricting the relative size of matrix singular values, we allow the weight matrix to change but restrict the overall impact which a layer can have on its neighbours. We also demonstrate that a similar effect can be achieved through a more biologically plausible and simple scheme of just capping the weights.

Keywords: Hierarchical predictive coding · Variational inference · Inference speed

1 Introduction

Predictive Coding (PC) is an increasingly influential theory in computational neuroscience, based on the hypothesis that the primary objective of the cortex is to minimize prediction error [6,19,26]. Prediction error represents the mismatch between predicted and actual observations. The concepts behind PC go back to Helmholtz's unconscious inference and the ideas of Kant [19]. There are also more recent roots in both machine learning [1,7] and neuroscience [22,24] which were then unified by Friston in a series of papers around 15 years ago [11–13]. In order to generate predictions, the brain instantiates a generative model of the world, producing sensory observations from latent variables. Typically, PC is assumed to be implemented in hierarchies, with each layer sending predictions down to the layer below it, although recent work has demonstrated its applicability in arbitrary graphs [25]. A key advantage of PC over backpropagation is that it

C. L. Buckley et al. (Eds.): IWAI 2022, CCIS 1721, pp. 1–15, 2023.
https://doi.org/10.1007/978-3-031-28719-0_1

requires only local updates. Despite this, recent work has shown that it can approximate backpropagation [21,27], and it is an active area of research in both machine learning and neuroscience.

A Predictive Coding Network (PCN) can be viewed as an example of an Energy Based Model (EBM), with the minimization of errors equating to minimization of the network's energy [12]. Minimizing energy by updating network node values corresponds to inference and perception, whereas reducing energy by updating the weights of the network corresponds to learning and improving the model of the world. The theory has close links with the concept of the Bayesian Brain [9,15,26] - the process of perception is implemented by setting a prior at the top layer being sent down the layers and then errors being sent back up the layers to create a new posterior given observations at the bottom of the network. This is done iteratively until the posterior "percept" at the top of the network and the incoming data at the bottom are in equilibrium which is when the energy of the network is minimized.

As the generative model is learned and the weights of the network are updated, the ability of the network to infer the correct latent variable (or label) should, in theory, improve. However, we have observed that, once an optimal amount of training has occurred, PCNs appear to then deteriorate in classification performance, with inference having to be run for increasingly many iterations to maintain a given level of performance. We are aware of only passing mentions of this issue in the literature [14,28], although it is often discussed informally. This paper provides an in-depth investigation and diagnosis of the problem, determines the reasons for it and then implements some techniques which can be used to avoid it. This allows us to stably train predictive coding networks for much longer numbers of epochs than previously possible without performance deterioration.

The remainder of this paper is set out as follows. Section 2 describes a typical PCN and gives a high level overview of the maths behind predictive coding. Section 3 analyses the reason for the degradation in performance, starting with a demonstration of the problem and then explains its causes at increasingly detailed levels of explanation, ultimately demonstrating that it is caused by a mismatch between the size of weights in different layers. Once we have identified this fundamental cause, Sect. 4 then demonstrates techniques for avoiding it such as weight regularisation or capping. Weight regularisation is a technique which is common in the world of machine learning to prevent overfitting [2]. However, we will demonstrate that this is not the problem faced in PCNs - our solutions are not designed to necessarily keep weights small, but rather to ensure that the relative impact of different weight layers stays optimal.

2   Predictive Coding Networks

PCNs can be trained
to operate in two dif-
ferent manners. Using
MNIST [16] as an
example, a "gener-
ative" PCN would
generate images from
labels in a single for-
ward sweep, but then
be faced with a diffi-
cult inversion task in
order to infer a label
from an image. Con-
versely, a "discrimi-
native" PCN would
be able to generate
labels from images
in a single forward
sweep but would have

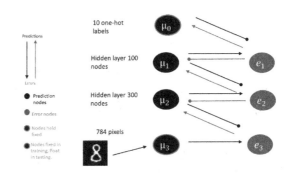

Fig. 1. Predictive Coding Network in test mode. The PCN learns to generate images from labels. At test time, the task of the network is to infer the correct label from a presented image. This is done by iteratively sending predictions down and errors back up the network, until the network reaches equilibrium.

difficulty producing an image for a given label [19]. Throughout this paper we use generative PCNs. Figure 1 shows a typical PCN. In training, it learns a genera-tive model which takes MNIST labels at the top of the network and generates MNIST images at the bottom. When testing the network's ability to classify MNIST images, an image is presented at the bottom of the network and the generative model is inverted using variational inference to infer a label. Deriva-tions of the maths involved can be found in [3,4,14,18] and we do not derive the update equations here, but simply give a brief overview and present the update equations which will be relevant later.

The network has N layers, where the top layer is labelled as layer 0 and the bottom layer as layer N-1. μ_n is a vector representing the node values of layer n, θ_n is a matrix giving the connection weights between layer n and layer $n+1$, f is an elementwise non-linear function and $\epsilon_{n+1} := \mu_{n+1} - f(\mu_n \theta_n)$ is the difference between the value of layer $n+1$ and the value predicted by layer n. Note that, in this implementation, $(\mu_n \theta_n)$ represents matrix multiplication of the node values and the weights, and therefore the prediction sent from layer n to layer $n+1$ is a non-linear function applied elementwise to a linear combination of layer n's node values. This is simply a specific instance of the more general case, where the prediction is produced by an arbitrary function f parameterised by θ_n, and is therefore given by $f(\mu_n, \theta_n)$.

Training the generative model of the network involves developing an auxiliary model (called the variational distribution) and minimizing the KL divergence between that auxiliary model and the true posterior. Variational free energy $\mathcal{F}$ is a measure which is closely related to this divergence and under certain assumptions it can be shown that:

$$\mathcal{F} \approx \left[\sum_{n}^{N} -\frac{1}{2}\epsilon_{n+1}^{T}\Sigma_{n+1}^{-1}\epsilon_{n+1} - \frac{1}{2}log(2\pi\,|\Sigma_{n+1}|) \right] \tag{1}$$

where Σ_n^{-1} is a term known as the precision, equal to the inverse of the variance of the nodes. It can also be shown that, in order to make the model a good fit for the data, it suffices to minimize $\mathcal{F}$. Therefore, to train the model, batches of label/image combinations are presented to the network and $\mathcal{F}$ is steadily reduced using the Expectation-Maximization approach [8,17], which alternately applies gradient descent on $\mathcal{F}$ with respect to node values (μ) on a fast timescale and weight values (θ) on a slower timescale. The process is repeated over multiple batches, with $\mathcal{F}$ steadily decreasing and the network's generative model improving. The gradients can be easily derived from Eq. (1). To update the nodes of the hidden layers, the gradient is given by:

$$\frac{d\mathcal{F}}{d\mu_n} = \epsilon_{n+1}\,\Sigma_{n+1}^{-1}\,\theta_n^T\,f'(\mu_n\theta_n) - \epsilon_n\,\Sigma_n^{-1} \tag{2}$$

After the node values have been updated, $\mathcal{F}$ is then further minimized by updating the weights using:

$$\frac{d\mathcal{F}}{d\theta_n} = \epsilon_{n+1}\,\Sigma_{n+1}^{-1}\,\mu_n^T\,f'(\mu_n\theta_n) \tag{3}$$

Training is carried out in a supervised manner and involves presentation of images and their corresponding labels. Therefore the bottom and top layers of the network are not updated during training and are held fixed with an MNIST image and a one-hot representation of the image's MNIST label respectively. Only the hidden layer nodes are updated. During testing, the task is to infer a label for a presented image and therefore the nodes of the top layer are allowed to update. Using the gradient for node updates, gradient descent is run through a number of iterations, ideally until the nodes of the network are at equilibrium. The inferred label can then be read out from the top layer. Since this layer receives no predictions from above, its gradient is slightly different from the other layers and is truncated to:

$$\frac{d\mathcal{F}}{d\mu_0} = \epsilon_1\,\Sigma_1^{-1}\,\theta_0^T\,f'(\mu_0\theta_0) \tag{4}$$

So far, we have described a supervised training regime. It is also possible to train in an unsupervised manner, in which case the top layer nodes are not held fixed during training and are updated in the same way as they are in testing. In this scenario, the top layer nodes will not be trained to converge to one-hot labels when inferring an image, but will still try to extract latent variables (see Appendix D for more discussion).

It is also possible to derive update equations for the precisions [3,12,19], but in our experiments we hold these fixed as the identity matrix and do not update them. Details of the network and the gradient descent techniques used are given in Appendix A.

3   Analysis

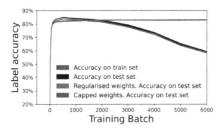

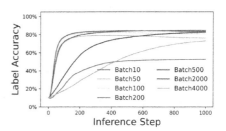

Fig. 2. Deteriorating Inference Accuracy. After training progresses beyond a certain stage, the ability of a PCN to infer the correct latent variable ("label") decreases. This is observed for both test and training set (black and blue lines) and therefore cannot be attributed to overfitting. This paper demonstrates that we can prevent this issue by simply capping size of weights (green line) or by regularising the weights so that the mean weight size on each layer remains constant (red line). Each line shows average of 3 networks, with standard error shown. (Color figure online)

Fig. 3. For different levels of training, development of label accuracy during iterative inference. During the early stages of training (roughly batches 10 to 200), the number of inference iterations required in testing reduces. However, after a certain amount of training, the asymptotic accuracy no longer improves, and the time taken to reach that asymptote worsens. This plot shows results from a single run - different runs vary slightly, but follow the same pattern. (Color figure online)

Figure 2 demonstrates a problem which is encountered when training generative PCNs for inference/classification tasks. In order to assess how well the network infers labels of MNIST images, we train the network over a large number of minibatches and regularly test the accuracy of label inference against a test dataset (black line), using 1000 iterations of inference. The network quickly improves with training, but accuracy then appears to deteriorate once training progresses beyond a certain point. The remainder of the paper demonstrates that this is caused by a mismatch between the way weights in different layers develop during training. We then show that simply capping the size of weights or regularising the weights so that the mean weight size on each layer remains constant prevents the problem and stabilizes training.

At any step in the iterative inference, it is possible to take a one-hot read-out from the inferred labels at the top layer. We can therefore construct a trajectory showing how accuracy develops with number of inference iterations. Figure 3 shows how this label accuracy develops for a network given different amounts of training. For example, the red line in the figure shows the trajectory of label accuracy for a network which has received 10 mini-batches of training. The plot demonstrates that, as training progresses through the early training batches, test accuracy improves both in terms of the asymptotic value achieved and in terms of how quickly the inference process reaches that asymptote. But, after a certain amount of training the inference process slows down. At batch 200, asymptotic performance is achieved after approximately 100 iterations. But if the network is trained for 4000 batches, the accuracy is still improving after 1000 test iterations. It is important to note that, if the inference process were allowed to run longer, the same asymptotic value would be achieved. Thus, if we infer the label after a set number of test iterations, performance will seem to deteriorate as the network gets better trained. This explains why Fig. 2 appears to show a deterioration in network performance since, following common practice, we stopped inference after a fixed number of steps (1000).

We have thus discovered the immediate cause of the performance deterioration - as we increase training, we need more test iterations to infer the correct label. To gain a deeper understanding of the phenomenon, however, we need to understand why this is the case. As an energy based model, the process of inference involves updating nodes until the network reaches equilibrium. Once this has been achieved, $\frac{d\mathcal{F}}{d\mu}$ for each node will be zero. It is instructive to examine whether any specific part of the network is the cause of the increase in the time to stabilise across epochs. Figure 4 shows that, as training progresses, the equilibrium $\mathcal{F}$ for each layer (which is proportional to the mean square error between layers) gets lower. Also, the error nodes on layers 2 and 3 stabilise quickly and even slightly speed up as training progresses. However, $\mathcal{F}$ on layer 1 (which is a measure of the errors between the node values at layers 0 and 1) takes an increasingly long time to settle down to equilibrium. Further investigation shows that this is caused by the nodes on the top layer taking increasingly long to stabilise, whereas hidden layer nodes all settle quickly (more details in Appendix B).

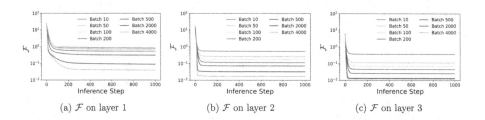

(a) $\mathcal{F}$ on layer 1   (b) $\mathcal{F}$ on layer 2   (c) $\mathcal{F}$ on layer 3

Fig. 4. For different levels of training, development of $\mathcal{F}$ on each layer during inference. Figures show the average $\mathcal{F}$ per node, for each layer. As training develops, $\mathcal{F}$ on all layers asymptote at lower values. Layers 2 and 3 also increase the speed which which they stabilise. However, on layer 1 (which represents the difference between the one-hot labels on the top layer and the nodes on the layer below), the network becomes slower to stabilise as training progresses. This plot shows results from a single run - different runs vary slightly, but follow the same pattern.

Summarising what this means for our 4 layer network as it tries to infer the label for an image: layer 3 at the bottom is fixed with the image, layer 2 quickly reaches equilibrium with layer 3, layer 1 quickly reaches equilibrium with layer 2, layer 0 is not at equilibrium with layer 1 and is still updating after many iterations. Crucially, it is this top layer which is used to read out the inferred label - thus causing the apparent reduction in label accuracy during training.

We now examine the reason why it is only the top layer which takes an increasing amount of time to stabilise. Recall that the node values are updated with Euler integration using $\frac{d\mathcal{F}}{d\mu}$ from Eqs. (2) and (4) as the gradient. As we are using fixed precision matrices, we can ignore the precision terms. Also, to help gain an intuition for these gradients, we ignore the non-linear activation function for now and assume the connections are simply linear. This allows us to see that the size of gradients are controlled by sizes of errors and weights:

$$\frac{d\mathcal{F}}{d\mu_0} \approx \epsilon_1 \, \theta_0^T \tag{5}$$

$$\frac{d\mathcal{F}}{d\mu_n} \approx \left[\epsilon_{n+1} \, \theta_n^T \; - \; \epsilon_n \right] \text{ for } n > 0 \tag{6}$$

We have seen already that $\mathcal{F}$, and therefore the errors at all layers, reduces in size as training progresses. On the other hand, Fig. 5 shows that there is a disparity between the weight sizes on the layers. As training progresses through the batches, the mean weight size reduces on the top layer but increases on the hidden layers. Applying this information to Eqs. (5) and (6) demonstrates why training causes the gradients on the top layer to reduce, but the hidden layers to increase: the gradient for the top layer is the product of two decreasing quantities, whereas on the hidden layers, diminishing errors are offset by increasing weights.

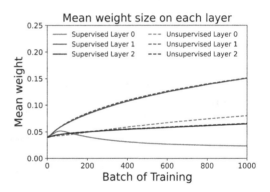

Fig. 5. Development with training of mean weight size for each weight layer. Solid lines show supervised mode, with top layer clamped in training - as training progresses, weight size reduces on the top layer but increases on the hidden layers. On the other hand, if the network is trained in unsupervised mode, all 3 weight layers increase in size. (Color figure online)

This leaves one final level of understanding to be dealt with - why do the weights at the top layer get smaller as training proceeds through the batches but the weights of hidden layers increase? To understand this, we now turn to the update equations for the weights which are applied after each batch of training (Eq. (3). Again, we ignore the precision and non-linearity in order to gain intuition and can see that the update equations for each weight layer θ_n comprise:

$$\forall n \ : \ \frac{d\mathcal{F}}{d\theta_n} \approx \epsilon_{n+1} \, \mu_n^T \ = \ (\mu_{n+1} - \mu_n \theta_n) \, \mu_n^T \ = \ \mu_{n+1} \, \mu_n^T - \mu_n \, \mu_n^T \, \theta_n \qquad (7)$$

This gradient is made up of a Hebbian component and an anti-Hebbian regularisation term and is reminiscent of the Oja learning rule [23] which was proposed as a technique to prevent exploding weights in Hebbian learning.

Although the equation for the weight gradient is the same for each layer, there is one crucial difference between the layers: when learning $\frac{d\mathcal{F}}{d\theta_n}$ in the case $n = 0$, the nodes μ_n are fixed, whereas the nodes μ_n are not fixed for other values of n. This interferes with the regularisation of those weight layers. As a result, the weight matrices grow in size when $n > 0$.

We have thus seen that accuracy of label inference deteriorates after a certain level of training because the different weight layers are being subject to different regularisation regimes. The top layer nodes are fixed during training, keeping the weight updates small, and this has the ultimate effect of causing top layer nodes to update more slowly as training progresses. The weights of the hidden layers are updated according to the same equation but with the crucial difference that the input nodes to the weight matrix are not clamped, reducing the anti-Hebbian impact of the regularisation term.

4   Techniques to Prevent Deterioration in Inference Accuracy

In order to prevent this deterioration we examine two possible approaches. One approach is to ensure that the top weight layer is treated in the same way as the hidden layers and is not regularised. Alternatively we can ensure that the hidden layers are regularised in the same way as the top layer. We address these two techniques in turn.

If we unclamp the top layer in training then we are training the PCN in unsupervised mode. Because the top layer of nodes is no longer clamped, the top layer of weights will be treated in the same way as the other layers and the implicit regularisation is reduced. This is demonstrated by the dotted red line in Fig. 5, which shows that, in unsupervised mode, the top layer weights now continue to increase as training continues just like the middle layers. In addition to monitoring mean weight size of each layer, we can obtain additional intuition as to what is happening by monitoring the development of the singular values of each weight matrix. This is briefly discussed in Appendix C.

Because training is now unsupervised, the top layer will no longer categorise the MNIST labels according to a one-hot label and therefore we cannot measure speed of label inference as we did in Fig. 3.[1] However, we can analyse the speed with which each layer reduces $\mathcal{F}$ (as we did in Fig. 4 for supervised mode). We find that there is no deterioration in time to reach equilibrium in any of the layers (chart not shown) and in fact, like the other hidden layers, the top layer converges slightly quicker as training progresses. We thus find that there is no longer a mismatch between the speeds at which the top layer and the hidden layers reach equilibrium - by having no prior forced on it, the top layer effectively becomes just another hidden layer.

But what if we want to have a supervised network? In that case, we want to maintain inference speed by ensuring that the weights on all layers are similarly regularised. A straightforward way of doing this is to simply ensure that the mean size of the weights for each layer stays constant throughout training. This is done by updating the weights as usual using $\frac{d\mathcal{F}}{d\theta}$ and then simply regularising them according to the formula

$$\forall i \; : \; \theta_i \; = \; \theta_i * \frac{\text{Target mean size}}{\Sigma_i |\theta_i|}, \tag{8}$$

where target mean size is a hyperparameter we set as described in Appendix A). The norm of the weight layers are now no longer changing with respect to each other and so, as predicted by our investigations, this prevents deterioration in label inference accuracy - shown by the red line in Fig. 2. A possible challenge to this approach is that this method of regularisation will no longer depend on purely local updates (although there is much neuroscience literature on the topic of homeostatic synaptic scaling - see for example [5, 29]). We used the same regularizing factor on each layer (see Appendix A), and we could find no other configuration of regularizing factor which worked better in terms of classification performance or speed of inference.

An even simpler, and possibly more biologically plausible, method of preventing exploding weights in supervised learning is to impose a simple cap on each weight. Empirically, we find that this method also maintains accuracy as shown by the green line in Fig. 2. Results shown were generated using a weight cap size of 0.1. It should be noted that this method is sensitive to the cap size used (although one could argue that evolution in the brain could select the correct cap size). Also, it may eventually lead to binary weight distribution and declining performance, although we have not investigated this.

5 Discussion

We have provided a detailed analysis of the dynamics of inference in predictive coding, and have shown that there is a tendency for that inference to slow down as training progresses. [20] separates the total energy of an Energy Based Model

---

[1] Note that this does not mean the top layer is no longer useful in terms of representing an image's label - energy minimization will still produce a representation at the top layer which separates the images by their characteristics - see Appendix D.

into the supervised loss (which depends on the errors at the top layer) and the internal energy (which corresponds to the energy of the hidden layers). We have shown that it is only the supervised loss which suffers from a slow-down in inference. As a result, this pathology does not exist in unsupervised training. We have also demonstrated that, even in supervised training, the decline can be prevented if the weights are constrained to ensure any weight regularisation is consistently applied across all layers. This is not something that happens automatically in the PCN framework without precisions.

In our implementation we have set all precisions to the identity matrix. We are aware of little research on the impact of precisions on inference dynamics, with most mentions of precisions pointing to them as an attention mechanism and a way of controlling the equilibrium point of the network [10]. But it can be seen from the update Eqs. (2), (4) and (3), that precisions also act as an adaptive weighting of the learning rate, both in the fast update of nodes and the slower update of weights. Therefore, as well as influencing direction of gradient updates, they should also have a significant impact on speed of update. Future work therefore needs to address the extent to which the phenomenon we observe is avoided if well-learned precisions are implemented. Having said that, we have experimented with different manually enforced relative precisions at each layer and found little benefit in terms of avoiding inference speed degradation. However, our testing simply implemented constant precisions for each layer - it is entirely possible that, by deriving a full covariance matrix, and allowing precisions to change with time, the phenomenon disappears. If precisions do prove to be a solution, then this paper will have at least pointed out the potential pitfalls of implementing predictive coding without them, and provided some techniques for coping with their omission.

Acknowledgements. PK is supported by the Sussex Neuroscience 4-year PhD Programme. CLB is supported by BBRSC grant number BB/P022197/1. BM is supervised by Rafal Bogacz who is supported by the BBSRC number BB/S006338/1 and MRC grant number MC_UU_00003/1.

A   Network Details

Network size: 4 layer
Number of nodes on each layer: 10, 100, 300, 784. In both training and testing, the 784 bottom layer nodes were fixed to the MNIST image. In supervised training mode, the top layer nodes were fixed as a one-hot representation of the MNIST label.
Non-linear function: tanh
Bias used: yes
Training set size: full MNIST training set of 60,000 images, in batches of 640. Thus the full training set is used after 93 batches.
Testing set size: full MNIST test set of 10,000 images
Learning parameters used in weight update of EM process: Learning Rate = 1e−3, Adam

Learning parameters used in node update of EM process: Learning Rate = 0.025, SGD
Number of SGD iterations in training: 50
Number of SGD iterations in testing: 1000.
Random node initialisation: Except where fixed, all nodes were initialized with a random values selected from $\mathcal{N}(0.5, 0.05)$
Weight regularisation: The weight regularisation technique used holds the L1 norm of each weight matrix constant. Rather than assigning a specific value, the algorithm measured the L1 norm of the matrix at initialisation (before any training took place) and then maintained that norm after each set of weight updates. The weights were randomly initialised using $\mathcal{N}(0, 0.05)$, giving a mean individual weight size of approximately 0.04 on all layers.

B   Development of Nodes

(See Fig. 6).

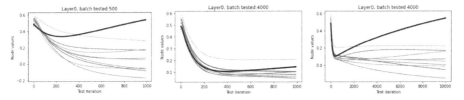

(a) After 500 batches of training   (b) After 4000 batches of training   (c) After 4000 batches of training - with more test iterations

Fig. 6. Nodes on top layer take longer to approach equilibrium after many epochs of training. For a PCN trained in supervised mode, the figures show how the 10 one-hot nodes in the top layer develop during test inference. The black lines represent the node which corresponds with the presented MNIST image's label (and therefore should be the maximum of the 10 nodes). Figure (a) shows the situation relatively early in training; after less than the 1000 test iterations, the system is inferring the correct label. But Figure (b) shows the situation when training has run much longer; the nodes are now much slower to update and therefore the system infers the wrong label. However, if the system were allowed to carry out inference over more iterations, the correct label would be inferred (Figure c).

C   Development of SVDs

(See Fig. 7).

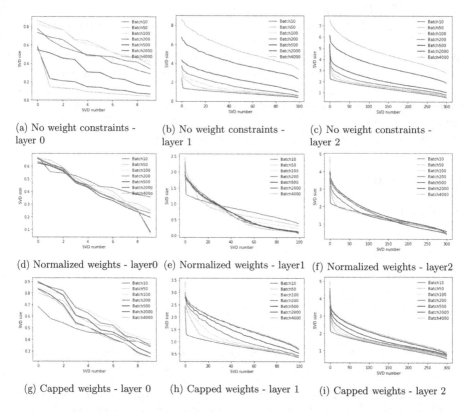

(a) No weight constraints - layer 0

(b) No weight constraints - layer 1

(c) No weight constraints - layer 2

(d) Normalized weights - layer0

(e) Normalized weights - layer1

(f) Normalized weights - layer2

(g) Capped weights - layer 0

(h) Capped weights - layer 1

(i) Capped weights - layer 2

Fig. 7. Singular value decomposition of weight matrices. If we do not constrain the weights (Figures a–c), the SVDs of layer 0 increase in early stages of training, and then decrease. But layers 1 and 2 steadily increase. In early stages of training, the distribution of SVDs changes, but after peak inference accuracy has been achieved at around training batch 500, subsequent changes in the weight matrix can be viewed largely as a parallel increase in all SVDs. By introducing normalisation (Figures d–f), the network learning is still able to redistribute the shape of the weight matrix if the gradients try to do this. But if the gradients are having no effect on the shape of the matrix then changes are not applied. Similar behaviour is observed for capped weights (Figures g–i). Monitoring the shape of SVDs could help identify when there is nothing to be gained from further training, although this would probably also require monitoring the change in singular vectors.

D   Unsupervised PCN Still Separates Top Layer into Labels

(See Fig. 8).

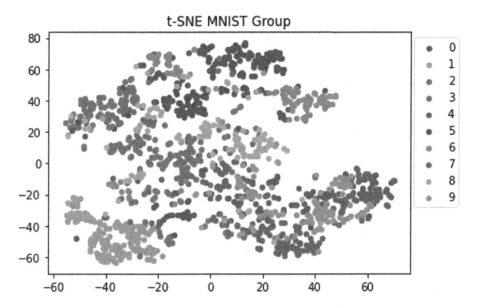

Fig. 8. Unsupervised PCN. Because the network is trained in an unsupervised manner, the top layer nodes do not contain a one-hot estimate of the image label. The nodes still contain a representation of the image, but it is simply in a basis that the network has created, rather than one which has been forced on it using supervised learning. tSNE analysis of top layer in testing demonstrates that unsupervised PCN still separates images according to label, despite lack of labels in training. Each dot shows the tSNE representation for an MNIST image - different colours represent images with different labels. (Color figure online)

References

1. Beal, M.J.: Variational algorithms for approximate Bayesian inference. University of London, University College London (United Kingdom) (2003)
2. Bishop, C.M., Nasrabadi, N.M.: Pattern Recognition and Machine Learning, vol. 4. Springer, Heidelberg (2006)
3. Bogacz, R.: A tutorial on the free-energy framework for modelling perception and learning. J. Math. Psychol. **76**, 198–211 (2017)
4. Buckley, C.L., Chang, S.K., McGregor, S., Seth, A.K.: The free energy principle for action and perception: a mathematical review (2017)
5. Carandini, M., Heeger, D.J.: Normalization as a canonical neural computation. Nat. Rev. Neurosci. **13**(1), 51–62 (2012)

6. Clark, A.: Whatever next? Predictive brains, situated agents, and the future of cognitive science. Behav. Brain Sci. **36**(3), 181–204 (2013)

7. Dayan, P., Hinton, G.E., Neal, R.M., Zemel, R.S.: The Helmholtz machine. Neural Comput. **7**(5), 889–904 (1995)

8. Dempster, A.P., Laird, N.M., Rubin, D.B.: Maximum likelihood from incomplete data via the EM algorithm. J. Roy. Stat. Soc.: Ser. B (Methodol.) **39**(1), 1–22 (1977)

9. Doya, K., Ishii, S., Pouget, A., Rao, R.P.: Bayesian Brain: Probabilistic Approaches to Neural Coding. MIT Press, Cambridge (2007)

10. Feldman, H., Friston, K.J.: Attention, uncertainty, and free-energy. Front. Hum. Neurosci. **4**, 215 (2010)

11. Friston, K.: Learning and inference in the brain. Neural Netw. **16**(9), 1325–1352 (2003)

12. Friston, K.: A theory of cortical responses. Philos. Trans. R. Soc. B: Biol. Sci. **360**(1456), 815–836 (2005)

13. Friston, K.: Hierarchical models in the brain. PLoS Comput. Biol. **4**(11), e1000211 (2008)

14. Kinghorn, P.F., Millidge, B., Buckley, C.L.: Habitual and reflective control in hierarchical predictive coding. In: Kamp, M., et al. (eds.) ECML PKDD 2021, pp. 830–842. Springer, Cham (2021). https://doi.org/10.1007/978-3-030-93736-2_59

15. Knill, D.C., Pouget, A.: The Bayesian brain: the role of uncertainty in neural coding and computation. Trends Neurosci. **27**(12), 712–719 (2004)

16. LeCun, Y., Cortes, C.: MNIST handwritten digit database (2010). http://yann.lecun.com/exdb/mnist/

17. MacKay, D.J., Mac Kay, D.J.: Information Theory, Inference and Learning Algorithms. Cambridge University Press, Cambridge (2003)

18. Millidge, B.: Combining active inference and hierarchical predictive coding: a tutorial introduction and case study. PsyArXiv (2019)

19. Millidge, B., Seth, A., Buckley, C.L.: Predictive coding: a theoretical and experimental review. arXiv preprint arXiv:2107.12979 (2021)

20. Millidge, B., Song, Y., Salvatori, T., Lukasiewicz, T., Bogacz, R.: Backpropagation at the infinitesimal inference limit of energy-based models: unifying predictive coding, equilibrium propagation, and contrastive hebbian learning. arXiv preprint arXiv:2206.02629 (2022)

21. Millidge, B., Tschantz, A., Buckley, C.L.: Predictive coding approximates backprop along arbitrary computation graphs. Neural Comput. **34**(6), 1329–1368 (2022)

22. Mumford, D.: On the computational architecture of the neocortex. Biol. Cybern. **66**(3), 241–251 (1992)

23. Oja, E.: Oja learning rule. Scholarpedia **3**(3), 3612 (2008). https://doi.org/10.4249/scholarpedia.3612. Revision #91607

24. Rao, R.P., Ballard, D.H.: Predictive coding in the visual cortex: a functional interpretation of some extra-classical receptive-field effects. Nat. Neurosci. **2**(1), 79–87 (1999)

25. Salvatori, T., Pinchetti, L., Millidge, B., Song, Y., Bogacz, R., Lukasiewicz, T.: Learning on arbitrary graph topologies via predictive coding. arXiv preprint arXiv:2201.13180 (2022)

26. Seth, A.K.: The cybernetic Bayesian brain. In: Open mind. Open MIND. Frankfurt am Main: MIND Group (2014)

27. Song, Y., Lukasiewicz, T., Xu, Z., Bogacz, R.: Can the brain do backpropagation?–exact implementation of backpropagation in predictive coding networks. In:

Advances in Neural Information Processing Systems, vol. 33, pp. 22566–22579 (2020)

28. Tschantz, A., Millidge, B., Seth, A.K., Buckley, C.L.: Hybrid predictive coding: inferring, fast and slow. arXiv preprint arXiv:2204.02169 (2022)

29. Turrigiano, G.G., Leslie, K.R., Desai, N.S., Rutherford, L.C., Nelson, S.B.: Activity-dependent scaling of quantal amplitude in neocortical neurons. Nature **391**(6670), 892–896 (1998)

Interpreting Systems as Solving POMDPs: A Step Towards a Formal Understanding of Agency

Martin Biehl[1]([⊠])[iD] and Nathaniel Virgo[2][iD]

[1] Cross Labs, Cross Compass, Tokyo 104-0045, Japan
`martin.biehl@cross-compass.com`
[2] Earth-Life Science Institute, Tokyo Institute of Technology,
Tokyo 152-8550, Japan

Abstract. Under what circumstances can a system be said to have beliefs and goals, and how do such agency-related features relate to its physical state? Recent work has proposed a notion of *interpretation map*, a function that maps the state of a system to a probability distribution representing its beliefs about an external world. Such a map is not completely arbitrary, as the beliefs it attributes to the system must evolve over time in a manner that is consistent with Bayes' theorem, and consequently the dynamics of a system constrain its possible interpretations. Here we build on this approach, proposing a notion of interpretation not just in terms of beliefs but in terms of goals and actions. To do this we make use of the existing theory of partially observable Markov decision processes (POMDPs): we say that a system can be interpreted as a solution to a POMDP if it not only admits an interpretation map describing its beliefs about the hidden state of a POMDP but also takes actions that are optimal according to its belief state. An agent is then a system together with an interpretation of this system as a POMDP solution. Although POMDPs are not the only possible formulation of what it means to have a goal, this nevertheless represents a step towards a more general formal definition of what it means for a system to be an agent.

Keywords: Agency · POMDP · Bayesian filtering · Bayesian inference

1 Introduction

This work is a contribution to the general question of when a physical system can justifiably be seen as an agent. We are still far from answering this question in full generality but employ here a set of limiting assumptions/conceptual commitments that allow us to provide an *example* of the kind of answer we are looking for.

This project was made possible through the support of Grant 62229 from the John Templeton Foundation. The opinions expressed in this publication are those of the authors and do not necessarily reflect the views of the John Templeton Foundation. Work on this project was also supported by a grant from GoodAI.

C. L. Buckley et al. (Eds.): IWAI 2022, CCIS 1721, pp. 16–31, 2023.
https://doi.org/10.1007/978-3-031-28719-0_2

The basic idea is inspired by but different from Dennett's proposal to use so-called stances [4], which says we should interpret a system as an agent if taking the *intentional stance* improves our predictions of its behavior beyond those obtained by the *physical stance* (or the design stance, but we ignore this stance here). Taking the physical stance means using the dynamical laws of the (microscopic) physical constituents of the system. Taking the intentional stance means ignoring the dynamics of the physical constituents of the system and instead interpreting it as a rational agent with beliefs and desires. (We content ourselves with only ascribing *goals* instead of desires.) A quantitative method to perform this comparison of stances can be found in [12].

In contrast to using a comparison of prediction performance of different stances we propose to decide whether a system can be interpreted as an agent by checking whether its physical dynamics are *consistent* with an interpretation as a rational agent with beliefs and goals. In other words, assuming that we know what happens in the system on the physical level (admittedly a strong assumption), we propose to check whether we can consistently ascribe meaning to its physical states, such that they appear to implement a process of belief updating and decision making.

A formal example definition of what it means for an interpretation to be consistent was recently published in [16]. This establishes a notion of consistent interpretation as a Bayesian *reasoner*, meaning something that receives inputs and uses them to make inferences about some hidden variable, but does not take actions or pursue a goal.

Briefly, such an interpretation consists of a map from the physical/internal states of the system to Bayesian beliefs about hidden states (that is, probability distributions over them), as well as a model describing how the hidden states determine the next hidden state and the input to the system. To be consistent, if the internal state at time t is mapped to some belief, then the internal state at time $t + 1$ must map to the Bayesian posterior of that belief, given the input that was received in between the two time steps.

In other words, the internal state parameterizes beliefs and the system updates the parameters in a way that makes the parameterized belief change according to Bayes law. A Bayesian reasoner is not an agent however. It lacks both goals and rationality since it neither has a goal nor actions that it could rationally take to bring the goal about.

Here we build on the notion of consistent interpretations of [16] and show how it can be extended to also include the attribution of goals and rationality.

For this we employ the class of problems called partially observable Markov decision processes (POMDPs), which are well suited to our purpose. These provide hidden states to parameterize beliefs over, a notion of a goal, and a notion of what it means to act optimally, and thus rationally, with respect to this goal. Note that both the hidden states and the goal (which will be represented by rewards) are not assumed to have a physical realization. They are part of the interpretation and therefore only need to exist in the mathematical sense. Informally, the hidden state is assumed by the agent to exist, but need not match a state of the true external world.

We will see that given a pair of a physical system (as modelled by a stochastic Moore machine) and a POMDP it can in principle be checked whether the system does indeed parameterize beliefs over the hidden states and act optimally with respect to the goal and its beliefs (Definition 5). We then say the system can be interpreted as solving the POMDP, and we propose to call the pair of system and POMDP an agent. This constitutes an example of a formal definition of a *rational agent with beliefs and goals*.

To get there however we need to make some conceptual commitments/ assumptions that restrict the scope of our definition. Note that we do not make these commitments because we believe they are particularly realistic or useful for the description of real world agents like living organisms, but only because they make it possible to be relatively precise. We suspect that each of these choices has alternatives that lead to other notions of agents. Furthermore, we do not argue that all agents are rational, nor that they all have beliefs and goals. These are properties of the particular notion of agent we define here, but there are certainly other notions of agent that one might want to consider.

The first commitment is with respect to the notion of system. Generally, the question of which physical systems are agents may require us to clarify how we obtain a candidate physical system from a causally closed universe and what the type of the resulting candidate physical system is. This can be done by defining what it means to be an individual and/or identifying some kind of boundary. Steps in this direction have been made in the context of cellular automata e.g. by [1,2] and in the context of stochastic differential equations by [5,7].

We here restrict our scope by assuming that the candidate physical system is a stochastic Moore machine (Definition 2). A stochastic Moore machine has inputs, a dynamic and possibly stochastic internal state, and outputs that deterministically depend on the internal state only. This is far from the most general types of system that could be considered, but it is general enough to represent the digital computers controlling most artificial agents at present. It it also similar to a time and space discretized version of the dynamics of the internal state of the literature on the free energy principle (FEP) [7].

Already at this point the reader may expect that the inputs of the Moore machine will play the role of sensor values and the outputs that of actions and this will indeed be the case. Furthermore, the role of the "physical constituents" or physical state (of Dennett's physical stance) will be played by the internal state of the machine and this state will be equipped with a kind of consistent Bayesian interpretation. In other words, it will be parameterizing/determining probabilistic beliefs. This is similar to the role of internal states in the FEP.

For our formal notion of beliefs we commit to probability distributions that are updated in accordance with Bayes law.

The third commitment is with respect to a formal notion of goals and rationality. As already mentioned, for those we employ POMDPs. These provide both a formal notion of goals via expected reward maximization and a formal notion of rational behavior via their optimal policy.

Combining these commitments we want to express when exactly a system can be interpreted as a rational agent with beliefs and goals.

Rational agents take the optimal actions with respect to their goals and beliefs. The convenient feature of POMDPs for our purposes is that the optimal policies are usually expressed as functions of probabilistic beliefs about the hidden state of the POMDP. For this to work, the probabilistic beliefs must be updated correctly according to Bayesian principles. It then turns out that these standard solutions for POMDPs can be turned into stochastic Moore machines whose states are the (correctly updated) probabilistic beliefs themselves and whose outputs are the optimal actions.

This has two consequences. One is that it seems justified to interpret such stochastic Moore machines as rational agents that have beliefs and goals. Another is that there are stochastic Moore machines that solve POMDPs. Accordingly, our definition of stochastic Moore machines that solve POMDPs (Definition 5) applies to these machines.

In addition to such machines, however, we want to include machines whose states only *parameterize* (and are not equal to) the probabilistic beliefs over hidden states and who output optimal actions.[1] We achieve this by employing an adapted notion of a consistent interpretation (Definition 3). A stochastic Moore machine can then be interpreted as solving a POMDP if it has this kind of consistent interpretation with respect to the hidden state dynamics of the POMDP and outputs the optimal policy.

We also show that the machines obeying our definition are optimal in the same sense as the machines whose states are the correctly updated beliefs, so we find it justified to interpret those machines as rational agents with beliefs and goals as well.

Before we go on to the technical part we want to highlight a few more aspects. The first is that the existence of a consistent interpretation (either in terms of filtering or in terms of agents) only depends on the stochastic Moore machine that's being interpreted, and not on any properties of its environment. This is because a consistent interpretation requires an agent's beliefs and goals to be *consistent*, and this is different from asking whether they are *correct*. An agent may have the wrong model, in that it doesn't correspond correctly to the true environment. Its conclusions in this case will be wrong, but its reasoning can still be consistent; see [16] for further discussion of this point. In the case of POMDP interpretations this means that the agent's actions only need to be optimal according to its model of the environment, but they might be suboptimal according to the true environment.

This differs from the perspective taken in the original FEP literature concerned with the question of when a system of stochastic differential equations contain an agent performing approximate Bayesian inference [3,5–7,14].[2]

---

[1] These machines are probably equivalent to the sufficient information state processes in [9, definition 2] but establishing this is beyond the scope of this work.

[2] The FEP literature includes both publications on how to construct agents that solve problems (e.g. [8]) and publications on when a system of stochastic differential equations contain an agent performing approximate Bayesian inference. Only the latter literature addresses a question comparable to the one addressed in the present manuscript.

This literature also interprets a system as modelling hidden state dynamics, but there the model is derived from the dynamics of the actual environment (the so called "external states"), and hence cannot differ from it. We consider it helpful to be able to make a clear distinction between the agent's model of its environment and its true environment. The case where the model is derived from the true environment is an interesting special case of this, but our framework covers the general case as well. To our knowledge, the possibility of choosing the model independently from the actual environment in a FEP-like theory was first proposed in [16], and has since also appeared in a setting closer to the original FEP one [13].

We will see here (Definition 3) that the independence of model from actual environment extends to actions in some sense. Even a machine without any outputs can have a consistent interpretation modelling an influence of the internal state on the hidden state dynamics even though it can't have an influence on the actual environment. Such "actions" remain confined to the interpretation.

Another aspect of using consistent interpretations of the internal state and thus the analogue of the physical state/the physical constituents of the system is that it automatically comes with a notion of coarse-graining of the internal state. Since interpretations map the internal state to beliefs but don't need to do so injectively they can include coarse-graining of the state.

Also note, all our current notions of interpretation in terms of Bayesian beliefs require exact Bayesian updating. This means approximate versions of Bayesian inference or filtering are outside of the scope. This limits the scope of our example definition in comparison with the FEP which, as mentioned, also uses beliefs parameterized by internal states but considers approximate inference. On the other hand this keeps the involved concepts simpler.

Finally, we want to mention that [11] recently proposed an agent discovery algorithm. This algorithm is based on a definition of agents that takes into account the creation process of the system. An agent discovery algorithm based on the approach presented here would take as input a machine (Definition 1) or a stochastic Moore machine (Definition 2) and try to find a POMDP interpretation (Definition 5). The creation process of the machine (system) would not be taken into account. This is one distinction between our notion of an agent and that of [11]. A more detailed comparison would be interesting but is beyond the scope of this work.

The rest of this manuscript presents the necessary formal definitions that allow us to precisely state our example of an agent definition.

2   Interpreting Stochastic Moore Machines

Throughout the manuscript we write $P\mathcal{X}$ for the set of all finitely supported probability distributions over a set $\mathcal{X}$. This ensures that all probability distributions we consider only have a finite set of outcomes that occur with non-zero probability. We can then avoid measure theoretic language and technicalities. For two sets $\mathcal{X}, \mathcal{Y}$ a Markov kernel is a function $\zeta : \mathcal{X} \to P\mathcal{Y}$. We write $\zeta(y|x)$

for the probability of $y \in \mathcal{Y}$ according to the probability distribution $\zeta(x) \in P\mathcal{Y}$. If we have a function $f : \mathcal{X} \to \mathcal{Y}$ we sometimes write $\delta_f : \mathcal{X} \to P\mathcal{Y}$ for the Markov kernel with $\delta_{f(x)}(y)$ (which is 1 if $y = f(x)$ and 0 else) then defining the probability of y given x.

We give the following definition, which is the same as the one used in [16], but specialised to the case where update functions map to the set of finitely supported probability distributions and not to the space of all probability distributions.

Definition 1. *A* machine *is a tuple* $(\mathcal{M}, \mathcal{I}, \mu)$ *consisting of a set* $\mathcal{M}$ *called* internal state space*; a set* $\mathcal{I}$ *called* input space*; and a Markov kernel* $\mu : \mathcal{I} \times \mathcal{M} \to P\mathcal{M}$ *called* machine kernel*, taking an input* $i \in \mathcal{I}$ *and a current machine state* $m \in \mathcal{M}$ *to a probability distribution* $\mu(i, m) \in P\mathcal{M}$ *over machine states.*

The idea is that at any given time the machine has a state $m \in \mathcal{M}$. At each time step it receives an input $i \in \mathcal{I}$, and updates stochastically to a new state, according to a probability distribution specified by the machine kernel. If we add a function that specifies an output given the machine state we get the definition of a stochastic Moore machine.

Definition 2. *A* stochastic Moore machine *is a tuple* $(\mathcal{M}, \mathcal{I}, \mathcal{O}, \mu, \omega)$ *consisting of a machine with internal state space* $\mathcal{M}$*, input space* $\mathcal{I}$*, and machine kernel* $\mu : \mathcal{I} \times \mathcal{M} \to P\mathcal{M}$*; a set* $\mathcal{O}$ *called the* output space*; and a function* $\omega : \mathcal{M} \to \mathcal{O}$ *called* expose function *taking any machine state* $m \in \mathcal{M}$ *to an output* $\omega(m) \in \mathcal{O}$*.*

Note that the expose function is an ordinary function and not stochastic.

We need to adapt the definition of a consistent Bayesian filtering interpretation [16, Definition 2]. For our purposes here we need to include models of dynamic hidden states that can be influenced. In particular we need to interpret a machine as modelling the dynamics of a hidden state that the machine itself can influence. This suggests that the interpretation includes a model of how the state of the machine influences the hidden state. We here call such influences "actions" and the function that takes states to actions *action kernel*.

Definition 3. *Given a machine with state space* $\mathcal{M}$*, input space* $\mathcal{I}$ *and machine kernel* $\mu : \mathcal{I} \times \mathcal{M} \to P\mathcal{M}$*, a* consistent Bayesian influenced filtering interpretation $(\mathcal{H}, \mathcal{A}, \psi, \alpha, \kappa)$ *consists of a set* $\mathcal{H}$ *called the* hidden state space*; a set* $\mathcal{A}$ *called the* action space*; a Markov kernel* $\psi : \mathcal{M} \to P\mathcal{H}$ *called* interpretation map *mapping machine states to probability distributions over the hidden state space; a function* $\alpha : \mathcal{M} \to \mathcal{A}$ *called* action function *mapping machine states to actions[3]; and a Markov kernel* $\kappa : \mathcal{H} \times \mathcal{A} \to P(\mathcal{H} \times \mathcal{I})$ *called the* model kernel *mapping pairs* (h, a) *of hidden states and actions to probability distributions* $\kappa(h, a)$ *over pairs* (h', i) *of next hidden states and an input.*

These components have to obey the following equation. First, in string diagram notation (see appendix A of [16] for an introduction to string diagrams for probability in a similar context to the current paper):

---

[3] We choose actions to be deterministic functions of the machine state because the stochastic Moore machines considered here also have deterministic outputs. Other choices may be more suitable in other cases.

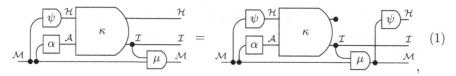

Second, in more standard notation, we must have for each $m \in \mathcal{M}$, $h' \in \mathcal{H}$, $i \in \mathcal{I}$, and $m' \in \mathcal{M}$:

$$\left(\sum_{h \in \mathcal{H}} \sum_{a \in \mathcal{A}} \kappa(h', i | h, a) \psi(h|m) \delta_{\alpha(m)}(a) \right) \mu(m' | i, m) =$$

$$\psi(h'|m') \left(\sum_{h \in \mathcal{H}} \sum_{a \in \mathcal{A}} \sum_{h'' \in \mathcal{H}} \kappa(h'', i | h, a) \psi(h|m) \delta_{\alpha(m)}(a) \right) \mu(m' | i, m). \tag{2}$$

In Appendix A we show how to turn Eq. (2) into a more familiar form.

Note that we defined consistent Bayesian influenced filtering interpretations for machines that have no actual output but that it also applies to those with outputs. If we want an interpretation of a machine with outputs we may choose the action space as the output space and the action kernel as the output kernel, but we don't have to. Interpretations can still be consistent.

Also note that when $\mathcal{A}$ is a space with only one element we recover the original definition of a consistent Bayesian filtering interpretation from [16].

3   Interpreting Stochastic Moore Machines as Solving POMDPs

Definition 4. *A partially observable Markov decision process (POMDP) can be defined as a tuple $(\mathcal{H}, \mathcal{A}, \mathcal{S}, \kappa, r)$ consisting of a set $\mathcal{H}$ called the* hidden state space; *a set $\mathcal{A}$ called the* action space; *a set $\mathcal{S}$ called the* sensor space; *a Markov kernel $\kappa : \mathcal{H} \times \mathcal{A} \to P(\mathcal{H} \times \mathcal{S})$ called the* transition kernel *taking a hidden state h and action a to a probability distribution over next hidden states and sensor values; and a function $r : \mathcal{H} \times \mathcal{A} \to \mathbb{R}$ called the* reward function *returning a real valued reward depending on the hidden state and an action.*

To solve a POMDP we have to choose a policy (as defined below) that maximizes the expected cumulative reward either for a finite horizon or discounted with an infinite horizon. We only deal with the latter case here.

POMDPs are commonly solved in two steps. First since the hidden state is unknown, probability distributions $b \in P\mathcal{H}$ (called belief states) over the hidden state are introduced and an updating function $f : P\mathcal{H} \times \mathcal{A} \times \mathcal{S} \to P\mathcal{H}$ for these belief states is defined. This updating is directly derived from Bayes rule [10]:

$$b'(h') = f(b, a, s)(h') := Pr(h' | b, a, s) := \frac{\sum_{h \in \mathcal{H}} \kappa(h', s | h, a) b(h)}{\sum_{\bar{h}, \bar{h}' \in \mathcal{H}} \kappa(\bar{h}', s | \bar{h}, a) b(\bar{h})}. \tag{3}$$

(Note that an assumption is that the denominator is greater than zero.) Then an optimal policy $\pi^* : P\mathcal{H} \to \mathcal{A}$ mapping those belief states to actions is derived from a so-called *belief state MDP* (see Appendix D for details). The optimal policy can be expressed using an optimal value function $V^* : P\mathcal{H} \to \mathbb{R}$ that solves the following *Bellman equation* [9]:

$$V^*(b) = \max_{a \in \mathcal{A}} \left(\sum_{h \in \mathcal{H}} b(h) r(h, a) + \gamma \sum_{\substack{s \in \mathcal{S} \\ h, h' \in \mathcal{H}}} \kappa(h', s | h, a) b(h) V^*(f(b, a, s)) \right). \quad (4)$$

The optimal policy is then [9]:

$$\pi^*(b) = \arg\max_{a \in \mathcal{A}} \left(\sum_{h \in \mathcal{H}} b(h) r(h, a) + \gamma \sum_{\substack{s \in \mathcal{S} \\ h, h' \in \mathcal{H}}} \kappa(h', s | h, a) b(h) V^*(f(b, a, s)) \right). \quad (5)$$

Note that the belief state update function f determines optimal value function and policy.

Define now $f_{\pi^*}(b, s) := f(b, \pi^*(b), s)$. Then note that if we consider $P\mathcal{H}$ a state space, $\mathcal{S}$ an input space, $\mathcal{A}$ an output space, $\delta_{f_{\pi^*}} : P\mathcal{H} \times \mathcal{S} \to PP\mathcal{H}$ a machine kernel, and $\pi^* : P\mathcal{H} \to \mathcal{A}$ an expose kernel, we get a stochastic Moore machine.[4]

This machine solves the POMDP and can be directly interpreted as a rational agent with beliefs and a goal. The beliefs are just the belief states themselves, the goal is expected cumulative reward maximization, and the optimal policy ensures it acts rationally with respect to the goal.

Our definition of interpretations of stochastic Moore machines as solutions to POMDPs includes this example and extends it to machines whose states aren't probability distributions/belief states directly but instead are parameters of such belief states that get (possibly stochastically) updated consistently.

We now state this main definition and then a proposition that ensures that our definition only applies to stochastic Moore machines that parameterize beliefs correctly as required by Eq. (3). This ensures that the optimal policy obtained via Eq. (5) is also the optimal policy for the states of the machine.

Definition 5. *Given a stochastic Moore machine $(\mathcal{M}, \mathcal{I}, \mathcal{O}, \mu, \omega)$, a consistent interpretation as a solution to a POMDP is given by a POMDP $(\mathcal{H}, \mathcal{O}, \mathcal{I}, \nu, \phi, r)$ and an interpretation map $\psi : \mathcal{M} \to P\mathcal{H}$ such that (i) $(\mathcal{H}, \mathcal{O}, \psi, \omega, \kappa)$ is a consistent Bayesian influenced filtering interpretation of the machine part $(\mathcal{M}, \mathcal{I}, \mu)$ of the stochastic Moore machine; and (ii) the machine expose function $\omega : \mathcal{M} \to \mathcal{O}$ (which coincides with the action function in the interpretation) maps any machine state m to the action $\pi^*(\psi(m))$ specified by the optimal POMDP policy for the belief $\psi(m)$ associated to machine state m by the interpretation. Formally:*

$$\omega(m) = \pi^*(\psi(m)). \quad (6)$$

---

[4] If the denominator in Eq. (3) is zero for some value $s \in \mathcal{S}$ then define e.g. $f_{\pi^*}(b, s) = b$.

Note that the machine never gets to observe the rewards of the POMDP we use to interpret it. An example of a stochastic Moore machine together with an interpretation of it as a solution to a POMDP is given in Appendix C.

Proposition 1. *Consider a stochastic Moore machine* $(\mathcal{M}, \mathcal{I}, \mathcal{O}, \mu, \omega)$, *together with a consistent interpretation as a solution to a POMDP, given by the POMDP* $(\mathcal{H}, \mathcal{O}, \mathcal{I}, \kappa, r)$ *and Markov kernel* $\psi : \mathcal{M} \to P\mathcal{H}$. *Suppose it is given an input* $i \in \mathcal{I}$, *and that this input has a positive probability according to the interpretation. (That is, Eq. (14) is obeyed.) Then the parameterized distributions* $\psi(m)$ *update as required by the belief state update equation (Eq. (3)) whenever* $a = \pi^*(b)$ *i.e. whenever the action is equal to the optimal action. More formally, for any* $m, m' \in \mathcal{M}$ *with* $\mu(m'|i, m) > 0$ *and* $i \in \mathcal{I}$ *that can occur according to the POMDP transition and sensor kernels, we have for all* $h' \in \mathcal{H}$

$$\psi(h'|m') = f(\psi(m), \pi^*(\psi(m)), i)(h'). \tag{7}$$

Proof. See Appendix B.

With this we can see that if V^* is the optimal value function for belief states $b \in P\mathcal{H}$ of Eq. (4), then $\bar{V}^*(m) := V^*(\psi(m))$ is an optimal value function on the machine's state space with optimal policy $\omega(m) = \pi^*(\psi(m))$.

4   Conclusion

We proposed a definition of when an stochastic Moore machine can be interpreted as solving a partially observable Markov decision process (POMDP). We showed that standard solutions of POMDPs have counterpart machines that this definition applies to. Our definition employs a newly adapted version of a consistent interpretation. We showed that with this our definition includes additional machines whose state spaces are parameters of probabilistic beliefs and not such beliefs directly. We suspect these machines are closely related to information state processes [9] but the precise relation is not yet known to us.

A   Consistency in More Familiar Form

One way to turn Eq. (2) into a probably more familiar form is to introduce some abbreviations and look at some special cases. We follow a similar strategy to [16]. Let

$$\psi_{\mathcal{H},\mathcal{I}}(h', i|m) := \sum_{h \in \mathcal{H}} \sum_{a \in \mathcal{A}} \kappa(h', i|h, a)\psi(h|m)\delta_{\alpha(m)}(a) \tag{8}$$

and

$$\psi_{\mathcal{I}}(i|m) := \sum_{h' \in \mathcal{H}} \psi_{\mathcal{H},\mathcal{I}}(h', i|m). \tag{9}$$

Then consider the case of a deterministic machine and choose the $m' \in \mathcal{M}$ that actually occurs for a given input $i \in \mathcal{I}$ such that $\mu(m'|i, m) = 1$ or abusing notation $m' = m'(i, m)$. Then we get from Eq. (2):

$$\psi_{\mathcal{H},\mathcal{I}}(h', i|m) = \psi(h'|\mu(i, m))\psi_{\mathcal{I}}(i|m). \tag{10}$$

If we then also consider an input $i \in \mathcal{I}$ that is *subjectively possible* as defined in [16] which here means that $\psi_{\mathcal{I}}(i|m) > 0$ we get

$$\psi(h'|m'(i, m)) = \frac{\psi_{\mathcal{H},\mathcal{I}}(h', i|m)}{\psi_{\mathcal{I}}(i|m)}. \tag{11}$$

This makes it more apparent that in the interpretation the updated machine state $m' = m'(i, m)$ parameterizes a belief $\psi(h'|m'(i, m))$ which is equal to the posterior distribution over the hidden state given input i. In the non-deterministic case, note that when $\mu(m'|i, m) = 0$ the consistency equation imposes no condition, which makes sense since that means the machine state m' can never occur. When $\mu(m'|i, m) > 0$ we can divide Eq. (2) by this to also get Eq. (10). The subsequent argument for $m' = m'(i, m)$ then must hold not only for this one possible next state but instead for every m' with $\mu(m'|i, m)$. So in this case (if s is subjectively possible) any of the possible next states will parameterize a belief $\psi(h'|m')$ equal to the posterior.

B   Proof of Proposition 1

For the readers's convenience we recall the proposition:

Proposition 2. *Consider a stochastic Moore machine $(\mathcal{M}, \mathcal{I}, \mathcal{O}, \mu, \omega)$, together with a consistent interpretation as a solution to a POMDP, given by the POMDP $(\mathcal{H}, \mathcal{O}, \mathcal{I}, \kappa, r)$ and Markov kernel $\psi : \mathcal{M} \to P\mathcal{H}$. Suppose it is given an input $i \in \mathcal{I}$, and that this input has a positive probability according to the interpretation. (That is, Eq. (14) is obeyed.) Then the parameterized distributions $\psi(m)$ update as required by the belief state update equation (Eq. (3)) whenever $a = \pi^*(b)$ i.e. whenever the action is equal to the optimal action. More formally, for any $m, m' \in \mathcal{M}$ with $\mu(m'|i, m) > 0$ and $i \in \mathcal{I}$ that can occur according to the POMDP transition and sensor kernels, we have for all $h' \in \mathcal{H}$*

$$\psi(h'|m') = f(\psi(m), \pi^*(\psi(m)), i)(h'). \tag{12}$$

Proof. By assumption the machine part $(\mathcal{M}, \mathcal{I}, \mu)$ of the stochastic Moore machine has a consistent Bayesian influenced filtering interpretation $(\mathcal{H}, \mathcal{O}, \psi, \omega, \kappa)$.

This means that the belief $\psi(m)$ parameterized by the stochastic Moore machine obeys Eq. (2). This means that for every possible next state m' (i.e. $\mu(m'|s, m) > 0$) we have

$$\sum_{h \in \mathcal{H}} \sum_{a \in \mathcal{A}} \kappa(h', i | h, a) \psi(h|m) \delta_{\omega(m)}(a)$$

$$= \psi(h'|m') \left(\sum_{h \in \mathcal{H}} \sum_{a \in \mathcal{A}} \sum_{h'' \in \mathcal{H}} \kappa(h'', i | h, a) \psi(h|m) \delta_{\omega(m)}(a) \right) \tag{13}$$

and for every subjectively possible input, that is, for every input $i \in \mathcal{I}$ with

$$\sum_{h \in \mathcal{H}} \sum_{a \in \mathcal{A}} \sum_{h'' \in \mathcal{H}} \kappa(h'', i | h, a) \psi(h|m) \delta_{\omega(m)}(a) > 0 \tag{14}$$

(see below for a note on why this assumption is reasonable) we will have:

$$\psi(h'|m') = \frac{\sum_{h \in \mathcal{H}} \sum_{a \in \mathcal{A}} \kappa(h', i | h, a) \psi(h|m) \delta_{\omega(m)}(a)}{\sum_{h \in \mathcal{H}} \sum_{a \in \mathcal{A}} \sum_{h'' \in \mathcal{H}} \kappa(h'', i | h, a) \psi(h|m) \delta_{\omega(m)}(a)} \tag{15}$$

$$= \frac{\sum_{h \in \mathcal{H}} \kappa(h', i | h, \omega(m)) \psi(h|m)}{\sum_{h \in \mathcal{H}} \sum_{h'' \in \mathcal{H}} \kappa(h'', i | h, \omega(m)) \psi(h|m)}. \tag{16}$$

Now consider the update function for which the optimal policy is found Eq. (3):

$$f(b, a, s)(h') := \frac{\sum_{h \in \mathcal{H}} \kappa(h', s | h, a) b(h)}{\sum_{\bar{h}, \bar{h}' \in \mathcal{H}} \kappa(\bar{h}', s | \bar{h}, a) b(\bar{h})} \tag{17}$$

and plug in the belief $b = \psi(m)$ parameterized by the machine state, the optimal action $\pi^*(\psi(m))$ specified for that belief by the optimal policy π^*, and the $s = i$:

$$f(\psi(m), \pi^*(m), i)(h') := \frac{\sum_{h \in \mathcal{H}} \kappa(h', i | h, \pi^*(\psi(m))) \psi(m)(h)}{\sum_{\bar{h}, \bar{h}' \in \mathcal{H}} \kappa(\bar{h}', i | \bar{h}, \pi^*(\psi(m))) \psi(m)(\bar{h})}. \tag{18}$$

Also introduce κ and write $\psi(h|m)$ for $\psi(m)(h)$ as usual

$$f(\psi(m), \pi^*(m), i)(h') := \frac{\sum_{h \in \mathcal{H}} \kappa(h', i | h, \pi^*(\psi(m))) \psi(h|m)}{\sum_{\bar{h}, \bar{h}' \in \mathcal{H}} \kappa(\bar{h}', i | \bar{h}, \pi^*(\psi(m))) \psi(\bar{h}|m)} \tag{19}$$

$$= \psi(h'|m'). \tag{20}$$

Which is what we wanted to prove.

Note that if Eq. (14) is not true and the probability of an input i is impossible according to the POMDP transition function, the kernel ψ, and the optimal policy ω then Eq. (13) puts no constraint on the machine kernel μ since both sides are zero. So the behavior of the stochastic Moore machine in this case is arbitrary. This makes sense since according to the POMDP that we use to interpret the machine this input is impossible, so our interpretation should tell us nothing about this situation.

C   Sondik's Example

We now consider the example from [15]. This has a known optimal solution. We constructed a stochastic Moore machine from this solution which has an interpretation as a solution to Sondik's POMDP. This proves existence of stochastic Moore machines with such interpretations.

Consider the following stochastic Moore machine:

- State space $\mathcal{M} := [0, 1]$. (This state will be interpreted as the belief probability of the hidden state being equal to 1.)
- input space $\mathcal{I} = \{1, 2\}$
- machine kernel $\mu : \mathcal{I} \times \mathcal{M} \to P\mathcal{M}$ defined by deterministic function $g : \mathcal{I} \times \mathcal{M} \to \mathcal{M}$:

$$\mu(m'|s, m) := \delta_{g(s,m)}(m') \tag{21}$$

where

$$g(S = 1, m) := \begin{cases} \frac{15}{6m+20} - \frac{1}{2} & \text{if } 0 \leq m \leq 0.1188 \\ \frac{9}{5} - \frac{72}{5m+60} & \text{if } 0.1188 \leq m \leq 1. \end{cases} \tag{22}$$

and

$$g(S = 2, m) := \begin{cases} 2 + \frac{20}{3m-15} & \text{if } 0 \leq m \leq 0.1188 \\ -\frac{1}{5} - \frac{12}{5m-40} & \text{if } 0.1188 \leq m \leq 1. \end{cases} \tag{23}$$

- output space $\mathcal{O} := \{1, 2\}$
- expose kernel $\omega : \mathcal{M} \to \mathcal{O}$ defined by

$$\omega(m) := \begin{cases} 1 \text{ if } 0 \leq m < 0.1188 \\ 2 \text{ if } 0.1188 \leq m \leq 1. \end{cases} \tag{24}$$

A consistent interpretation as a solution to a POMDP for this stochastic Moore machine is given by

- The POMDP with
  - state space $\mathcal{H} := \{1, 2\}$
  - action space equal to the output space $\mathcal{O}$ of the machine above
  - sensor space equal to the input space $\mathcal{I}$ of the machine above
  - model kernel $\kappa : \mathcal{H} \times \mathcal{O} \to \mathcal{H} \times \mathcal{I}$ defined by

$$\kappa(h', s|h, a) := \nu(h'|h, a)\phi(s|h', a) \tag{25}$$

  where $\nu : \mathcal{H} \times \mathcal{O} \to P\mathcal{H}$ and $\phi : \mathcal{H} \times \mathcal{O} \to P\mathcal{I}$ are shown in Table 1
  - reward function $r : \mathcal{H} \times \mathcal{O} \to \mathbb{R}$ also shown in Table 1.
- Markov kernel $\psi : \mathcal{M} \to P\mathcal{H}$ given by:

$$\psi(h|m) := m^{\delta_1(h)}(1 - m)^{\delta_2(h)}. \tag{26}$$

Table 1. Sondik's POMDP data.

Action $a \in \mathcal{O}$	$\nu(h'\|h, A = a)$	$\phi(s\|h', A = a)$	$r(h, A = a)$
1	$\begin{pmatrix} 1/5 & 1/2 \\ 4/5 & 1/2 \end{pmatrix}$	$\begin{pmatrix} 1/5 & 3/5 \\ 4/5 & 2/5 \end{pmatrix}$	$\begin{pmatrix} 4 \\ -4 \end{pmatrix}$
2	$\begin{pmatrix} 1/2 & 2/5 \\ 1/2 & 3/5 \end{pmatrix}$	$\begin{pmatrix} 9/10 & 2/5 \\ 1/10 & 3/5 \end{pmatrix}$	$\begin{pmatrix} 0 \\ -3 \end{pmatrix}$

To verify this we have to check that $(\mathcal{H}, \mathcal{O}, \psi, \omega, \kappa)$ is a consistent Bayesian influenced filtering interpretation of the machine $(\mathcal{M}, \mathcal{I}, \mu)$. For this we need to check Eq. (2) with $\delta_{\alpha(m)}(a) := \delta_{\omega(m)}(a)$. So for each each $m \in [0,1]$, $h' \in \{1,2\}$, $i \in \{1,2\}$, and $m' \in [0,1]$ we need to check:

$$\left(\sum_{h \in \mathcal{H}} \sum_{a \in \mathcal{A}} \kappa(h', i\|h, a) \psi(h\|m) \delta_{\omega(m)}(a) \right) \mu(m'\|i, m)$$

$$= \psi(h'\|m') \left(\sum_{h \in \mathcal{H}} \sum_{a \in \mathcal{A}} \sum_{h'' \in \mathcal{H}} \kappa(h'', i\|h, a) \psi(h\|m) \delta_{\omega(m)}(a) \right) \mu(m'\|i, m). \tag{27}$$

This is tedious to check but true. We would usually also have to show that ω is indeed the optimal policy for Sondik's POMDP but this is shown in [15].

D   POMDPs and Belief State MDPs

Here we give some more details about belief state MDPs and the optimal value function and policy of Eqs. (4) and (5). There is no original content in this section and it follows closely the expositions in [9,10].

We first define an MDP and its solution and then discuss then add some details about the belief state MDP associated to a POMDP.

Definition 6. *A Markov decision process (MDP) can be defined as a tuple $(\mathcal{X}, \mathcal{A}, \nu, r)$ consisting of a set $\mathcal{X}$ called the* state space, *a set $\mathcal{A}$ called the* action space, *a Markov kernel $\nu : \mathcal{X} \times \mathcal{A} \to P(\mathcal{X})$ called the* transition kernel, *and a reward function $r : \mathcal{X} \times \mathcal{A} \to \mathbb{R}$. Here, the transition kernel takes a state $x \in \mathcal{X}$ and an action $a \in \mathcal{A}$ to a probability distribution $\nu(x, a)$ over next states and the reward function returns a real-valued instantaneous reward $r(x, a)$ depending on the hidden state and an action.*

A solution to a given MDP is a control policy. As the goal of the MDP we here choose the maximization of expected cumulative discounted reward for an infinite time horizon (an alternative would be to consider finite time horizons). This means an optimal policy maximizes

$$\mathbb{E}\left[\sum_{t=1}^{\infty} \gamma^{t-1} r(x_t, a_t) \right]. \tag{28}$$

where $0 < \gamma < 1$ is a parameter called the discount factor. This specifies the goal.

To express the optimal policy explicitly we can use the optimal value function $V^* : \mathcal{X} \to \mathbb{R}$. This is the solution to the Bellman equation [10]:

$$V^*(x) = \max_{a \in \mathcal{A}} \left(r(x, a) + \gamma \sum_{x' \in \mathcal{X}} \nu(x'|a, x) V^*(x') \right). \tag{29}$$

The optimal policy is then the function $\pi^* : \mathcal{X} \to \mathcal{A}$ that greedily maximizes the optimal value function [10]:

$$\pi^*(x) = \arg\max_{a \in \mathcal{A}} \left(r(x, a) + \gamma \sum_{x' \in \mathcal{X}} \nu(x'|a, x) V^*(x') \right). \tag{30}$$

D.1   Belief State MDP

The belief state MDP for a POMDP (see Definition 4) is defined using the belief state update function of Eq. (3). We first define this function again here with an additional intermediate step:

$$f(b, a, s)(h') := Pr(h'|b, a, s) \tag{31}$$

$$= \frac{Pr(h', s|b, a)}{Pr(s|b, a)} \tag{32}$$

$$= \frac{\sum_{h \in \mathcal{H}} \kappa(h', s|h, a) b(h)}{\sum_{\bar{h}, \bar{h}' \in \mathcal{H}} \kappa(\bar{h}', s|\bar{h}, a) b(\bar{h})}. \tag{33}$$

The function $f(b, a, s)$ returns the posterior belief over hidden states h given prior belief $b \in P\mathcal{H}$, an action $a \in \mathcal{A}$ and observation $s \in \mathcal{S}$. The Markov kernel $\delta_f : P\mathcal{H} \times \mathcal{S} \times \mathcal{A} \to PP\mathcal{H}$ associated to this function can be seen as a probability of the next belief state b' given current belief state b, action a and sensor value s:

$$Pr(b'|b, a, s) = \delta_{f(b,a,s)}(b'). \tag{34}$$

Intuitively, the belief state MDP has as its transition kernel the probability $Pr(b'|b, a)$ expected over all next sensor values of the next belief state b' given that the current belief state is b the action is a and beliefs get updated according to the rules of probability, so

$$Pr(b'|b, a) = \sum_s Pr(b'|b, a, s) Pr(s|b, a) \tag{35}$$

$$= \sum_{s \in \mathcal{S}} \delta_{f(b,a,s)}(b') \sum_{h, h' \in \mathcal{H}} \kappa(h', s|h, a) b(h). \tag{36}$$

This gives some intuition behind the definition of belief state MDPs.

Definition 7. *Given a POMDP* $(\mathcal{H}, \mathcal{A}, \mathcal{S}, \kappa, r)$ *the associated belief state Markov decision process (belief state MDP) is the MDP* $(P\mathcal{H}, \mathcal{A}, \beta, \rho)$ *where*

- *the state space* $P\mathcal{H}$ *is the space of probability distributions beliefs over the hidden state of the POMDP. We write* $b(h)$ *for the probability of a hidden state* $h \in \mathcal{H}$ *according to belief* $b \in P\mathcal{H}$.
- *the action space* $\mathcal{A}$ *is the same as for the underlying POMDP*
- *the transition kernel* $\kappa : P\mathcal{H} \times A \rightarrow P\mathcal{H}$ *is defined as [10, Section 3.4]*

$$\beta(b'|b, a) := \sum_{s \in \mathcal{S}} \delta_{f(b,a,s)}(b') \sum_{h,h' \in \mathcal{H}} \kappa(h', s|h, a)b(h). \tag{37}$$

- *the reward function* $\rho : P\mathcal{H} \times \mathcal{A} \rightarrow \mathbb{R}$ *is defined as*

$$\rho(b, a) := \sum_{h \in \mathcal{H}} b(h)r(h, a). \tag{38}$$

So the reward for action a under belief b is equal to the expectation under belief b of the original POMDP reward of that action a.

D.2   Optimal Belief-MDP Policy

Using the belief MDP we can express the optimal policy for the POMDP.

The optimal policy can be expressed in terms of the *optimal value function of the belief MDP*. This is the solution to the equation [9]

$$V^*(b) = \max_{a \in \mathcal{A}} \left(\rho(b, a) + \gamma \sum_{b' \in P\mathcal{H}} \beta(b'|a, b)V^*(b') \right) \tag{39}$$

$$V^*(b) = \max_{a \in \mathcal{A}} \left(\rho(b, a) + \gamma \sum_{b' \in P\mathcal{H}} \sum_{s \in \mathcal{S}} \delta_{f(b,a,s)}(b') \sum_{h,h' \in \mathcal{H}} \kappa(h', s|h, a)b(h)V^*(b') \right) \tag{40}$$

$$V^*(b) = \max_{a \in \mathcal{A}} \left(\rho(b, a) + \gamma \sum_{s \in \mathcal{S}} \sum_{h,h' \in \mathcal{H}} \kappa(h', s|h, a)b(h)V^*(f(b, a, s)) \right). \tag{41}$$

This is the expression we used in Eq. (4). The optimal policy for the belief MDP is then [9]:

$$\pi^*(b) = \arg\max_{a \in \mathcal{A}} \left(\rho(b, a) + \gamma \sum_{s \in \mathcal{S}} \sum_{h,h' \in \mathcal{H}} \kappa(h', s|h, a)b(h)V^*(f(b, a, s)) \right). \tag{42}$$

This is the expression we used in Eq. (5).

References

1. Beer, R.D.: The cognitive domain of a glider in the game of life. Artif. Life **20**(2), 183–206 (2014). https://doi.org/10.1162/ARTL_a_00125
2. Biehl, M., Ikegami, T., Polani, D.: Towards information based spatiotemporal patterns as a foundation for agent representation in dynamical systems. In: Proceedings of the Artificial Life Conference 2016, pp. 722–729. The MIT Press (2016). https://doi.org/10.7551/978-0-262-33936-0-ch115. https://mitpress.mit.edu/sites/default/files/titles/content/conf/alife16/ch115.html
3. Da Costa, L., Friston, K., Heins, C., Pavliotis, G.A.: Bayesian mechanics for stationary processes. arXiv:2106.13830 [math-ph, physics:nlin, q-bio] (2021)
4. Dennett, D.C.: True believers: the intentional strategy and why it works. In: Heath, A.F. (ed.) Scientific Explanation: Papers Based on Herbert Spencer Lectures Given in the University of Oxford, pp. 53–75. Clarendon Press (1981)
5. Friston, K.: Life as we know it. J. R. Soc. Interface **10**(86) (2013). https://doi.org/10.1098/rsif.2013.0475. http://rsif.royalsocietypublishing.org/content/10/86/20130475
6. Friston, K.: A free energy principle for a particular physics. arXiv:1906.10184 [q-bio] (2019)
7. Friston, K., et al.: The free energy principle made simpler but not too simple (2022). https://doi.org/10.48550/arXiv.2201.06387 [cond-mat, physics:nlin, physics:physics, q-bio]
8. Friston, K., Rigoli, F., Ognibene, D., Mathys, C., Fitzgerald, T., Pezzulo, G.: Active inference and epistemic value. Cogn. Neurosci. **6**(4), 187–214 (2015). https://doi.org/10.1080/17588928.2015.1020053
9. Hauskrecht, M.: Value-function approximations for partially observable Markov decision processes. J. Artif. Intell. Res. **13**, 33–94 (2000). https://doi.org/10.1613/jair.678. http://arxiv.org/abs/1106.0234 [cs]
10. Kaelbling, L.P., Littman, M.L., Cassandra, A.R.: Planning and acting in partially observable stochastic domains. Artif. Intell. **101**(1–2), 99–134 (1998). https://doi.org/10.1016/S0004-3702(98)00023-X. http://www.sciencedirect.com/science/article/pii/S000437029800023X
11. Kenton, Z., Kumar, R., Farquhar, S., Richens, J., MacDermott, M., Everitt, T.: Discovering agents (2022). https://doi.org/10.48550/arXiv.2208.08345 [cs]
12. Orseau, L., McGill, S.M., Legg, S.: Agents and devices: a relative definition of agency. arXiv:1805.12387 [cs, stat] (2018)
13. Parr, T.: Inferential dynamics: comment on: How particular is the physics of the free energy principle? by Aguilera et al. Phys. Life Rev. **42**, 1–3 (2022). https://doi.org/10.1016/j.plrev.2022.05.006. https://www.sciencedirect.com/science/article/pii/S1571064522000276
14. Parr, T., Da Costa, L., Friston, K.: Markov blankets, information geometry and stochastic thermodynamics. Philos. Trans. R. Soc. A: Math. Phys. Eng. Sci. **378**(2164), 20190159 (2020). https://doi.org/10.1098/rsta.2019.0159. https://royalsocietypublishing.org/doi/full/10.1098/rsta.2019.0159
15. Sondik, E.J.: The optimal control of partially observable Markov processes over the infinite horizon: discounted costs. Oper. Res. **26**(2), 282–304 (1978). http://www.jstor.org/stable/169635
16. Virgo, N., Biehl, M., McGregor, S.: Interpreting dynamical systems as Bayesian reasoners. arXiv:2112.13523 [cs, q-bio] (2021)

Disentangling Shape and Pose for Object-Centric Deep Active Inference Models

Stefano Ferraro[✉], Toon Van de Maele, Pietro Mazzaglia, Tim Verbelen, and Bart Dhoedt

IDLab, Department of Information Technology, Ghent University - imec, Ghent, Belgium
stefano.ferraro@ugent.be

Abstract. Active inference is a first principles approach for understanding the brain in particular, and sentient agents in general, with the single imperative of minimizing free energy. As such, it provides a computational account for modelling artificial intelligent agents, by defining the agent's generative model and inferring the model parameters, actions and hidden state beliefs. However, the exact specification of the generative model and the hidden state space structure is left to the experimenter, whose design choices influence the resulting behaviour of the agent. Recently, deep learning methods have been proposed to learn a hidden state space structure purely from data, alleviating the experimenter from this tedious design task, but resulting in an entangled, non-interpretable state space. In this paper, we hypothesize that such a learnt, entangled state space does not necessarily yield the best model in terms of free energy, and that enforcing different factors in the state space can yield a lower model complexity. In particular, we consider the problem of 3D object representation, and focus on different instances of the ShapeNet dataset. We propose a model that factorizes object shape, pose and category, while still learning a representation for each factor using a deep neural network. We show that models, with best disentanglement properties, perform best when adopted by an active agent in reaching preferred observations.

Keywords: Active inference · Object perception · Deep learning · Disentanglement

1 Introduction

In our daily lives, we manipulate and interact with hundreds of objects without even thinking. In doing so, we make inferences about an object's identity, location in space, 3D structure, look and feel. In short, we learn a generative model of how objects come about [24]. Robots however still lack this kind of intuition,

C. L. Buckley et al. (Eds.): IWAI 2022, CCIS 1721, pp. 32–49, 2023.
https://doi.org/10.1007/978-3-031-28719-0_3

and struggle to consistently manipulate a wide variety of objects [2]. Therefore, in this work, we focus on building object-centric generative models to equip robots with the ability to reason about shape and pose of different object categories, and generalize to novel instances of these categories.

Active inference offers a first principles approach for learning and acting using a generative model, by minimizing (expected) free energy. Recently, deep learning techniques were proposed to learn such generative models from high dimensional sensor data [7,27,33], which paves the way to more complex application areas such as robot perception [14]. In particular, Van de Maele et al. [16,18] introduced object-centric, deep active inference models that enable an agent to infer the pose and identity of a particular object instance. However, this model was restricted to identify unique object instances, i.e. "this sugar box versus that particular tomato soup can", instead of more general object categories, i.e. "mugs versus bottles". This severely limits generalization, as it requires to learn a novel model for each particular object instance, i.e. for each particular mug.

In this paper, we further extend upon this line of work, by learning object-centric models not by object instance, but by object category. This allows the agent to reduce the number of required object-centric models, as well as to generalize to novel instances of known object categories. Of course, this requires the agent to not only infer object pose and identity, but also the different shapes that comprise this category. An important research question is then how to define and factorize the generative model, i.e. do we need to explicitly split the different latent factors in our model (i.e. shape and pose), or can a latent structure be learnt purely from data, and to what extent is this learnt latent structure factorized?

In the brain, there is also evidence for disentangled representations. For instance, processing visual inputs in primates consists of two pathways: the ventral or "what" pathway, which is involved with object identification and recognition, and the dorsal or "where" pathway, which processes an object's spatial location [22]. Similarly, Hawkins et al. hypothesize that cortical columns in the neocortex represent an object model, capturing their pose in a local reference frame, encoded by cortical grid cells [8]. This fuels the idea of treating object pose as a first class citizen when learning an object-centric generative model.

In this paper, we present a novel method for learning object-centric models for distinct object categories, that promotes a disentangled representation for shape and pose. We demonstrate how such models can be used for inferring actions that move an agent towards a preferred observation. We show that a better pose-shape disentanglement indeed seems to improve performance, yet further research in this direction is required. In the remainder of the paper we first give an overview on related work, after which we present our method. We present some results on object categories of the ShapeNet database [3], and conclude the paper with a thorough discussion.

2 Related Work

Object-Centric Models. Many techniques have been proposed for representing 3D objects using deep neural networks, working with 2D renders [5], 3D voxel representations [32], point clouds [15] or implicit signed distance function representations [20,21,23,28]. However, none of these take "action" into account, i.e. there is no agent that can pick its next viewpoint.

Disentangled Representations. Disentangling the hidden factors of variation of a dataset is an long sought feature for representation learning [1]. This can be encouraged during training by restricting the capacity of the information bottleneck [9], by penalizing the total correlation of the latent variables [4,11], or by matching moments of a factorized prior [13]. It has been shown that disentangled representations yield better performance on down-stream tasks, enabling quicker learning using fewer examples [29].

Deep Active Inference. Parameterizing generative models using deep neural networks for active inference has been coined "deep active inference" [30]. This enables active inference applications on high-dimensional observations such as pixel inputs [7,27,33]. In this paper, we propose a novel model which encourages a disentangled latent space, and we compare with other deep active inference models such as [17,33]. For a more extensive review, see [19].

3 Object-Centric Deep Active Inference Models

In active inference, an agent acts and learns in order to minimize an upper bound on the negative log evidence of its observations, given its generative model of the world i.e. the free energy. In this section, we first formally introduce the different generative models considered for our agents for representing 3D objects. Next we discuss how we instantiate and train these generative models using deep neural networks, and how we encourage the model to disentangle shape and pose.

Generative Model. We consider the same setup as [18], in which an agent receives pixel observations o of a 3D object rendered from a certain camera viewpoint v, and as an action a can move the camera to a novel viewpoint. The action space is restricted to viewpoints that look at the object, such that the object is always in the center of the observation.

Figure 1 depicts different possible choices of generative model to equip the agent with. The first (1a) considers a generic partially observable Markov decision process (POMDP), in which a hidden state s_t encodes all information at timestep t to generate observation o_t. Action a_t determines together with the current state s_t how the model transitions to a new state s_{t+1}. This a model can be implemented as a variational autoencoder (VAE) [12,25], as shown in [7,33]. A second option (1b) is to exploit the environment setup, and assume we can also observe the camera viewpoint v_t. Now the agent needs to infer the object shape s which stays fixed over time. This resembles the architecture of a generative query network (GQN), which is trained to predict novel viewpoints of a given a

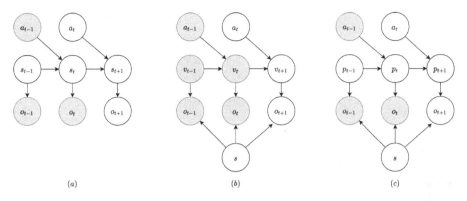

(a)   (b)   (c)

Fig. 1. Different generative models for object-centric representations, blue nodes are observed. (a) A generic POMDP model with a hidden state s_t that is transitioned through actions and which generates the observations. (b) The hidden state s encodes the appearance of the object, while actions transition the camera viewpoint v which is assumed to be observable. (c) Similar as (b), but without access to the camera viewpoint, which in this case has to be inferred as a separate pose latent variable p_t. (Color figure online)

scene [6,17]. Finally, in (1c), we propose our model, in which we have the same structure as (1b), but without access to the ground truth viewpoint. In this case, the model needs to learn a hidden latent representation of the object pose in view p_t. This also allows the model to learn a different pose representation than a 3D pose in SO(3), which might be more suited. We call this model a VAEsp, as it is trained in similar vein as (1a), but with a disentangled shape and pose latent.

VAEsp. Our model is parameterized by three deep neural networks: an encoder q_ϕ, a transition model p_χ, and a decoder p_ψ, as shown in Fig. 2. Observations o^i of object instance i are processed by the encoder q_ϕ, that outputs a belief over a pose latent $q_\phi(p_t^i|o_t^i)$ and a shape latent $q_\phi(s_t^i|o_t^i)$. From the pose distribution a sample p_t^i is drawn and fed to the transition model p_χ, paired with an action a_t. The output is a belief $p_\chi(p_{t+1}^i|p_t^i, a_t)$. From the transitioned belief a sample p_{t+1}^i is again drawn which is paired with a shape latent sample s^i and input to the decoder $p_\psi(o_t^i|p_t^i, s^i)$. The output of the decoding process is again an image $\hat{o}_{t+1}^i$. These models are jointly trained end-to-end by minimizing free energy, or equivalently, maximizing the evidence lower bound [18]. More details on the model architecture and training hyperparameters can be found in Appendix A.

Enforcing Disentanglement. In order to encourage the model to encode object shape features in the shape latent, while encoding object pose in the pose latent, we only offer the pose latent p_t as input to the transition model, whereas the decoder uses both the shape and pose. Similar to [10], in order

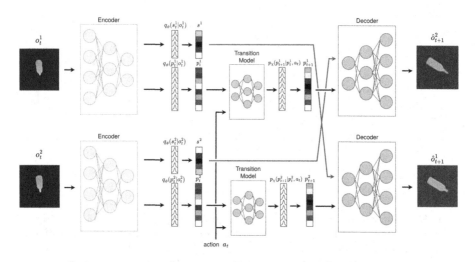

Fig. 2. The proposed VAEsp architecture consists of three deep neural networks: an encoder q_ϕ, a transition model p_χ, and a decoder p_ψ. By swapping the shape latent samples, we enforce the model to disentangle shape and pose during training.

to further disentangle, we randomly swap the shape latent code for two object instances at train time while keeping the same latent pose, refer to Fig. 2.

4   Experiments

We train our model on a subset of the ShapeNet dataset [3]. In particular, we use renders of 15 instances of the 'mug', 'bottle','bowl' and 'can' categories, train a separate model for each category, and evaluate on unseen object instances. We compare our VAEsp approach against a VAE model [33] that has equal amount of latent dimensions, but without a shape and pose split, and a GQN-like model [17], which has access to the ground truth camera viewpoint.

We evaluate the performance of the three considered generative models. First we look at the reconstruction and prediction quality of the models for unseen object instances. Next we investigate how good an agent can move the camera to match a preferred observation by minimizing expected free energy. Finally, we investigate the disentanglement of the resulting latent space.

One-Step Prediction. First, we evaluate all models on prediction quality over a test set of 500 observations of unseen objects in unseen poses. We provide each model with an initial observation which is encoded into a latent state. Next we sample a random action, predict the next latent state using the transition model, for which we reconstruct the observation and compare with a ground truth. We report both pixel-wise mean squared error (MSE) and structural similarity (SSIM) [31] in Table 1. In terms of MSE results are comparable for all the proposed architectures. In terms of SSIM however, VAEsp shows better performance for 'bottle' and 'can' category. Performance for 'bowl' category are

Table 1. One-step prediction errors, averaged over the entire test set. MSE (lower the better) and SSIM (higher the better) are considered.

		bottle	bowl	can	mug
MSE ⇓	GQN	0.473 ± 0.0874	0.487 ± 0.141	0.707 ± 0.1029	0.656 ± 0.0918
	VAE	0.471 ± 0.0824	0.486 ± 0.1487	0.693 ± 0.1103	0.646 ± 0.0886
	VAEsp	0.480 ± 0.0879	0.485 ± 0.1486	0.702 ± 0.1108	0.626 ± 0.0915
SSIM ⇑	GQN	0.748 ± 0.0428	0.814 ± 0.0233	0.868 ± 0.0203	0.824 ± 0.0279
	VAE	0.828 ± 0.0238	**0.907 ± 0.0178**	0.844 ± 0.0361	**0.874 ± 0.0323**
	VAEsp	**0.854 ± 0.0190**	0.902 ± 0.0291	**0.880 ± 0.0176**	0.814 ± 0.0348

comparable to the best performing VAE model. For the 'mug' category, the negative gap over the VAE model is consistent. Qualitative results for all models are shown in Appendix B.

Reaching Preferred Viewpoints. Next, we consider an active agent that is tasked to reach a preferred observation that was provided in advance. To do so, the agent uses the generative model to encode both the preferred and initial observation and then uses Monte Carlo sampling to evaluate the expected free energy for 10000 potential actions, after which the action with the lowest expected free energy is executed. The expected free energy formulation is computed as the negative log probability of the latent representation with respect to the distribution over the preferred state, acquired through encoding the preferred observation. This is similar to the setup adopted by Van de Maele et al. [18], with the important difference that now the preferred observation is an image of a *different* object instance.

To evaluate the performance, we compute the pixel-wise mean squared error (MSE) between a render of the target object in the preferred pose, and the render of the environment after executing the chosen action after the initial observation. The results are shown in Table 2. VAEsp performs on par with the other approaches for 'bowl' and 'mug', but significantly outperforms the GQN on the 'bottle' and 'can' categories, reflected by p-values of 0.009 and 0.001 for these respective objects. The p-values for the comparison with the VAE are 0.167

Table 2. MSE for the reached pose through the minimization of expected free energy. For each category, 50 meshes are evaluated, where for each object a random pose is sampled from a different object as preferred pose, and the agent should reach this pose.

	bottle	bowl	can	mug
GQN	0.0833 ± 0.0580	0.0888 ± 0.0594	0.0806 ± 0.0547	0.1250 ± 0.0681
VAE	0.0698 ± 0.0564	**0.0795 ± 0.0599**	0.0608 ± 0.0560	0.1247 ± 0.0656
VAEsp	**0.0557 ± 0.0404**	0.0799 ± 0.0737	**0.0487 ± 0.0381**	**0.1212 ± 0.0572**

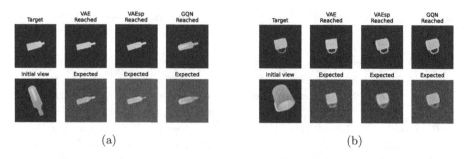

(a)   (b)

Fig. 3. Two examples of the experiment on reaching preferred viewpoints for a 'bottle' (a) and a 'mug' (b). First column shows the target view (top) and initial view given to the agent (bottom). Next, for the three models we show the actual reached view (top), versus the imagined expected view of the model (bottom).

and 0.220, which are not significant. A qualitative evaluation is shown in Fig. 3. Here we show the preferred target view, the initial view of the environment, as well as the final views reached by each of the agents, as well as what each model was imagining. Despite the target view being from a different object instance, the agent is able to find a matching viewpoint.

Disentangled Latent Space. Finally, we evaluate the disentanglement of shape and pose for the proposed architecture. Given that our VAEsp model outperforms the other models on 'bottle' and 'can', but not on 'bowl' and 'mug', we hypothesize that our model is able to better disentangle shape and pose for the first categories, but not for the latter. To evaluate this, we plot the distribution of each latent dimension when encoding 50 random shapes in a fixed pose, versus 50 random poses for a fixed shape, as shown on Fig. 4. We see that indeed the VAEsp model has a much more disentangled latent space for 'bottle' compared to 'mug', which supports our hypothesis. Hence, it will be interesting to further experiment to find a correlation between latent space disentanglement and model performance. Moreover, we could work on even better enforcing disentanglement when training a VAEsp model, for example by adding additional regularization losses [4,11]. Also note that the GQN does not outperform the other models, although this one has access to the ground truth pose factor. This might be due to the fact that an $SO(3)$ representation of pose is not optimal for the model to process, and it still encodes (entangled) pose information in the resulting latent space, as illustrated by violin plots for GQN models in Appendix C. Figure 5 qualitatively illustrates the shape and pose disentanglement for our best performing model (bottle). We plot reconstructions of latent codes consisting of the shape latent of the first column, combined with the pose latent of the first row.

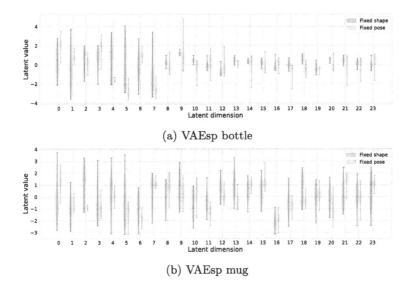

(a) VAEsp bottle

(b) VAEsp mug

Fig. 4. Violin plots representing the distribution over the latent dimension when keeping either the pose or shape fixed. For the bottle model (a) the pose latent dimensions (0–7) vary when only varying the pose, whereas the shape latent dimensions (8–23) don't vary with the pose. For the mug model (b) we see the shape and pose latent are much more entangled.

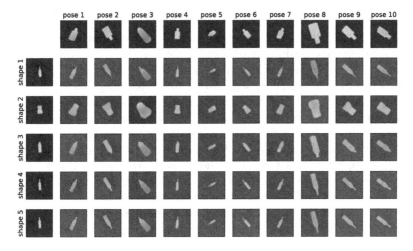

Fig. 5. Qualitative experimentation for the bottle category. Images are reconstructed from the different pairings of the pose latent and shape latent of the first row and column respectively.

5   Conclusion

In this paper, we proposed a novel deep active inference model for learning object-centric representations of object categories. In particular, we encourage the model to have a disentangled pose and shape latent code. We show that the better our model disentangles shape and pose, the better the results are on prediction, reconstruction as well as action selection towards a preferred observation. As future work, we will further our study on the impact of disentanglement, and how to better enforce disentanglement in our model. We believe that this line of work is important for robotic manipulation tasks, i.e. where a robot learns to pick up a cup by the handle, and can then generalize to pick up any cup by reaching to the handle.

A   Model and Training Details

This paper compares three generative models for representing the shape and pose of an object. Each of the models has a latent distribution of 24 dimensions, parameterized as a Gaussian distribution and has a similar amount of total trainable parameters.

VAE: The VAE baseline is a traditional variational autoencoder. The encoder consists of 6 convolutional layers with a kernel size of 3, a stride of 2 and padding of 1. The features for each layer are doubled every time, starting with 4 for the first layer. After each convolution, a LeakyReLU activation function is applied to the data. Finally, two linear layers are used on the flattened output from the convolutional pipeline, to directly predict the mean and log variance of the latent distribution. The decoder architecture is a mirrored version of the encoder. It consists of 6 convolutional layers with kernel size 3, padding 1 and stride 1. The layers have 32, 8, 16, 32 and 64 output features respectively. After each layer the LeakyReLU activation function is applied. The data is doubled in spatial resolution before each such layer through bi-linear upsampling, yielding a 120 by 120 image as final output. A transition model is used to predict the expected latent after applying an action. This model is parameterized through a fully connected neural network, consisting of three linear layers, where the output features are 64, 128 and 128 respectively. The input is the concatenation of a latent sample, and a 7D representation of the action (coordinate and orientation quaternion). The output of this layer is then again through two linear layers transformed in the predicted mean and log variance of the latent distribution. This model has 474.737 trainable parameters.

GQN: The GQN baseline only consists of an encoder and a decoder. As the model is conditioned on the absolute pose of the next viewpoint, there is no need for a transition model. The encoder is parameterized exactly the same as the encoder of the VAE baseline. The decoder is now conditioned on both a latent sample and the 7D representation of the absolute viewpoint (coordinate and orientation quaternion). These are first concatenated and transformed through a linear layer with 128 output features. This is then used as a latent code for the decoder, which is parameterized the same as the decoder used in the VAE baseline. In total, the GQN has 361.281 trainable parameters.

VAEsp: Similar to the VAE baseline, the VAEsp consists of an encoder, decoder and transition model. The encoder is also a convolutional neural network, parameterized the same as the encoder of the VAE, except that instead of two linear layers predicting the parameters of the latent distribution, this model contains 4 linear layers. Two linear layers with 16 output features are used to predict the mean and log variance of the shape latent distribution, and two linear layers with 8 output features are used to predict the mean and log variance of the pose latent distribution. In the decoder, a sample from the pose and shape latent distributions are concatenated and decoded through a convolutional neural network, parameterized exactly the same as the decoder from the VAE baseline. The transition model, only transitions the pose latent, as we make the assumption that the object shape does not change over time. The transition model is parameterized the same as the transition model of the VAE, with the exception that the input is the concatenation of the 8D pose latent vector and the 7D action, in contrast to the 24D latent in the VAE. The VAEsp model has 464.449 trainable parameters.

All models are trained using a constrained loss, where Lagrangian optimizers are used to weigh the separate terms [26]. During training, we tuned the reconstruction tolerance for each object empirically. Respectively to 'bottle', 'bowl', 'can' and 'mug' categories, MSE tolerances are: 350, 250, 280 and 520. Regularization terms are considered for each latent element. For all models, the Adam optimizer was used to minimize the objective.

B   Additional Qualitative Results

(See Fig. 6).

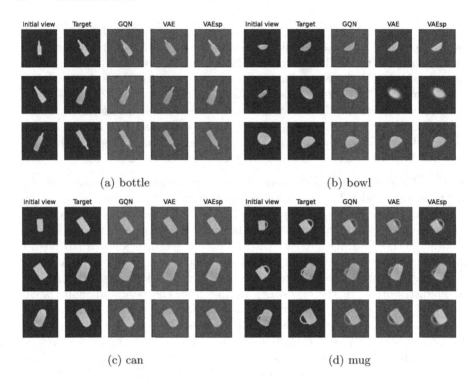

Fig. 6. One-step prediction for different object categories.

C   Latent Disentanglement

In Figs. 7, 8, 9 and 10, we show the distribution over the latent values when encoding observation where a single input feature changes. The blue violin plots represent the distribution over the latent values for observations where the shape is kept fixed, and renders from different poses are fed through the encoder. The orange violin plots represent the distribution over the latent values for observations where the pose is kept fixed, and renders from different shapes within the object class are encoded through the encoder models.

In these figures, we can clearly see that the encoding learnt by the VAE is not disentangled for any of the objects as the latent dimensions vary for both the fixed shape and pose cases. With the GQN, we would expect that the latent dimensions would remain static for the fixed shape case, as the pose is an explicit external signal for the decoder, however we can see that for a fixed shape, the variation over the latent value still varies a lot, in similar fashion as for the fixed pose. We conclude that the encoding of the GQN is also not disentangled. For the VAEsp model, we can see that in Figs. 7 and 8, the first eight dimensions are used for the encoding of the pose, as the orange violins are much denser distributed for the fixed pose case. However, in Figs. 9 and 10, we see that the model still shows a lot of variety for the latent codes describing the non-varying feature of the input. This result also strokes with our other experiments where for these objects both reconstruction as well as the move to perform worse.

In this paper, we investigated the disentanglement for the different considered object classes. We see that our approach does not yield a disentangled representation each time. Further investigation and research will focus on better enforcing this disentanglement.

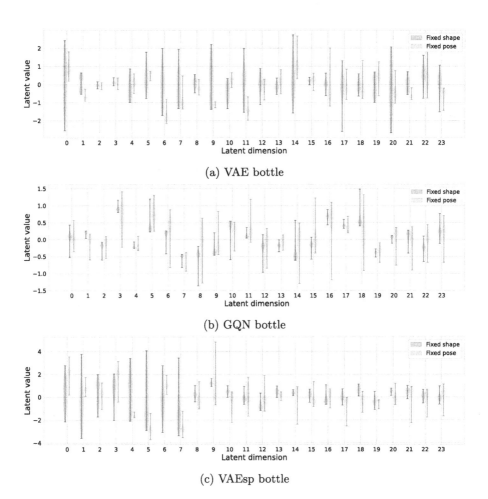

(a) VAE bottle

(b) GQN bottle

(c) VAEsp bottle

Fig. 7. Distribution of the latent values for the different models (VAE, GQN and VAEsp) for objects from the "bottle" class. In this experiment, 50 renders from a fixed object shape with a varying pose (fixed shape, marked in blue) are encoded. The orange violin plots represent the distribution over the latent values for 50 renders from the same object pose, with a varying object shape. (Color figure online)

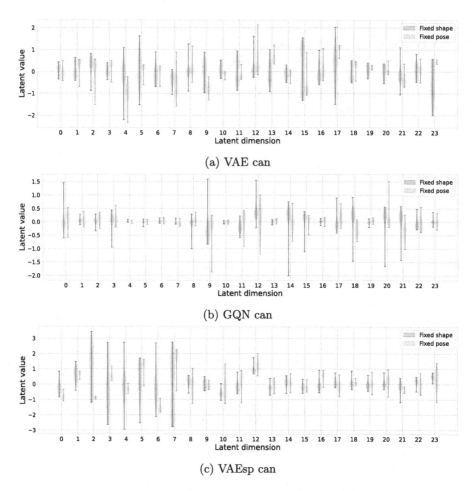

Fig. 8. Distribution of the latent values for the different models (VAE, GQN and VAEsp) for objects from the "can" class. In this experiment, 50 renders from a fixed object shape with a varying pose (fixed shape, marked in blue) are encoded. The orange violin plots represent the distribution over the latent values for 50 renders from the same object pose, with a varying object shape. (Color figure online)

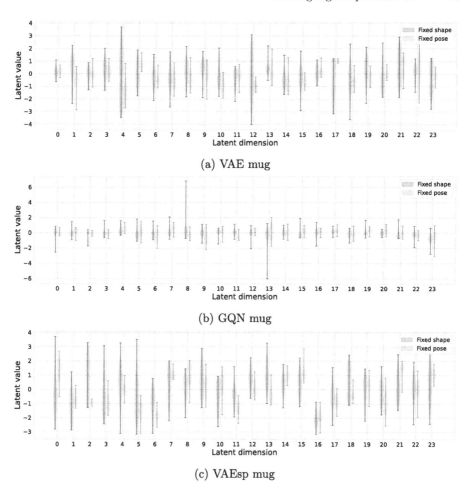

(a) VAE mug

(b) GQN mug

(c) VAEsp mug

Fig. 9. Distribution of the latent values for the different models (VAE, GQN and VAEsp) for objects from the "mug" class. In this experiment, 50 renders from a fixed object shape with a varying pose (fixed shape, marked in blue) are encoded. The orange violin plots represent the distribution over the latent values for 50 renders from the same object pose, with a varying object shape. (Color figure online)

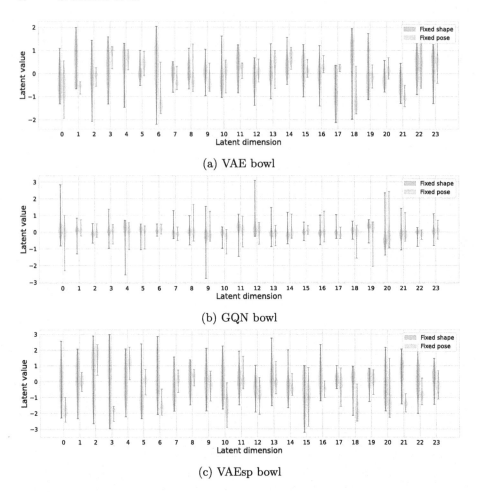

(a) VAE bowl

(b) GQN bowl

(c) VAEsp bowl

Fig. 10. Distribution of the latent values for the different models (VAE, GQN and VAEsp) for objects from the "bowl" class. In this experiment, 50 renders from a fixed object shape with a varying pose (fixed shape, marked in blue) are encoded. The orange violin plots represent the distribution over the latent values for 50 renders from the same object pose, with a varying object shape. (Color figure online)

References

1. Bengio, Y., Courville, A., Vincent, P.: Representation learning: a review and new perspectives. IEEE Trans. Pattern Anal. Mach. Intell. **35**, 1798–1828 (2013). https://doi.org/10.1109/TPAMI.2013.50
2. Billard, A., Kragic, D.: Trends and challenges in robot manipulation. Science **364**, eaat8414 (2019). https://doi.org/10.1126/science.aat8414

3. Chang, A.X., et al.: ShapeNet: an information-rich 3D model repository. Technical report. arXiv:1512.03012 [cs.GR], Stanford University – Princeton University – Toyota Technological Institute at Chicago (2015)
4. Chen, R.T.Q., Li, X., Grosse, R., Duvenaud, D.: Isolating sources of disentanglement in VAEs. In: Proceedings of the 32nd International Conference on Neural Information Processing Systems, NIPS 2018, pp. 2615–2625. Curran Associates Inc., Red Hook (2018)
5. Dosovitskiy, A., Springenberg, J.T., Tatarchenko, M., Brox, T.: Learning to generate chairs, tables and cars with convolutional networks. IEEE Trans. Pattern Anal. Mach. Intell. **39**(4), 692–705 (2017). https://doi.org/10.1109/TPAMI.2016.2567384
6. Eslami, S.M.A., et al.: Neural scene representation and rendering. Science **360**(6394), 1204–1210 (2018). https://doi.org/10.1126/science.aar6170. https://www.science.org/doi/10.1126/science.aar6170
7. Fountas, Z., Sajid, N., Mediano, P., Friston, K.: Deep active inference agents using Monte-Carlo methods. In: Larochelle, H., Ranzato, M., Hadsell, R., Balcan, M.F., Lin, H. (eds.) Advances in Neural Information Processing Systems, vol. 33, pp. 11662–11675. Curran Associates, Inc. (2020)
8. Hawkins, J., Ahmad, S., Cui, Y.: A theory of how columns in the neocortex enable learning the structure of the world. Front. Neural Circuits **11**, 81 (2017). https://doi.org/10.3389/fncir.2017.00081. http://journal.frontiersin.org/article/10.3389/fncir.2017.00081/full
9. Higgins, I., et al.: Beta-VAE: learning basic visual concepts with a constrained variational framework. In: 5th International Conference on Learning Representations, ICLR 2017, Toulon, France, 24–26 April 2017, Conference Track Proceedings (2017)
10. Huang, X., Liu, M.Y., Belongie, S., Kautz, J.: Multimodal unsupervised image-to-image translation. In: Proceedings of the European Conference on Computer Vision (ECCV) (2018)
11. Kim, H., Mnih, A.: Disentangling by factorising. In: Dy, J., Krause, A. (eds.) Proceedings of the 35th International Conference on Machine Learning. Proceedings of Machine Learning Research, vol. 80, pp. 2649–2658. PMLR (2018)
12. Kingma, D.P., Welling, M.: Auto-encoding variational bayes. arXiv:1312.6114 [cs, stat] (2014)
13. Kumar, A., Sattigeri, P., Balakrishnan, A.: Variational inference of disentangled latent concepts from unlabeled observations. In: 6th International Conference on Learning Representations, ICLR 2018, Vancouver, BC, Canada, 30 April–3 May 2018, Conference Track Proceedings (2018)
14. Lanillos, P., et al.: Active inference in robotics and artificial agents: survey and challenges (2021)
15. Lin, C.H., Kong, C., Lucey, S.: Learning efficient point cloud generation for dense 3D object reconstruction. In: Proceedings of the Thirty-Second AAAI Conference on Artificial Intelligence and Thirtieth Innovative Applications of Artificial Intelligence Conference and Eighth AAAI Symposium on Educational Advances in Artificial Intelligence, AAAI 2018/IAAI 2018/EAAI 2018. AAAI Press (2018)
16. Van de Maele, T., Verbelen, T., Catal, O., Dhoedt, B.: Disentangling what and where for 3D object-centric representations through active inference. arXiv:2108.11762 [cs] (2021)

17. Van de Maele, T., Verbelen, T., Çatal, O., De Boom, C., Dhoedt, B.: Active vision for robot manipulators using the free energy principle. Front. Neuro-robotics **15**, 642780 (2021). https://doi.org/10.3389/fnbot.2021.642780. https://www.frontiersin.org/articles/10.3389/fnbot.2021.642780/full

18. Van de Maele, T., Verbelen, T., Çatal, O., Dhoedt, B.: Embodied object representation learning and recognition. Front. Neurorobotics **16** (2022). https://doi.org/10.3389/fnbot.2022.840658. https://www.frontiersin.org/article/10.3389/fnbot.2022.840658

19. Mazzaglia, P., Verbelen, T., Çatal, O., Dhoedt, B.: The free energy principle for perception and action: a deep learning perspective. Entropy **24**(2) (2022). https://doi.org/10.3390/e24020301. https://www.mdpi.com/1099-4300/24/2/301

20. Mescheder, L., Oechsle, M., Niemeyer, M., Nowozin, S., Geiger, A.: Occupancy networks: learning 3D reconstruction in function space. arXiv:1812.03828 [cs] (2019)

21. Mildenhall, B., Srinivasan, P.P., Tancik, M., Barron, J.T., Ramamoorthi, R., Ng, R.: NeRF: representing scenes as neural radiance fields for view synthesis. arXiv:2003.08934 [cs] (2020)

22. Mishkin, M., Ungerleider, L.G., Macko, K.A.: Object vision and spatial vision: two cortical pathways. Trends Neurosci. **6**, 414–417 (1983)

23. Park, J.J., Florence, P., Straub, J., Newcombe, R., Lovegrove, S.: DeepSDF: learning continuous signed distance functions for shape representation. arXiv:1901.05103 [cs] (2019)

24. Parr, T., Sajid, N., Da Costa, L., Mirza, M.B., Friston, K.J.: Generative models for active vision. Front. Neurorobotics **15**, 651432 (2021). https://doi.org/10.3389/fnbot.2021.651432. https://www.frontiersin.org/articles/10.3389/fnbot.2021.651432/full

25. Rezende, D.J., Mohamed, S., Wierstra, D.: Stochastic backpropagation and approximate inference in deep generative models. arXiv:1401.4082 [cs, stat] (2014)

26. Rezende, D.J., Viola, F.: Taming VAEs. arXiv:1810.00597 [cs, stat] (2018)

27. Sancaktar, C., van Gerven, M.A.J., Lanillos, P.: End-to-end pixel-based deep active inference for body perception and action. In: 2020 Joint IEEE 10th International Conference on Development and Learning and Epigenetic Robotics (ICDL-EpiRob) (2020). https://doi.org/10.1109/icdl-epirob48136.2020.9278105

28. Sitzmann, V., Martel, J.N.P., Bergman, A.W., Lindell, D.B., Wetzstein, G.: SIREN: implicit neural representations with periodic activation functions. arXiv:2006.09661 [cs, eess] (2020)

29. van Steenkiste, S., Locatello, F., Schmidhuber, J., Bachem, O.: Are disentangled representations helpful for abstract visual reasoning? In: Wallach, H., Larochelle, H., Beygelzimer, A., d'Alché-Buc, F., Fox, E., Garnett, R. (eds.) Advances in Neural Information Processing Systems, vol. 32. Curran Associates, Inc. (2019). https://proceedings.neurips.cc/paper/2019/file/bc3c4a6331a8a9950945a1aa8c95ab8a-Paper.pdf

30. Ueltzhöffer, K.: Deep active inference. Biol. Cybern. **112**(6), 547–573 (2018). https://doi.org/10.1007/s00422-018-0785-7

31. Wang, Z., Bovik, A., Sheikh, H., Simoncelli, E.: Image quality assessment: from error visibility to structural similarity. IEEE Trans. Image Process. **13**(4), 600–612 (2004). https://doi.org/10.1109/TIP.2003.819861

32. Wu, J., Zhang, C., Xue, T., Freeman, B., Tenenbaum, J.: Learning a probabilistic latent space of object shapes via 3D generative-adversarial modeling. In: Lee, D., Sugiyama, M., Luxburg, U., Guyon, I., Garnett, R. (eds.) Advances in Neural Information Processing Systems, vol. 29. Curran Associates, Inc. (2016). https://proceedings.neurips.cc/paper/2016/file/44f683a84163b3523afe57c2e008bc 8c-Paper.pdf

33. Çatal, O., Wauthier, S., De Boom, C., Verbelen, T., Dhoedt, B.: Learning generative state space models for active inference. Front. Comput. Neurosci. **14**, 574372 (2020). https://doi.org/10.3389/fncom.2020.574372. https://www.frontiersin.org/ articles/10.3389/fncom.2020.574372/full

Object-Based Active Inference

Ruben S. van Bergen[✉] and Pablo Lanillos

Department of Artificial Intelligence and Donders Institute for Brain,
Cognition and Behavior, Radboud University, Nijmegen, The Netherlands
{ruben.vanbergen,pablo.lanillos}@donders.ru.nl

Abstract. The world consists of objects: distinct entities possessing
independent properties and dynamics. For agents to interact with the
world intelligently, they must translate sensory inputs into the bound-
together features that describe each object. These object-based representations form a natural basis for planning behavior. Active inference (AIF)
is an influential unifying account of perception and action, but existing
AIF models have not leveraged this important inductive bias. To remedy
this, we introduce 'object-based active inference' (OBAI), marrying AIF
with recent deep object-based neural networks. OBAI represents distinct
objects with separate variational beliefs, and uses selective attention to
route inputs to their corresponding object slots. Object representations
are endowed with independent action-based dynamics. The dynamics and
generative model are learned from experience with a simple environment
(active multi-dSprites). We show that OBAI learns to correctly segment
the action-perturbed objects from video input, and to manipulate these
objects towards arbitrary goals.

Keywords: Multi-object representation learning · Active inference

1 Introduction

Intelligent agents are not passive entities that observe the world and learn its
causality. They learn the relationship of action and effect by interacting with
the world, in order to fulfil their goals [1]. In higher-order intelligence, such as
exhibited by primates, these interactions very often take place at the level of
objects [2,3]. Whether picking a ripe fruit from a tree branch, kicking a football,
or taking a drink from a glass of water; all require reasoning and planning in
terms of objects. Objects, thus, are natural building blocks for representing the
world and planning interactions with it.

While there have been recent advances in unsupervised multi-object representation learning and inference [4,5], to the best of the authors knowledge,
no existing work has addressed how to leverage the resulting representations
for generating actions. In addition, object perception itself could benefit from
being placed in an active loop, as carefully selected actions could resolve ambiguity about object properties (including their segmentations - i.e., which inputs
belong to which objects). Meanwhile, state-of-the-art behavior-based learning

C. L. Buckley et al. (Eds.): IWAI 2022, CCIS 1721, pp. 50–64, 2023.
https://doi.org/10.1007/978-3-031-28719-0_4

(control), such as model-free reinforcement learning [6] uses complex encoding of high-dimensional pixel inputs without taking advantage of objects as an inductive bias (though see [7,8].

To bridge the gap between these different lines of work, we here introduce 'object-based active inference' (OBAI, pronounced /ə'beɪ/), a new framework that combines deep, object-based neural networks [4] and active inference [9,10]. Our proposed neural architecture functions like a Bayesian filter that iteratively refines perceptual representations. Through selective attention, sensory inputs are routed to high-level object modules (or *slots* [5]) that encode each object as a separated probability distribution, whose evolution over time is constrained by an internal model of action-dependent object dynamics. These object representations are highly compact and abstract, thus enabling efficient unrolling of possible futures in order to select optimal actions in a tractable manner. Furthermore, we introduce a closed-form procedure to learn preferences or goals in the network's latent space.

As a proof-of-concept, we evaluate our proposed framework on an active version of the multi-dSprites dataset, developed for this work (See Fig. 1a). Our preliminary results show that OBAI is able to: *i*) learn to segment and represent objects *ii*) learn the action-dependent, object-based dynamics of the environment; and *iii*) plan in the latent space – obviating the need to imagine detailed pixel-level outcomes in order to generate behavior. This work is a first step towards building more complex object-based active inference systems that can perform more cognitively challenging tasks on naturalistic input.

2   Methods

2.1   Object-Structured Generative Model

We extend the IODINE architecture proposed in [4] for object representation learning, to incorporate dynamics. Zablotskaia et al. [11] previously developed a similar extension to IODINE, in which object dynamics were modeled implicitly, through LSTM units operating one level below the latent-space representation. Here, we instead implement the dynamics directly in the latent space, and allow these dynamics to be influenced by actions on the part of the agent.

Like IODINE, our framework relies on iterative amortized inference [12] (IAI) on an object-structured generative model. This model describes images of up to K objects with a Normal mixture density (illustrated in Fig. 1):

$$p(o_i|\{\mathbf{s}^{(k)}\}_{k \in 1:K}, m_i) = \sum_k [m_i = k] \mathcal{N}\left(g_i(\mathbf{s}^{(k)}), \sigma_o^2\right) \qquad (1)$$

where o_i is the value of the i-th image pixel, $\mathbf{s}^{(k)}$ is the state of the k-th object, $g_i(\bullet)$ is a decoder function (implemented as a deep neural network (DNN)) that translates an object state to a predicted mean value at pixel i, σ_o^2 is the variability of pixels around their mean values and, crucially, m_i is a categorical variable that

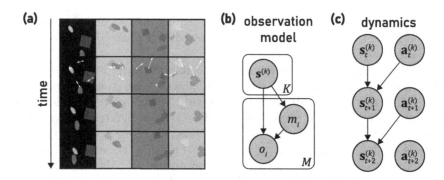

Fig. 1. Environment and generative model. (a) Active multi-dSprites. Non-zero accelerations in action fields (in the 2nd frame) are indicated as white arrows originating at the accelerated grid location. (b) Object-structured generative model for a single image (time indices omitted here for clarity of exposition). (c) Dynamics model in state-space for a single object k, shown for three time points.

indicates which object (out of a possible K choices) pixel i belongs to[1]. Note that the same decoder function is shared between objects. The pixel assignments themselves also depend on the object states:

$$p(m_i|\{\mathbf{s}^{(k)}\}_{k\in 1:K}) = \mathrm{Cat}\left(\mathrm{Softmax}\left(\{\pi_i(\mathbf{s}^{(k)})\}_{k\in 1:K}\right)\right) \qquad (2)$$

where $\pi_i(\bullet)$ is another DNN that maps an object state to a log-probability at pixel i, which (up to a constant of addition) defines the probability that the pixel belongs to that object. Marginalized over the assignment probabilities, the pixel likelihoods are given by:

$$p(o_i|\{\mathbf{s}^{(k)}\}_{k\in 1:K}) = \sum_k \hat{m}_{ik}\mathcal{N}\left(g_i(\mathbf{s}^{(k)}), \sigma_o^2\right) \qquad (3)$$

$$\hat{m}_{ik} = p(m_i = k|\{\mathbf{s}^{(k)}\}_{k\in 1:K}) \qquad (4)$$

During inference, the soft pixel assignments $\{\hat{m}_{ik}\}$ introduce dynamics akin to selective attention, as each object slot is increasingly able to focus on those pixels that are relevant to that object.

2.2 Incorporating Action-Dependent Dynamics

So far, this formulation is identical to the generative model in IODINE. We now extend this with an action-based dynamics model. We want to endow objects with (approximately) linear dynamics, and to allow actions that accelerate the objects. First, we redefine the state of an object at time point t in generalized coordinates, i.e. $\mathbf{s}_t^\dagger = \begin{bmatrix} \mathbf{s}_t \\ \mathbf{s}_t' \end{bmatrix}$, where $\mathbf{s}'$ refers to the first-order derivative of the state. The action-dependent state dynamics are then given by:

---

[1] Note the use of Iverson-bracket notation; the bracket term is binary and evaluates to 1 iff the expression inside the brackets is true.

$$\mathbf{s}_t'^{(k)} = \mathbf{s}_{t-1}'^{(k)} + \mathbf{D}\mathbf{a}_{t-1}^{(k)} + \sigma_s \epsilon_1 \tag{5}$$

$$\mathbf{s}_t^{(k)} = \mathbf{s}_{t-1}^{(k)} + \mathbf{s}_t'^{(k)} + \sigma_s \epsilon_2 \tag{6}$$

where $\mathbf{a}_t^{(k)}$ is the action on object k at time t. This action is a 2-D vector that specifies the acceleration on the object in pixel coordinates. Multiplication by $\mathbf{D}$ (which is learned during training) transforms the pixel-space acceleration to its effect in the latent space[2]. Equations 5–6 thus define the object dynamics model $p(\mathbf{s}_t^{\dagger(k)}|\mathbf{s}_{t-1}^{\dagger(k)}, \mathbf{a}_{t-1}^{(k)})$.

We established that $\mathbf{a}_t^{(k)}$ is the action on object k in the model at a given time. However, note that the correspondence between objects represented by the model, and the true objects in the (simulated) environment, is unknown.[3] To solve this correspondence problem, we introduce the idea of *action fields*. An action field $\boldsymbol{\Psi} = [\boldsymbol{\psi}_1, ..., \boldsymbol{\psi}_M]^T$ is an $[M \times 2]$ matrix (with M the number of pixels in an image or video frame), such that the i-th row in this matrix ($\boldsymbol{\psi}_i$) specifies the (x,y)-acceleration applied at pixel i. In principle, a different acceleration can be applied at each pixel coordinate (in practice, we apply accelerations sparsely). These pixel-wise accelerations affect objects through the rule that each object receives the sum of all accelerations that occur at its visible pixels:

$$\mathbf{a}_t^{(k)} = \sum_i [m_i = k]\boldsymbol{\psi}_i + \sigma_\psi \epsilon_3 \tag{7}$$

where we include a small amount of Normally distributed noise in order to make this relationship amenable to variational inference. This definition of actions in pixel-space is unambiguous and allows the model to interact with the environment.

2.3   Inference

On the generative model laid out in the previous section, we perform *iterative amortized inference* (IAI). IAI generalizes variational autoencoders (VAEs), which perform inference in a single feedforward pass, to architectures which use several iterations (implemented in a recurrent network) to minimize the Evidence Lower Bound (ELBO). As in VAEs, the final result is a set of variational

---

[2] Since the network will be trained unsupervised, we do not know in advance the nature of the latent space representation that will emerge. In particular, we do not know in what format (or even if) the network will come to represent the positions of the objects.

[3] In particular, since the representation across objects slots in the network is permutation-invariant, their order is arbitrary – just as the order in the memory arrays that specify the environment is also arbitrary. Thus, the object-actions, as represented in the model, cannot be unambiguously mapped to objects in the environment that the agent interacts with. This problem is exacerbated if the network has not inferred the object properties and segmentations with perfect accuracy or certainty, and thus cannot accurately or unambiguously refer to a true object in the environment.

beliefs in the latent space of the network. In our case, this amounts to inferring $q(\{\mathbf{s}^{\dagger(k)}, \mathbf{a}^{(k)}\}_{k \in 1:K})$. We choose these beliefs to be independent Normal distributions. Inference and learning both minimize the following ELBO loss:

$$\mathcal{L} = -\sum_{t=0}^{T} \left[\mathcal{H}\left(q\left(\{\mathbf{s}_t^{\dagger(k)}, \mathbf{a}_t^{(k)}\}\right)\right) + E_{q(\{\mathbf{s}_t^{(k)}\})}[\log p(\mathbf{o}_t | \{\mathbf{s}_t^{(k)}\})] \right.$$

$$\left. + \sum_k E_{q(\mathbf{a}_t^{(k)})}[\log p(\mathbf{a}_t^{(k)} | \mathbf{\Psi}_t)] + \sum_k E_{q\left(\mathbf{s}_t^{\dagger(k)}, \mathbf{s}_{t-1}^{\dagger(k)}, \mathbf{a}_{t-1}^{(k)}\right)}[\log p(\mathbf{s}_t^{\dagger(k)} | \mathbf{s}_{t-1}^{\dagger(k)}, \mathbf{a}_{t-1}^{(k)})] \right] \quad (8)$$

for some time horizon T. Note that for $t = 0$, we define $p(\mathbf{s}_t^{\dagger(k)} | \mathbf{s}_{t-1}^{\dagger(k)}, \mathbf{a}_{t-1}^{(k)}) = p(\mathbf{s}^{\dagger(k)}) = \prod_{jk} \mathcal{N}(s_j^{\dagger(k)}; 0, 1)$, i.e. a fixed standard-Normal prior. To compute $E_{q(\mathbf{a}_t^{(k)})}[\log p(\mathbf{a}_t^{(k)} | \mathbf{\Psi}_t)]$, we employ a sampling procedure, described in Appendix D.

The IAI architecture consists of a decoder module that implements the generative model, and a refinement module which outputs updates to the parameters $\mathbf{\lambda}$ of the variational beliefs. Mirroring the decoder, the refinement module consists of K copies of the same network (sharing the same parameters), such that refinement network k outputs updates to $\mathbf{\lambda}^{(k)}$. Network architectures for the decoder and refinement modules are detailed in Appendix B. To perform inference across multiple video frames, we simply copy the refinement and decoder networks across frames as well as object slots. Importantly, as in [4], each refinement network instance also receives as input a stochastic estimate of the current gradient $\nabla_{\mathbf{\lambda}_t^{(k)}} \mathcal{L}$. Since the ELBO loss includes a temporal dependence term between time points, the inference dynamics in the network are automatically coupled between video frames, constraining the variational beliefs to be consistent with the dynamics model. To infer $q(\{\mathbf{a}^{(k)}\}_{k \in 1:K})$, we employ a separate (small) refinement network (again copied across objects slots; details in Appendix B.2).

2.4 Task and Training

We apply OBAI to a simple synthetic environment, developed for this work, which we term *active-dSprites*. This environment was created by re-engineering the original dSprites shapes [13], to allow these objects to be translated at will and by non-integer pixel offsets. The environment simulates these shapes moving along linear trajectories that can be perturbed through the action fields we introduced above. In the current work, OBAI was trained on pre-generated experience with this environment, with action fields sampled arbitrarily (i.e. not based on any intent on the part of the agent).

Specifically, we generated video sequences (4 frames, 64×64 pixels) from the active-dSprites environment, with 3 objects per video, in which we applied an action field only at a single time point (in the 2nd frame). Action fields were sparsely sampled such that (whenever possible) every object received exactly one non-zero acceleration at one of its pixels. Exactly one background pixel also received a non-zero acceleration, to encourage the model to learn not to assign background pixels to the segmentation masks of foreground objects. In practice, the appearance of the background was unaffected by these actions (conceptually, the background can be thought of as an infinitely large plane extending outside the image frame, and thus shifting it by any amount will not change its visual appearance in the image).

OBAI was trained on 50,000 pre-generated video sequences, to minimize the ELBO loss from Eq. 8. This loss was augmented to include the losses at intermediate iterations of the inference procedure, and this composite loss was backpropagated through time to compute parameter gradients. More details about the environment and training procedure can be found in Appendices A and C.

2.4.1   Learning Goals in the Latent Space

The active-dSprites environment was conceived to support cognitive tasks that require object-based reasoning. A natural task objective in active-dSprites is to move a certain object to a designated location. This type of objective is simple in and of itself, but the rules that determine which object must be moved where can be arbitrarily complex. For now, we restrict ourselves to the simple objective of moving all objects in a scene to a fixed location, and focus on how to encode this objective. We follow previous Active Inference work (e.g. [14, 15]) in conceptualizing goals as a preference distribution $\tilde{p}$. However, rather than defining this preference to be over observations, as is common (though see [16–18]), we instead opt to define it over latent states, i.e. $\tilde{p}(\{s^{\dagger(k)}\})$, which simplifies action selection (a full discussion of the merits of this choice is outside the scope of this paper).

Assuming that we can define a preference over the true state of the environment, s_{true} (e.g. the ground-truth object positions), the preference distribution in latent space can be obtained through the following importance-sampling procedure:

$$\tilde{p}(\mathbf{s}) \propto \sum_j p(\mathbf{s}|\mathbf{o}_j^*)u_j \approx \sum_j q(\mathbf{s}|\mathbf{o}_j^*)u_j \qquad (9)$$

$$\mathbf{o}_j^* \sim p(\mathbf{o}|\mathbf{s}_{\text{true}_j}^*), \quad u_j = \tilde{p}(\mathbf{s}_{\text{true}_j}^*), \quad \mathbf{s}_{\text{true}_j}^* \sim p(\mathbf{s}_{\text{true}}) \propto \text{Constant} \qquad (10)$$

This allows the latent-space preference to be estimated in closed form from a set of training examples, constructed by sampling true states uniformly from the environment and rendering videos from these states. Inference is performed on the resulting videos, and the latent-state preference is computed as the importance-weighted average of the inferred state-beliefs (alternatively, we can sample states directly from $\tilde{p}(\mathbf{s}_{\text{true}}^*)$, and let the importance weights drop out of the equation). In particular, if $\tilde{p}(\mathbf{s}_{\text{true}}^*)$ is Normal, then the preference in the latent space is also Normal:

$$q(\mathbf{s}|\mathbf{o}_j^*) = \mathcal{N}\left(\boldsymbol{\mu}(\mathbf{o}_j^*), \boldsymbol{\sigma}(\mathbf{o}_j^{*2})\right), \quad \tilde{p}(\mathbf{s}) = \mathcal{N}(\tilde{\boldsymbol{\mu}}, \tilde{\boldsymbol{\sigma}}^2) \qquad (11)$$

$$\tilde{\boldsymbol{\mu}} = \frac{1}{\sum_j u_j}\sum_j u_j\boldsymbol{\mu}(\mathbf{o}_j^*), \quad \tilde{\boldsymbol{\sigma}} = \sqrt{\frac{1}{\sum_j u_j}\sum_j u_j\left((\tilde{\boldsymbol{\mu}} - \boldsymbol{\mu}(\mathbf{o}_j^*))^2 + \boldsymbol{\sigma}(\mathbf{o}_j^*)^2\right)} \qquad (12)$$

2.4.2 Planning Actions

OBAI can plan actions aimed at bringing the environment more closely in line with its learned preferences. Specifically, we choose actions that minimize the Free Energy of the Expected Future (FEEF) [18]. When preferences are defined with respect to latent states, the FEEF of a policy (action sequence) $\pi = \left\{ [\mathbf{a}_1^{(k)}, \mathbf{a}_2^{(k)}, \dots, \mathbf{a}_T^{(k)}] \right\}_{k \in 1:K}$ is given by:

$$\mathcal{G}(\pi) = \sum_{\tau=1}^{T} \sum_{k} D_{KL} \left(q(\mathbf{s}_\tau^{(k)} | \pi) || \tilde{p}(\mathbf{s}) \right) \tag{13}$$

where $q(\mathbf{s}_\tau^{(k)} | \pi)$ is the policy-conditioned variational prior, obtained by propagating the most recent state beliefs through the dynamics model by the requisite number of time steps.

Given this objective, the optimal policy can be calculated in closed form for arbitrary planning horizons. In this work, as a first proof-of-principle, we only consider greedy, one-step-ahead planning. In this case, the optimal "policy" (single action per object) is given by:

$$\hat{\mathbf{a}}^{(k)} = (\mathbf{D}^T \mathbf{L} \mathbf{D})^{-1} \mathbf{D}^T \mathbf{L} (\tilde{\mu} - \mu_s^{(k)}) \tag{14}$$

where $\mathbf{L} = \mathrm{diag}(\tilde{\sigma}^{-2})$, and $\mu_s^{(k)}$ is the mean of the current state belief for object k.

3 Results

3.1 Object Segmentation and Reconstruction in Dynamic Scenes

We first evaluated OBAI on its inference and reconstruction capabilties, when presented with novel videos of moving objects (not seen during training). To evaluate this, we examined the quality of its object segmentations, and of the video frame reconstructions (Fig. 2 & Table 1). Segmentation quality was computed using the Adjusted Rand Index (ARI), as well as a modified version of this index that only considers (ground-truth) foreground pixels (FARI). Across a test set of 10,000 4-frame video sequences of 3 randomly sampled moving objects each, OBAI achieved an average ARI and FARI of 0.948 and 0.939, respectively (where 1 means perfect accuracy and 0 equals chance-level performance), and a MSE of 9.51×10^{-4} (note that pixel values were in the range of $[0, 1]$). For comparison, a re-implementation of IODINE, trained on 50,000 static images of 3 dSprite objects, achieved an ARI of 0.081, FARI of 0.856 and MSE of 1.63×10^{-3} (on a test set of identically sampled static images). The very low ARI score reflects the fact that IODINE has no built-in incentive to assign background pixels to their own object slot. OBAI, on the other hand, has to account for the effects of actions being applied to the background, which must not affect the dynamics of the foreground objects. Thus, for OBAI to accurately model the dynamics in the training data, it must learn not to assign background pixels to the segmentation masks of foreground objects, lest an action might be placed on one of these spurious pixels.

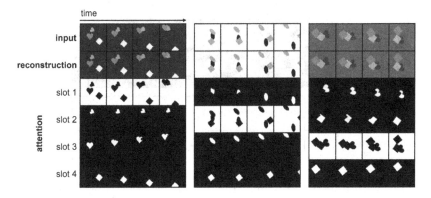

Fig. 2. Reconstruction and segmentation of videos of moving objects with actions. Three instances of segmentation from the test set of videos. The masks shows how each slot attends to the three objects and the background. The action field in this experiment is randomly generated for each instance at the beginning of the simulation.

Table 1. Quantitative segmentation and reconstruction results.

	ARI ($\uparrow$)	F-ARI ($\uparrow$)	MSE ($\downarrow$)
IODINE (static)	0.081	0.856	1.63×10^{-3}
OBAI (ours; 4 frames)	0.948	0.939	9.51×10^{-4}

3.2 Predicting the Future State of Objects

An advantage of our approach is that the network can predict future states of the world at the level of objects using the learned state dynamics. Figure 3 shows three examples of the network predicting future video frames. The first 4 video frames are used by the network to infer the state. Afterwards, we extrapolate the inferred state and dynamics of the last observed video frame into the future, and decode the thus-predicted latent states into predicted video frames. These predictions are highly accurate, with a MSE of 4×10^{-3}.

3.3 Goal-Directed Action Planning

Can OBAI learn and accomplish behavioral objectives? As a first foray into this question, we asked OBAI to learn fixed preference distributions defined in the true state-space of the environment, using the method described in Sect. 2.4.1. Specifically, we placed a Gaussian preference distribution on the location of the object and had the network learn the corresponding preference in its latent space from a set of 10,000 videos (annotated with the requisite importance weights). We then presented the network with static images of dSprite objects in random locations, and asked it to "imagine" the action that would bring the state of the environment into alignment with the learned preference. Finally, we applied this

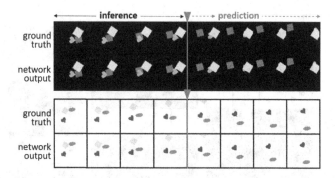

Fig. 3. Prediction of future video frames. Two instances of prediction from the test set of videos. We let the network perform inference on 4 consecutive frames, and then predict the future.

imagined action to the latent state, and decoded the image that would result from this. As illustrated in Fig. 4, the network is reliably able to imagine actions that would accomplish the goals we wanted it to learn.

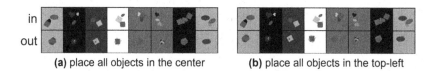

(a) place all objects in the center   **(b)** place all objects in the top-left

Fig. 4. Goal-directed action planning. We give the network an arbitrary input image with three objects (in) and it infers the action that will move the state towards the learned preference, and imagines the resulting image (out). In (a), $\tilde{p}(\mathbf{s}_{\text{true}})$ was biased towards the center of the image; in (b), it was biased more towards the top-left.

4 Conclusion

This work seeks to bridge an important gap in the field. On the one hand, computer vision research has developed object-based models, but these only perform passive inference. On the other hand, there is a wealth of research on behavioral learning (e.g. reinforcement learning and active inference), which has not generally leveraged objects as an inductive bias, built into the network architecture (cf. [7,8]). OBAI reconciles these two lines of research, by extending object-based visual inference models with action-based dynamics. We showed that OBAI can accurately track and predict the dynamics (and other properties) of simple objects whose movements are perturbed by actions – an important prerequisite for an agent to plan its own actions. In addition, we presented an efficient method for internalizing goals as a preference distribution over latent states, and showed that the agent can infer the actions necessary to accomplish these goals, at the same abstract level of reasoning. While our results are preliminary, they are an important proof-of-concept, establishing the potential of our approach. In future work, we aim to scale OBAI to more naturalistic environments, and more cognitively demanding tasks.

Acknowledgements. RSvB is supported by Human Brain Project Specific Grant Agreement 3 grant ID 643945539: "SPIKEFERENCE".

Appendix

A   Active-dSprites

Active-dSprites can be thought of as an "activated" version of the various multi-dSprites datasets that have been used in previous work on object-based visual inference (e.g. [4,5]). Not only does it include dynamics, but these dynamics can be acted on by an agent. Thus, active-dSprites is an interactive environment, rather than a dataset.

Objects in the active-dSprites environment are 2.5-D shapes (squares, ellipses and hearts): they have no depth dimension of their own, but can occlude each other within the depth dimension of the image. When an active-dSprites instance is intialized, object shapes, positions, sizes and colors are all sampled Uniformly at random. Initial velocities are drawn from a Normal distribution with mean 0 and standard deviation 4 (in units of pixels). Shape colors are sampled at discrete intervals spanning the full range of RGB-colors. Shapes are presented in random depth order against a solid background with a random grayscale color. Accelerations in action fields (at those locations that have been selected to incur a non-zero acceleration) are drawn from a Normal distribution with mean 0 and s.d. of 4.

B   Network Architectures

The OBAI architecture discussed in this paper consists of two separate IAI modules, each of which in turn contains a refinement and a decoder module. The first IAI module concerns the inference of the state beliefs $q(\{\mathbf{s}^{\dagger(k)}\})$ – we term this the *state inference module*. The second IAI module infers the object action beliefs $q(\{\mathbf{a}^{(k)}\})$, and we refer to this as the *action inference module*.

We ran inference for a total of $F \times 4$ iterations, where F is the number of frames in the input. Inference initially concerns just the first video frame, and beliefs for this frame are initialized to $\boldsymbol{\lambda}_0$, which is learned during training. After every 4 iterations, an additional frame is added to the inference window, and the beliefs for this new frame are initialized predictively, by extrapolating the object dynamics inferred up to that point. This procedure minimizes the risk that object representations are swapped between slots across frames, which can constitute a local minimum for the ELBO loss and leads to poor inference. We trained all IAI modules with $K=4$ object slots.

B.1   State Inference Module

This module used a latent dimension of 16. Note that, in the output of the refinement network, this number is doubled once as each latent belief is encoded by a mean and variance, and then doubled again as we represent (and infer) both the states and their first-order derivatives. In the decoder, the latent dimension is doubled only once, as the state derivatives do not enter into the reconstruction of a video frame. As in IODINE [4], we use a spatial broadcast decoder, meaning that the latent beliefs are copied along a spatial grid with the same dimensions as a video frame, and each latent vector is concatenated with the (x, y) coordinate of its grid location, before passing through a stack of transposed convolution layers. Decoder and refinement network architectures are summarized in the tables below. The refinement network takes in 16 image-sized inputs, which are identical to those used in IODONE [4], except that we omit the leave-one-out likelihoods. Vector-sized inputs join the network after the convolutional stage (which processes only the image-sized inputs), and consist of the variational parameters and (stochastic estimates of) their gradients.

Decoder

Type	Size/#Chan.	Act. func.	Comment
Input (λ)	32		
Broadcast	34		Appends coordinate channels
Conv^T 5×5	32	ELU	
Conv^T 5×5	32	ELU	
Conv^T 5×5	32	ELU	
Conv^T 5×5	32	ELU	
Conv^T 5×5	4		Outputs RGB + mask

Refinement Network

Type	Size/#Chan.	Act. func.	Comment
Linear	64		
LSTM	128	tanh	
Concat $[..., \lambda, \nabla_\lambda \mathcal{L}]$	256		Appends vector-sized inputs
Linear	128	ELU	
Flatten	800		
Conv 5×5	32	ELU	
Conv 5×5	32	ELU	
Conv 5×5	32	ELU	
Inputs	16		

B.2   Action Inference Module

The action inference module does not incorporate a decoder network, as the quality of the action beliefs is computed by evaluating Eq. 7 and plugging this into the ELBO loss from Eq. 8. While this requires some additional tricks (see Appendix D), no neural network is required for this. This module does include a (shallow) refinement network, which is summarized in the table below. This network takes as input the current variational parameters $\boldsymbol{\lambda_a}^{(k)}$ (2 means and 2 variances), their gradients, and the 'expected object action', $\sum_i \hat{m}_{ik}\boldsymbol{\psi}_i$.

Refinement Network

Type	Size/#Chan.	Act. func.	Comment
Linear	4		
LSTM	32	tanh	
Inputs	10		

C   Training Procedure

The above network architecture was trained on pre-generated experience with the active-dSprites environment, as described in the main text. The training set comprised 50,000 videos of 4 frames each. An additional validation set of 10,000 videos was constructed using the same environment parameters as the training set, but using a different random seed. Training was performed using the ADAM optimizer [REF] with default parameters and an initial learning rate of 3×10^{-4}. This learning rate was reduced automatically by a factor 3 whenever the validation loss had not decreased in the last 10 training epochs, down to a minimum learning rate of 3×10^{-5}. Training was performed with a batch size of 64 (16 × 4 GPUs), and was deemed to have converged after 245 epochs.

C.1   Modified ELBO Loss

OBAI optimizes an ELBO loss for both learning and inference. The basic form of this loss is given by Eq. 8. In practice, we modify this loss in two ways (similar to previous work, e.g. [4]). First, we re-weight the reconstruction term in the ELBO loss as follows:

$$
\mathcal{L}_\beta = -\sum_{t=0}^{T} \left[\mathcal{H}\left(q\left(\{\mathbf{s}_t^{\dagger(k)}, \mathbf{a}_t^{(k)}\}\right)\right) + \beta E_{q(\{\mathbf{s}_t^{(k)}\})}[\log p(\mathbf{o}_t|\{\mathbf{s}_t^{(k)}\})] \right.
$$

$$
\left. + \sum_k E_{q(\mathbf{a}_t^{(k)})}[\log p(\mathbf{a}_t^{(k)}|\boldsymbol{\Psi}_t)] + \sum_k E_{q\left(\mathbf{s}_t^{\dagger(k)}, \mathbf{s}_{t-1}^{\dagger(k)}, \mathbf{a}_{t-1}^{(k)}\right)}[\log p(\mathbf{s}_t^{\dagger(k)}|\mathbf{s}_{t-1}^{\dagger(k)}, \mathbf{a}_{t-1}^{(k)})] \right]
$$

$$(15)$$

Second, we train the network to minimize not just the loss at the end of the inference iterations through the network, but a composite loss that also includes the loss after earlier iterations. Let $\mathcal{L}_\beta^{(n)}$ be the loss after n inference iterations, then the composite loss is given by:

$$\mathcal{L}_{\text{comp}} = \sum_{n=1}^{N_{\text{iter}}} \frac{n}{N_{\text{iter}}} \mathcal{L}_\beta^{(n)} \tag{16}$$

C.2   Hyperparameters

OBAI includes a total of 4 hyperparameters: (1) the loss-reweighting coefficient β (see above); (2) the variance of the pixels around their predicted values, σ_o^2; (3) the variance of the noise in the latent space dynamics, σ_s^2; and (4) the variance of the noise in the object actions, σ_ψ^2. The results described in the current work were achieved with the following settings:

Param.	Value
β	5.0
σ_o	0.3
σ_s	0.1
σ_ψ	0.3

D   Computing $E_{q(\mathbf{a}^{(k)})}[\log p(\mathbf{a}^{(k)}|\mathbf{\Psi})]$

The expectation under $q(\mathbf{a}^{(k)})$ of $\log p(\mathbf{a}^{(k)}|\mathbf{\Psi})$, which appears in the ELBO loss (Eq. 8), cannot be computed in closed form, because the latter log probability requires us to marginalize over all possible configurations of the pixel-to-object assignments, and to do so inside of the logarithm. That is:

$$\log p(\mathbf{a}^{(k)}|\mathbf{\Psi}) = \sum_{\mathbf{m}} \log \left(p(\mathbf{a}^{(k)}|\mathbf{\Psi}, \mathbf{m}) p(\mathbf{m}|\{\mathbf{s}^{(k)}\}) \right) \tag{17}$$

$$= \log \left(E_{p(\mathbf{m}|\{\mathbf{s}^{(k)}\})}[p(\mathbf{a}^{(k)}|\mathbf{\Psi}, \mathbf{m})] \right) \tag{18}$$

However, note that within the ELBO loss, we want to maximize the expected value of this quantity (as its negative appears in the ELBO, which we want to minimize). From Jensen's inequality, we have:

$$E_{p(\mathbf{m}|\{\mathbf{s}^{(k)}\})}[\log p(\mathbf{a}^{(k)}|\mathbf{\Psi}, \mathbf{m})] \leq \log \left(E_{p(\mathbf{m}|\{\mathbf{s}^{(k)}\})}[p(\mathbf{a}^{(k)}|\mathbf{\Psi}, \mathbf{m})] \right) \tag{19}$$

Therefore, the l.h.s. of this equation provides a lower bound on the quantity we want to maximize. Thus, we can approximate our goal by maximizing this lower bound instead. This is convenient, because this lower bound, and its expectation under $q(\mathbf{a}^{(k)})$ can be approximated through sampling:

$$E_{q(\mathbf{a}^{(k)})}\left[E_{p(\mathbf{m}|\{\mathbf{s}^{(k)}\})}[\log p(\mathbf{a}^{(k)}|\boldsymbol{\Psi}, \mathbf{m})]\right] \approx \frac{1}{N_{\text{samples}}} \sum_j \log p(\mathbf{a}_j^{(k)*}|\boldsymbol{\Psi}, \mathbf{m}_j^*) \quad (20)$$

$$= \frac{1}{N_{\text{samples}}} \sum_j \log \mathcal{N}\left(\mathbf{a}_j^{(k)*}; \sum_i \hat{m}_{jk}^{*(i)} \boldsymbol{\psi}_i, \sigma_\psi^2 \mathbf{I}\right) \quad (21)$$

$$\hat{\mathbf{m}}_j^{*(i)} \sim p(m_i|\{\mathbf{s}^{(k)}\}), \quad \mathbf{a}_j^{(k)*} \sim q(\mathbf{a}^{(k)}), \quad \mathbf{s}_j^{(k)*} \sim q(\mathbf{s}^{(k)}) \quad (22)$$

where we slightly abuse notation in the sampling of the pixel assignments, as a vector is sampled from a distribution over a categorical variable. The reason this results in a vector is because this sampling step uses the Gumbel-Softmax trick [19], which is a differentiable method for sampling categorical variables as "approximately one-hot" vectors. Thus, for every pixel i, we sample a vector $\hat{\mathbf{m}}_j^{*(i)}$, such that the k-th entry of this vector, $\hat{m}_{jk}^{*(i)}$, denotes the "soft-binary" condition of whether pixel i belongs to object k. In practice, we use $N_{\text{samples}} = 1$, based on the intuition that this will still yield a good approximation over many training instances, and that we rely on the refinement network to learn to infer good beliefs. The Gumbel-Softmax sampling method depends on a temperature τ, which we gradually reduce across training epochs, so that the samples gradually better approximate the ideal one-hot vectors.

It is worth noting that, as the entropy of $p(\mathbf{m}|\{\mathbf{s}^{(k)}\})$ decreases (i.e. as object slots "become more certain" about which pixels are theirs), the bound in Eq. 19 becomes tighter. In the limit as the entropy becomes 0, the network is perfectly certain about the pixel assignments, and so the distribution collapses to a point mass. The expectation then becomes trivial, and so the two sides of Eq. 19 become equal. Sampling the pixel assignments is equally trivial in this case, as the distribution has collapsed to permit only a single value for each assignment. In short, at this extreme point, the procedure becomes entirely deterministic. In our data, we typically observe very low entropy for $p(\mathbf{m}|\{\mathbf{s}^{(k)}\})$, and so we likely operate in a regime close to the deterministic one, where the approximation is very accurate.

References

1. Lanillos, P., Dean-Leon, E., Cheng, G.: Yielding self-perception in robots through sensorimotor contingencies. IEEE Trans. Cogn. Dev. Syst. **9**(2), 100–112 (2016)
2. Kourtzi, Z., Connor, C.E.: Neural representations for object perception: structure, category, and adaptive coding. Annu. Rev. Neurosci. **34**, 45–67 (2011)
3. Peters, B., Kriegeskorte, N.: Capturing the objects of vision with neural networks. Nat. Hum. Behav. **5**, 1127–1144 (2021). https://doi.org/10.1038/s41562-021-01194-6
4. Greff, K., et al.: Multi-object representation learning with iterative variational inference. In: International Conference on Machine Learning, pp. 2424–2433. PMLR (2019)

5. Locatello, F., et al.: Object-centric learning with slot attention. In: Advances in Neural Information Processing Systems, vol. 33, pp. 11525–11538 (2020)
6. Mnih, V., et al.: Human-level control through deep reinforcement learning. Nature **518**(7540), 529–533 (2015)
7. Veerapaneni, R., et al.: Entity abstraction in visual model-based reinforcement learning (2019). http://arxiv.org/abs/1910.12827
8. Watters, N., Matthey, L., Bosnjak, M., Burgess, C.P., Lerchner, A.: Cobra: data-efficient model-based RL through unsupervised object discovery and curiosity-driven exploration (2019). http://arxiv.org/abs/1905.09275
9. Parr, T., Pezzulo, G., Friston, K.J.: Active Inference: The Free Energy Principle in Mind, Brain, and Behavior. MIT Press, Cambridge (2022)
10. Lanillos, P., et al.: Active inference in robotics and artificial agents: survey and challenges. arXiv preprint arXiv:2112.01871 (2021)
11. Zablotskaia, P., Dominici, E.A., Sigal, L., Lehrmann, A.M.: Unsupervised video decomposition using spatio-temporal iterative inference (2020). http://arxiv.org/abs/2006.14727
12. Marino, J., Yue, Y., Mandt, S.: Iterative amortized inference. In: 35th International Conference on Machine Learning, ICML 2018, vol. 8, pp. 5444–5462 (2018). http://arxiv.org/abs/1807.09356
13. Matthey, L., Higgins, I., Hassabis, D., Lerchner, A.: dSprites: disentanglement testing sprites dataset (2017). https://github.com/deepmind/dsprites-dataset/
14. Active inference and epistemic value. Cogn. Neurosci. **6**, 187–214 (2015). https://doi.org/10.1080/17588928.2015.1020053
15. Sajid, N., Ball, P.J., Parr, T., Friston, K.J.: Active inference: demystified and compared. Neural Comput. **33**, 674–712 (2021). https://doi.org/10.1162/neco_a_01357. https://direct.mit.edu/neco/article/33/3/674-712/97486
16. Friston, K.: A free energy principle for a particular physics (2019). http://arxiv.org/abs/1906.10184
17. Da Costa, L., Parr, T., Sajid, N., Veselic, S., Neacsu, V., Friston, K.: Active inference on discrete state-spaces: a synthesis. J. Math. Psychol. **99**, 102447 (2020)
18. Millidge, B., Tschantz, A., Buckley, C.L.: Whence the expected free energy? (2020). http://arxiv.org/abs/2004.08128
19. Jang, E., Gu, S., Poole, B.: Categorical reparameterization with Gumbel-Softmax (2016). http://arxiv.org/abs/1611.01144

Knitting a Markov Blanket is Hard When You are Out-of-Equilibrium: Two Examples in Canonical Nonequilibrium Models

Miguel Aguilera[1]([✉]), Ángel Poc-López[2], Conor Heins[3], and Christopher L. Buckley[1]

[1] School of Engineering and Informatics, University of Sussex, Falmer, Brighton, UK
`sci@maguilera.net`, `C.L.Buckley@sussex.ac.uk`
[2] ISAAC Lab, I3A Engineering Research Institute of Aragon, University of Zaragoza, Zaragoza, Spain
[3] Department of Collective Behaviour, Max Planck Institute of Animal Behavior, Konstanz, Germany
`cheins@ab.mpg.de`

Abstract. Bayesian theories of biological and brain function speculate that Markov blankets (a conditional independence separating a system from external states) play a key role for facilitating inference-like behaviour in living systems. Although it has been suggested that Markov blankets are commonplace in sparsely connected, nonequilibrium complex systems, this has not been studied in detail. Here, we show in two different examples (a pair of coupled Lorenz systems and a nonequilibrium Ising model) that sparse connectivity does not guarantee Markov blankets in the steady-state density of nonequilibrium systems. Conversely, in the nonequilibrium Ising model explored, the more distant from equilibrium the system appears to be correlated with the distance from displaying a Markov blanket. These result suggests that further assumptions might be needed in order to assume the presence of Markov blankets in the kind of nonequilibrium processes describing the activity of living systems.

Keywords: Markov blankets · Nonequilibrium dynamics · Bayesian inference · Lorenz attractor · Ising model

1 Introduction

In statistical inference, a Markov blanket describes a subset of variables containing all the required information to infer the state of another subset. Identifying a Markov blanket reduces the computational complexity of inferring generative models of some variables to capturing dependencies with blanket states. Specifically, a Markov blanket describes a set of variables (the 'blanket') separating two other sets of variables, that become independent conditioned on the state of the

© The Author(s), under exclusive license to Springer Nature Switzerland AG 2023
C. L. Buckley et al. (Eds.): IWAI 2022, CCIS 1721, pp. 65–74, 2023.
https://doi.org/10.1007/978-3-031-28719-0_5

blanket. If a system $\mathbf{s} = \{s_1, s_2, \ldots, s_N\}$ can be decomposed into three subsets $\mathbf{x}$, $\mathbf{b}$ and $\mathbf{y}$, $\mathbf{b}$ is a Markov blanket if it renders $\mathbf{x}$, $\mathbf{y}$ conditionally independent:

$$p(\mathbf{x}, \mathbf{y}|\mathbf{b}) = p(\mathbf{x}|\mathbf{b})p(\mathbf{y}|\mathbf{b}). \tag{1}$$

This property, also referred to as the global Markov condition [24], implies an absence of functional couplings between $\mathbf{x}$ and $\mathbf{y}$, given the blanket $\mathbf{b}$.

Beyond its role as a technical tool for statistical inference, Markov blankets are becoming a subject of discussion in Bayesian approaches to biological systems, specially in literature addressing the free energy principle (FEP). The FEP is a framework originating in theoretical neuroscience promoting a Bayesian perspective of the dynamics of self-organizing adaptive systems (including living organisms) [8,9,11]. The FEP claims that the internal states of certain systems can be described as if they were (approximately) inferring the hidden sources of sensory variations. Its foundational literature assumes that Markov blankets emerge from a sparse structural connectivity, decoupling internal states of a self-organizing system from its environmental milieu, (external states), via some interfacing states (Fig. 1) – e.g., the cell's membrane, or a combination of retinal and oculomotor states during vision. The assumption is that this sparse connectivity leads to a statistical decoupling of internal states conditioned on the blanket [8]. Although different versions of the theory address different aspects of the idea of a Markov blanket (e.g. its temporal evolution [21] or its role in paths outside a stationary density [11]), in the present article, we restrict our analysis of Markov blankets to the 'traditional' formulation of the FEP [9], where conditional independence relationships are expected to hold between states in the steady-state probability density that defines a stochastic system.

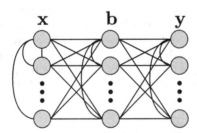

Fig. 1. Sparse structural connectivity. The FEP assumes that Markov blankets naturally arise (under some conditions) when internal and external states are not structurally connected [13]. All the models explored in this article will display this sparse connectivity pattern.

Acyclic Networks. Originally, Markov blankets were introduced in acyclic Bayesian networks [23], where they can be identified using a simple rule applied over the structural connections of the network (e.g. Fig. 2A). By this rule, the Markov blanket $\mathbf{b}$ of a subset $\mathbf{x}$ contains the parent nodes of $\mathbf{x}$, the children nodes of $\mathbf{x}$ and the parents of each child. This specific sparse structural connectivity is defined as the local Markov condition [24]. Originally, the FEP derived its intuitions about Markov blankets from acyclic models, considering the local Markov condition for a Markov blanket [8,9], suggesting that a boundary between system and environment arises naturally from this sparse structural connectivity as in directed acyclic graphs, without considering functional dynamics.

a) Bayesian directed acyclic graph

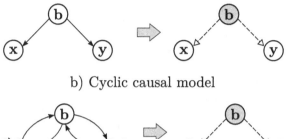

b) Cyclic causal model

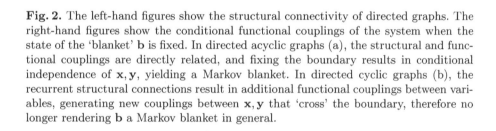

Fig. 2. The left-hand figures show the structural connectivity of directed graphs. The right-hand figures show the conditional functional couplings of the system when the state of the 'blanket' **b** is fixed. In directed acyclic graphs (a), the structural and functional couplings are directly related, and fixing the boundary results in conditional independence of **x**, **y**, yielding a Markov blanket. In directed cyclic graphs (b), the recurrent structural connections result in additional functional couplings between variables, generating new couplings between **x**, **y** that 'cross' the boundary, therefore no longer rendering **b** a Markov blanket in general.

Equilibrium Systems. More recent literature on the FEP justifies a similar equivalence of Markov blankets and structural connectivity under an asymptotic approximation to a weak-coupling equilibrium [14, see Eq. S8 in Supplementary Material]. Under this assumption, it has been predicted that Markov blankets will be commonplace in adaptive systems, e.g., in brain networks [13,17]. It is easy to observe that many instances of equilibrium systems will display Markov blankets under sparse, pairwise connectivity (Fig. 1). For example, consider any causal system described as a dynamical Markov chain in discrete time:

$$p(\mathbf{s}_t) = \sum_{\mathbf{s}_{t-1}} w(\mathbf{s}_t|\mathbf{s}_{t-1})p(\mathbf{s}_t|\mathbf{s}_{t-1}), \tag{2}$$

or its continuous state-time equivalent using a master equation:

$$\frac{dp(\mathbf{s}_t)}{dt} = \int_{\mathbf{s}'_t} w(\mathbf{s}_t|\mathbf{s}'_t)p(\mathbf{s}'_t)d\mathbf{s}'_t, \tag{3}$$

where w describes transition probabilities between states. Eventually, if the system converges to a global attractor, it will be described by a probability distribution

$$p(\mathbf{s}) = Z^{-1} \exp\left(-\beta E(\mathbf{s})\right) \tag{4}$$

where Z is a partition function. In thermodynamic equilibrium $E(\mathbf{s})$ will capture the Hamiltonian function of the system. Thermodynamic equilibrium implies a condition called 'detailed balance', which requires that, in steady state, transitions are time-symmetric, i.e., $w(\mathbf{s}|\mathbf{s}')p(\mathbf{s}') = w(\mathbf{s}'|\mathbf{s})p(\mathbf{s})$, resulting in

$$\frac{w(\mathbf{s}|\mathbf{s}')}{w(\mathbf{s}'|\mathbf{s})} \propto \exp\left(-\beta(E(\mathbf{s}) - E(\mathbf{s}'))\right). \tag{5}$$

If a system is described by the sparse connectivity structure in Fig. 1, then its energy can be decomposed into

$$E(\mathbf{s}) = E_{\text{int}}(\mathbf{x}, \mathbf{b}) + E_{\text{ext}}(\mathbf{b}, \mathbf{y}), \tag{6}$$

leading to a conditional indpendence

$$p(\mathbf{x}, \mathbf{y}|\mathbf{b}) = Z_{\mathbf{b}}^{-1} \exp\left(-\beta E_{\text{int}}(\mathbf{x}, \mathbf{b})\right) \cdot \exp\left(-\beta E_{\text{ext}}(\mathbf{b}, \mathbf{y})\right) = p(\mathbf{x}|\mathbf{b})p(\mathbf{y}|\mathbf{b}). \tag{7}$$

Recurrent, Nonequilibrium Systems. The most recent arguments in favour of why sparse coupling implies conditional independence follow from analysis of a stochastic system's coupling structure using a Helmholtz decomposition redescribing a continuous Langevin equation in terms of a gradient flow on the system's (log) stationary probability [7,15,22,28]. Briefly, a dynamical system described by a (Ito) stochastic differential equation:

$$\frac{d\mathbf{s}_t}{dt} = f(\mathbf{s}_t) + \varsigma(\mathbf{s}_t)\boldsymbol{\omega} \tag{8}$$

where f is the drift or deterministic part of the flow, ς is the diffusive or stochastic part (which can be state-dependent) and $\boldsymbol{\omega}$ a Wiener noise with covariance $2\Gamma(\mathbf{s}_t)$. The Helmholtz decomposition expresses f as follows [16, Equation 3]:

$$f(\mathbf{s}) = \underbrace{-\Gamma(\mathbf{s})\nabla E(\mathbf{s}) + \nabla \cdot \Gamma(\mathbf{s})}_{\text{dissipative}} + \underbrace{Q(\mathbf{s})\nabla E(\mathbf{s}) - \nabla \cdot Q(\mathbf{s})}_{\text{solenoidal}}, \tag{9}$$

expressing the total drift f as a gradient flow on the log of the stationary density $E(\mathbf{s}) \propto \log p(\mathbf{s})$. This decomposition involves two orthogonal gradient fields, a dissipative (or curl-free) term and a rotational (or divergence-free) term.

In a system subject only to dissipative forces, Eq. 9 is compatible with Eq. 5 for continuous-time systems. In contrast, a system driven by nonequilibrium dynamics will no longer show a direct correspondence between its dynamics and steady state distribution, thus a Markov blanket is not guaranteed from sparse connectivity. Given this difficulty [3,5], recent extensions of the FEP require additional conditions besides the absence of solenoidal couplings $Q(\mathbf{s})$ between internal and external states to guarantee a Markov blanket [13]. Nevertheless, a recent exploration of nonequilibrum linear systems showed that these extra conditions are unlikely to emerge without stringent restrictions of the parameter space [3]. In such linear systems, their cyclic, asymmetric structure propagates reverberant activity system-wide, generating couplings beyond their structural connectivity (e.g. Fig. 2B). As a consequence, for most parameter configurations of a system, the sparse connectivity of the local Markov condition does not result in a Markov blanket. That is, even if a system only interacts with the environment via a physical boundary, it will in general not display the conditional independence associated with a Markov blanket [3]. Recently, these arguments have been dismissed under the argument that living systems are poorly described by linear dynamics and thermodynamic equilibrium, and thus the scope of the FEP is focused on non-equilibrium systems [10]. Further work has argued Markov blankets may appear in high-dimensional state-spaces and spatially-localized interactions [16], under the assumption of a quadratic potential. The rest of this paper will explore how likely are Markov blankets to emerge for canonical nonlinear out-of-equilibrium models.

2   Results

To test empirically the extent to which Markov blankets can be expected out of equilibrium, we have performed conditional independence tests over two canonical non-linear systems: the Lorenz system and the asymmetric kinetic Ising model. Lorenz systems have long been studied due to their chaotic behaviour [20]. In contrast, asymmetric kinetic Ising models are recently becoming a popular tool to study non-equilibrium biological systems like neural networks [3,25].

Measure of Conditional Independence. Markov blanket conditional independence (Eq. 1) implies an absence of functional couplings between internal states $\mathbf{x}$ and external states $\mathbf{y}$ once the value of the blanket $\mathbf{b}$ is fixed. This condition is captured by the conditional mutual information being equal to zero:

$$I(\mathbf{x};\mathbf{y}|\mathbf{b}) = \sum_{\mathbf{x},\mathbf{b},\mathbf{y}} p(\mathbf{x},\mathbf{b},\mathbf{y}) \log \frac{p(\mathbf{x},\mathbf{y}|\mathbf{b})}{p(\mathbf{x}|\mathbf{b})p(\mathbf{y}|\mathbf{b})} \tag{10}$$

This conditional mutual information is equivalent to the Kullback Leibler divergence $D_{\mathrm{KL}}(p(\mathbf{x},\mathbf{y}|\mathbf{b})||p(\mathbf{x}|\mathbf{b})p(\mathbf{y}|\mathbf{b}))$, i.e. the dissimilarity between the joint and conditionally independent probability distributions. Thus, it is trivial to show that Eq. 1 holds only and only if $I(\mathbf{x};\mathbf{y}|\mathbf{b}) = 0$.

Pair of Coupled Lorenz Systems. In [12], the authors explore a system composed of two coupled Lorenz systems. The Lorenz system is a three-dimensional system of differential equations first studied by Edward Lorenz [20], displaying chaotic dynamics for certain parameter configurations. The system explored in [12] describes two three-dimensional systems that are coupled to each other through the states b_1 and b_2. The equations of motion for the full six-dimensional system are:

$$\frac{d}{dt}\begin{pmatrix} b_{1,t} \\ x_{1,t} \\ x_{2,t} \\ b_{2,t} \\ y_{1,t} \\ y_{2,t} \end{pmatrix} = \begin{pmatrix} \sigma(x_{1,t} - \chi b_{2,t} - (1-\chi)b_{1,t}) \\ \rho b_{1,t} - x_{1,t} - b_{1,t}x_{2,t} \\ b_{1,t}x_{1,t} - \beta x_{2,t} \\ \sigma(y_{1,t} - \chi b_{1,t} - (1-\chi)b_{2,t}) \\ \rho b_{2,t} - y_{1,t} - b_{2,t}y_{2,t} \\ b_{2,t}y_{1,t} - \beta y_{2,t} \end{pmatrix} \tag{11}$$

with $\sigma = 10$, $\beta = 8/3$, and $\rho = 32$. The coupling parameter is set to $\chi = 0.5$ (we will use $\chi = 0$ as reference of an uncoupled system) expecting the system to display nonequilibrium, chaotic dynamics. Even in the absence of random fluctuations, the chaotic nature of the system will result in a rich steady-state probability distribution $p(\mathbf{s}_t)$. In [12], authors show a Markov blanket conditional independence (Eq. 1) by approximating $p(\mathbf{s}_t)$ with a multivariate Gaussian (the so-called 'Laplace assumption' [21]). A careful analysis of the conditional mutual information $I(\mathbf{x}; \mathbf{y}|\mathbf{b})$ reveals that the system does not display a Markov blanket. In Fig. 3a we show the conditional mutual information $I(\mathbf{x}; \mathbf{y}|\mathbf{b})$ of the coupled Lorenz systems (solid line, $\chi = 0.5$), estimating over an ensemble of 10^7 trajectories from a random starting point (each variable $\mathcal{N}(0,1)$), and estimating its probability density using a histogram with 25 bins for each of the 6 dimensions. In comparison, the pair of decoupled Lorenz systems (dashed line, $\chi = 0$), shows near zero conditional mutual information only due to sampling noise). We note that the authors of [12] never claim that the true stochastic Lorenz system (or the coupled equivalent) has Markov blankets, only that their Laplace-approximated equivalents do.

Nonequilibrium Kinetic Ising Model. The asymmetric kinetic Ising model is a dynamical model with asymmetric couplings between binary spins $\mathbf{s}$ (with values ± 1) at times t and $t-1$, describing spin updates as:

$$w(s_{i,t}|\mathbf{s}_{t-1}) = \frac{\exp(s_{i,t}h_{i,t})}{2\cosh h_{i,t}} \tag{12}$$

$$h_{i,t} = \sum_j J_{ij}s_{j,t-1} \tag{13}$$

We define asynchronous dynamics in which, at each time step, only one spin is updated. In the case of symmetric couplings, $J_{ij} = J_{ji}$, the system converges to an equilibrium steady state, guaranteed by the detailed balance condition $p(\mathbf{s})$ maximum entropy distribution (Eq. 4), with $E(\mathbf{s}) = \sum_{ij} J_{ij}s_is_j$, displaying emerging phenomena like critical phase transitions maximizing information integration and transfer [1]. In the case of asymmetric couplings, the system converges to a nonequilibrium steady state distribution $p(\mathbf{s}_t)$, generally displaying a complex statistical structure with higher-order interactions [4]. In contrast with

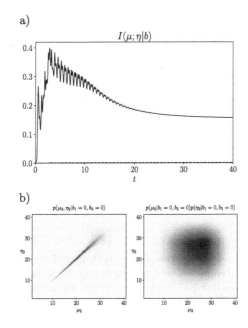

Fig. 3. Pair of coupled Lorenz systems. a) Conditional mutual information $I(\mathbf{x}; \mathbf{y}|\mathbf{b})$ of the coupled (solid line, $\chi = 0.5$) and decoupled (dashed line, $\chi = 0$) system, estimating using a 25 bin 6-dimensional histogram. b) Comparison of the joint and independent probability densities (estimated for a 100 bin bidimensional histogram) of variables x_2, y_2.

static equilibrium systems, asymmetries in $\mathbf{J}$ result in loops of oscillatory activity involving a nonequilibrium entropy production [2], corresponding to entropy dissipation in a steady-state irreversible process. In stochastic thermodynamics this is described as the divergence between forward and reverse trajectories [6,19], relating the system's time asymmetry with the entropy change of the reservoir. The entropy production σ_t at time t is then given as

$$\sigma_t = \sum_{\mathbf{s}_t, \mathbf{s}_{t-1}} p(\mathbf{s}_t, \mathbf{s}_{t-1}) \log \frac{w(\mathbf{s}_t|\mathbf{s}_{t-1})p(\mathbf{s}_{t-1})}{w(\mathbf{s}_{t-1}|\mathbf{s}_t)p(\mathbf{s}_t)}, \tag{14}$$

which is the Kullback-Leibler divergence between the forward and backward trajectories [18,26,27].

In the asymmetric Ising model, when couplings J_{ij} have a Gaussian distribution $\mathcal{N}(J_0/N, \Delta J^2/N)$ (an asymmetric equivalent of the Sherrington-Kirkpatrick model). In the thermodynamic limit the system generates out-of-equilibrium structures both in an order-disorder critical phase transition ($\Delta J < \Delta J_c$), and in a regime showing highly-deterministic disordered dynamics ($\Delta J > \Delta J_c$ and large β) [2,4]. Here, we will study in detail a network with just 6 nodes and random couplings with the connectivity in Fig. 1 where the probability distribution $p(\mathbf{s}_t)$ can be calculated exactly. We will use parameters correspond-

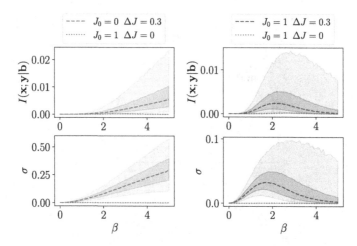

Fig. 4. Conditional mutual information $I(\mathbf{x};\mathbf{y}|\mathbf{b})$ (top) and entropy production σ (bottom) at different inverse temperatures (β) for a kinetic Ising model with Gaussian couplings and connectivity as in Fig. 1, including systems with disordered dynamics ($J_0 = 0, \Delta J = 0.3$, red curves), a nonequilibrium order-disorder transition ($J_0 = 1, \Delta J = 0.3$, blue curves), and an equilibrium transition ($J_0 = 1, \Delta J = 0$, green curves). Areas show the median, 25/75 and 5/95 percentiles for 10^4 configurations. (Color figure online)

ing to an order-disorder phase transition ($J_0 = 1, \Delta J = 0.3$) and a disordered dynamics ($J_0 = 0, \Delta J = 0.3$)). We will compare the results with the behaviour of the system in equilibrium, when disorder between couplings is removed ($J_0 = 1, \Delta J = 0$). The equilibrium system (equivalent to independent functional couplings as in Eq. 7), results in a Markov blanket with zero conditional mutual information $I(\mathbf{x};\mathbf{y}|\mathbf{b})$, as well as zero entropy production σ (Fig. 4, red line). Nonetheless, this is not the case when couplings are asymmetric (Fig. 4). Out of equilibrium, we observe how as the entropy production increases (i.e., the further the system is from equilibrium), the larger is the conditional mutual information $I(\mathbf{x};\mathbf{y}|\mathbf{b})$ (i.e., the further the system is from displaying a Markov blanket). This is particularly noticeable around the transition point in the order-disorder transition ($J_0 = 1, \Delta J = 0.3$), suggesting that Markov blankets might be specially challenging near nonequilibrium critical points.

Discussion. These results raise fundamental concerns about the frequent use of Markov blankets as an explanatory concept in studying the behaviour of biological systems. Our results however suggest that additional assumptions are needed for Markov blankets to arise under nonequilibrium conditions. In consequence, without further assumptions, it may not be possible to take for granted that biological systems operate in a regime where Markov blankets arise naturally. We shall note that the examples explored here have a reduced dimensionality (6 variables for both the Lorenz and asymmetric Ising systems). Previous work in the literature has suggested that high-dimensionality might be required to guarantee Markov blankets [12,16], but this remains a speculation. Further work could extend the type of analysis performed here to larger-dimensional systems.

References

1. Aguilera, M., Di Paolo, E.A.: Integrated information in the thermodynamic limit. Neural Netw. **114**, 136–146 (2019)
2. Aguilera, M., Igarashi, M., Shimazaki, H.: Nonequilibrium thermodynamics of the asymmetric Sherrington-Kirkpatrick model. arXiv preprint arXiv:2205.09886 (2022)
3. Aguilera, M., Millidge, B., Tschantz, A., Buckley, C.L.: How particular is the physics of the free energy principle? Phys. Life Rev. (2022). https://doi.org/10. 1016/j.plrev.2021.11.001
4. Aguilera, M., Moosavi, S.A., Shimazaki, H.: A unifying framework for mean-field theories of asymmetric kinetic Ising systems. Nat. Commun. **12**(1), 1–12 (2021)
5. Biehl, M., Pollock, F.A., Kanai, R.: A technical critique of some parts of the free energy principle. Entropy **23**(3), 293 (2021)
6. Crooks, G.E.: Nonequilibrium measurements of free energy differences for microscopically reversible Markovian systems. J. Stat. Phys. **90**(5), 1481–1487 (1998)
7. Eyink, G.L., Lebowitz, J.L., Spohn, H.: Hydrodynamics and fluctuations outside of local equilibrium: driven diffusive systems. J. Stat. Phys. **83**(3), 385–472 (1996)
8. Friston, K.: Life as we know it. J. R. Soc. Interface **10**(86), 20130475 (2013)
9. Friston, K.: A free energy principle for a particular physics. arXiv preprint arXiv:1906.10184 (2019)
10. Friston, K.: Very particular: comment on "how particular is the physics of the free energy principle?". Phys. Life Rev. **41**, 58–60 (2022)
11. Friston, K., et al.: The free energy principle made simpler but not too simple. arXiv preprint arXiv:2201.06387 (2022)
12. Friston, K., Heins, C., Ueltzhöffer, K., Da Costa, L., Parr, T.: Stochastic chaos and Markov blankets. Entropy **23**(9), 1220 (2021)
13. Friston, K.J., Da Costa, L., Parr, T.: Some interesting observations on the free energy principle. Entropy **23**(8), 1076 (2021)
14. Friston, K.J., et al.: Parcels and particles: Markov blankets in the brain. Netw. Neurosci. **5**(1), 211–251 (2021)
15. Graham, R.: Covariant formulation of non-equilibrium statistical thermodynamics. Zeitschrift für Phys. B Condens. Matt. **26**(4), 397–405 (1977)
16. Heins, C., Da Costa, L.: Sparse coupling and markov blankets: a comment on "how particular is the physics of the free energy principle?" By Aguilera, Millidge, Tschantz and Buckley. Phys. Life Rev. **42**, 33–39 (2022)
17. Hipólito, I., Ramstead, M.J., Convertino, L., Bhat, A., Friston, K., Parr, T.: Markov blankets in the brain. Neurosci. Biobehav. Rev. **125**, 88–97 (2021)
18. Ito, S., Oizumi, M., Amari, S.I.: Unified framework for the entropy production and the stochastic interaction based on information geometry. Phys. Rev. Res. **2**(3), 033048 (2020)
19. Jarzynski, C.: Hamiltonian derivation of a detailed fluctuation theorem. J. Stat. Phys. **98**(1), 77–102 (2000)
20. Lorenz, E.N.: Deterministic nonperiodic flow. J. Atmos. Sci. **20**(2), 130–141 (1963)
21. Parr, T., Da Costa, L., Heins, C., Ramstead, M.J.D., Friston, K.J.: Memory and Markov blankets. Entropy **23**(9), 1105 (2021)

22. Pavliotis, G.A.: Stochastic Processes and Applications: Diffusion Processes, the Fokker-Planck and Langevin Equations, vol. 60. Springer, Heidelberg (2014). https://doi.org/10.1007/978-1-4939-1323-7

23. Pearl, J.: Probabilistic Reasoning in Intelligent Systems: Networks of Plausible Inference. Morgan Kaufmann (1988)

24. Richardson, T.S., Spirtes, P., et al.: Automated discovery of linear feedback models. Carnegie Mellon [Department of Philosophy] (1996)

25. Roudi, Y., Dunn, B., Hertz, J.: Multi-neuronal activity and functional connectivity in cell assemblies. Curr. Opin. Neurobiol. **32**, 38–44 (2015)

26. Schnakenberg, J.: Network theory of microscopic and macroscopic behavior of master equation systems. Rev. Mod. Phys. **48**(4), 571–585 (1976). https://doi.org/10.1103/RevModPhys.48.571

27. Seifert, U.: Stochastic thermodynamics, fluctuation theorems and molecular machines. Rep. Prog. Phys. **75**(12), 126001 (2012)

28. Tomita, K., Tomita, H.: Irreversible circulation of fluctuation. Progress Theoret. Phys. **51**(6), 1731–1749 (1974)

Spin Glass Systems as Collective Active Inference

Conor Heins[1,2,3,4]([envelope]), Brennan Klein[1,6,7], Daphne Demekas[5,6], Miguel Aguilera[8,9], and Christopher L. Buckley[1,8,9]

[1] VERSES Research Lab, Los Angeles, CA, USA
`conor.heins@verses.io`
[2] Department of Collective Behaviour, Max Planck Institute of Animal Behavior, Konstanz, Germany
[3] Centre for the Advanced Study of Collective Behaviour, Konstanz, Germany
[4] Department of Biology, University of Konstanz, 78457 Konstanz, Germany
[5] Birkbeck Department of Film, Media and Cultural Studies, London, UK
[6] Network Science Institute, Northeastern University, Boston, MA, USA
[7] Laboratory for the Modeling of Biological and Socio-Technical Systems, Northeastern University, Boston, USA
[8] Sussex AI Group, Department of Informatics, University of Sussex, Brighton, UK
[9] Sackler Centre for Consciousness Science, University of Sussex, Brighton, UK

Abstract. An open question in the study of emergent behaviour in multi-agent Bayesian systems is the relationship, if any, between individual and collective inference. In this paper we explore the correspondence between generative models that exist at two distinct scales, using spin glass models as a sandbox system to investigate this question. We show that the collective dynamics of a specific type of active inference agent is equivalent to sampling from the stationary distribution of a spin glass system. A collective of specifically-designed active inference agents can thus be described as implementing a form of sampling-based inference (namely, from a Boltzmann machine) at the higher level. However, this equivalence is very fragile, breaking upon simple modifications to the generative models of the individual agents or the nature of their interactions. We discuss the implications of this correspondence and its fragility for the study of multiscale systems composed of Bayesian agents.

Keywords: Active inference · Boltzmann machines · Spin glass models

1 Introduction

Emergent phenomena in multi-agent systems are central to the study of self-organizing, complex systems, yet the relationship between individual properties and group-level phenomena remains opaque and lacks formal explanation. One principled approach to understanding such phenomena is offered by the Free Energy Principle and so-called 'multiscale active inference' [6,27]. Proponents of this multiscale approach propose that groups of individually-Bayesian agents

© The Author(s), under exclusive license to Springer Nature Switzerland AG 2023
C. L. Buckley et al. (Eds.): IWAI 2022, CCIS 1721, pp. 75–98, 2023.
https://doi.org/10.1007/978-3-031-28719-0_6

necessarily entail an emergent, higher-order Bayesian agent—in other words, systems are 'agents all the way down' by definition [13,17,25,27,28]. However, to date, there has been little theoretical or modeling work aimed at investigating whether this proposition is true in general or even demonstrating an existence proof in a specific system.

In this work we investigate this proposal by building a network of active inference agents that collectively implement a spin glass system. Spin glasses are a well-studied model class with a long history in statistical physics and equilibrium thermodynamics [5,12]. In the context of machine learning and computational neuroscience, spin glass systems can be tied to models of Bayesian inference and associative memory, particularly in the form of Hopfield networks and undirected graphical models like Boltzmann machines [16,33]. Boltzmann machines are a canonical example of an energy-based model in machine learning—they are defined by a global energy function and are analytically equivalent to a particular sort of spin glass model. The Boltzmann machine can straightforwardly be shown to be an inferential model by conditioning on the states of particular spins and sampling from the posterior distribution over the remaining spins' states [14,15]. In doing so, Boltzmann machines and spin glass systems can be described as performing Bayesian inference about the latent causes (spin configurations) of the 'data' (conditioned spins).

In this paper, we set out to investigate whether an inference machine (in our case, a Boltzmann machine) that exists at a 'higher-level', can be hierarchically decomposed into an ensemble of agents collectively performing active inference at a 'lower level.' We show a simple but rigorous equivalence between collective active inference and spin glass systems, providing the first steps for future quantitative study into the relationship between individual and collective generative models. We show that a group of active inference agents, equipped with a simple but very specific generative model, collectively sample from the stationary distribution of a spin glass system at the higher scale. This can be connected to a particular form of sampling known as Glauber dynamics [12,32]. When we further condition on the states of particular agents, then the system can be interpreted as collectively performing a form of sampling-based posterior inference over the configurations of the unconditioned agents, i.e., Boltzmann machine inference.

This paper is structured as follows: first, we specify the generative model that each spin site in a multi-agent spin glass system is equipped with, noting that the single agents are constructed explicitly such that their interactions at the lower-level lead to a spin glass system at the higher level. Then, we establish the equivalence between this multi-agent active inference process and Glauber dynamics, a scheme that samples from the stationary distribution of a spin glass system. We then generalize this result to sampling-based inference in Boltzmann machines by relaxing the homogeneous parameterization of each agent's generative models. We draw exact equivalences between the precisions of each agent's generative model and the parameters of a higher-level Boltzmann machine. We conclude by noting the fragility of the equivalence between multi-agent active inference and sampling from a collective spin glass system, and discuss the implications of our results for the study of emergent Bayesian inference in multiscale systems.

2   Generative Model for a Single Spin

We begin by constructing a generative model for a single Bayesian agent, which we imagine as a single spin in a collective of similar agents. From the perspective of this single 'focal agent' or spin, this generative model describes the agent's internal model of how the local environment generates its sensory data. Throughout this section we will take the perspective of this single focal agent, keeping in mind that any given agent is embedded in a larger system of other agents.

2.1   States and Observations

We begin by specifying the state-space of observations and hidden states that characterize the focal agent's 'world'. The focal agent's observations or sensations are comprised of a collection of binary spin states $\tilde{\sigma} = \{\sigma_j : j \in M\}$ where $\sigma_j = \pm 1$ and M is the set of neighbour spins that the focal agent has direct access to. In other words, the agent directly observes the spin states of neighbouring agents (but not its own).

The focal agent assumes these observed spin states $\tilde{\sigma}$ all depend on a single, binary latent variable z—the 'hidden spin state', which could also be interpreted as a coarse-grained 'average spin' of its neighbours. Having specified observations $\tilde{\sigma}$ and latent states z, the full generative model can be written as a joint distribution over observations and the hidden spin state, $P(\tilde{\sigma}, z)$. This in turn factorizes into a set of distributions that describe the likelihood over observations, given the latent state $P(\tilde{\sigma}|z)$, and prior beliefs about the latent state $P(z)$:

$$P(\tilde{\sigma}, z) = P(\tilde{\sigma}|z)P(z)$$

We parameterize the likelihood and prior distributions as Bernoulli distributions (expressed in a convenient exponential form):

$$P(\tilde{\sigma}|z; \gamma) = \prod_{j \in M} \frac{\exp(\gamma \sigma_j z)}{2 \cosh(\gamma z)} \qquad P(z; \zeta) = \frac{\exp(\zeta z)}{2 \cosh(\zeta z)}$$

$$\text{Likelihood} \qquad\qquad\qquad \text{Prior}$$

The likelihood factorizes into a product of independent Bernoulli distributions over each neighbouring spin σ_j. The full likelihood is parameterized with a single sensory precision parameter γ whose magnitude captures the focal agent's assumption about how reliably neighbouring spin states σ_j indicate the identity of the latent state z. A positive γ indicates that σ_j lends evidence to z being aligned with σ_j, whereas a negative γ means that σ_j lends evidence to z being opposite to σ_j.

The prior over z is also a Bernoulli distribution, parameterized by a precision ζ that acts as a 'bias' in the focal agent's prior belief about the value of z. When $\zeta > 0$, the focal agent believes the 'UP' ($z = +1$) state is more likely *a priori*, whereas $\zeta < 0$ indicates that the agent believes that $z = -1$ is more likely, with the magnitude of ζ reflecting the strength or confidence of this prior belief.

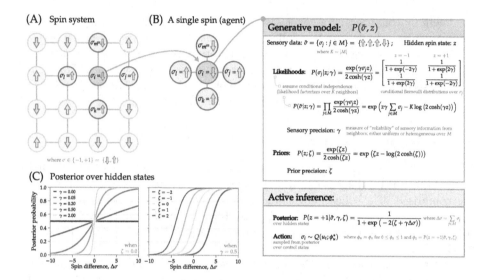

Fig. 1. Schematic illustration of individual and collective dynamics. (A) Example of a system of 16 spin sites connected via a 2-D lattice, each in a state of $\sigma \in \{-1, +1\}$ (green down-arrow or yellow up-arrow above), with a focal agent and its spin, $\sigma_i = -1$, highlighted in blue. **(B)** Generative model of a single spin. **(C)** The posterior belief over $z = +1$ as a function of the spin difference $\Delta\sigma$. Left: The steepness of the function is tuned by γ ($\zeta = 0.0$ shown). Right: The horizontal shift depends on ζ ($\gamma = 0.5$ shown). (Color figure online)

2.2 Bayesian Inference of Hidden States

Having specified a generative model, we now consider (from the perspective of a focal agent) the problem of estimating z, given observations $\tilde{\sigma} = \{\sigma_j : j \in M\}$ and generative model parameters γ, ζ. This is the problem of Bayesian inference, specifically the calculation of the posterior distribution over z. The conjugate-exponential form of the generative model means that the Bayesian posterior can be calculated exactly, and has a Bernoulli form that depends on $\tilde{\sigma}$ and z:

$$P(z|\tilde{\sigma}; \gamma, \zeta) = \frac{P(\tilde{\sigma}, z; \gamma, \zeta)}{P(\tilde{\sigma}; \gamma, \zeta)} = \frac{\exp\left(z(\zeta + \gamma \sum_j \sigma_j)\right)}{2\cosh\left(\zeta + \gamma \sum_j \sigma_j\right)} \tag{1}$$

If we fix the hidden state z to a particular value (e.g. $z = +1$), then we arrive at a simple expression for the posterior probability that the hidden spin state is in the 'UP' state, given the observations and the generative model parameters γ, ζ. The posterior belief expressed as the sum of sensory input $\sum_j \sigma_j$ assumes a logistic or sigmoid form. Hereafter we refer to the sum of observed spin states as the 'spin difference' $\Delta\sigma = \sum_j \sigma_j$, since this sum is equivalent to the difference in the number of positive ($\sigma_j = +1$) and negative ($\sigma_j = -1$) spins. Intuitively, the steepness and horizontal shift of this logistic function are determined by the likelihood and prior precisions:

$$P(z = +1|\tilde{\sigma}, \gamma, \zeta) = \frac{1}{1 + \exp\left(-2(\zeta + \gamma\Delta\sigma)\right)} \tag{2}$$

Figure 1C shows the effect of varying the likelihood and prior precisions on the posterior belief over z as a function of $\Delta\sigma$. We can also express the posterior as the equivalent Bernoulli distribution using the more common form, with parameter ϕ_z:

$$P(z|\tilde{\sigma}, \gamma, \zeta; \phi_z) = (1 - \phi_z)^{1 - \frac{z+1}{2}} \phi_z^{\frac{z+1}{2}}$$

$$\phi_z = \frac{1}{1 + \exp(-2(\zeta + \gamma\Delta\sigma))} \tag{3}$$

We now have a simple update rule that expresses how a focal agent updates its beliefs about z in response to observed spins $\tilde{\sigma}$. This sigmoid belief update has a clear, intuitive relationship to the parameters of the focal agent's generative model, with γ encoding the sensitivity of the belief to small changes in $\Delta\sigma$ and ζ encoding a 'bias' that skews the belief towards either -1 or $+1$. In the next section, we connect the generation of spins themselves to an active inference process, that leverages the Bayesian estimation problem of the current section to determine a focal agent's inference of its own spin state.

2.3   Active Inference of Spin States

Having addressed the issue of Bayesian inference or state estimation, we can now specify a mechanism by which agents generate their own spin states. These generated spin states will then serve as observations for the neighbours to whom the focal agent is connected. This turns into a problem of belief-guided action selection or decision-making. To enable agents to sample spin states as a function of their beliefs, we supplement each agent's current generative model with an extra random variable that corresponds to *control states*, and some forward model of how those control states determine observations. We use *active inference* to optimize a posterior belief over these control states [7,9,10]; an agent can then act to change its spin state by sampling from this posterior. By equipping each agent with a particular type of forward model of how its actions impact observations, we can formally tie the collective dynamics of active inference agents with this generative model to a sampling scheme from a spin glass model. Appendix B walks through the steps needed to add a representation of control states into the generative model introduced in the previous section, and perform active inference with respect to this augmented generative model.

Active inference agents entertain posterior beliefs not only about the hidden states of the world, but also about how their own actions affect the world. Posterior beliefs about actions are denoted $Q(u; \phi_u)$, where u is a random variable corresponding to actions and ϕ_u are the parameters of this belief. As opposed to the analytic posterior over hidden states z, $Q(u; \phi_u)$ is an approximate posterior, optimized using variational Bayesian inference [4]. In our focal agent's simple action model, control states have the same support as hidden states, i.e. $u = \pm1$. The value of u represents a possible spin action to take ('UP' vs. 'DOWN'). We parameterize $Q(u; \phi_u)$ as a Bernoulli distribution with parameter ϕ_u, which itself encodes the probability of taking the 'UP' $(+1)$ action:

$$Q(u_t; \phi_u) = (1 - \phi_z)^{1 - \frac{u_t+1}{2}} \phi_z^{\frac{u_t+1}{2}} \tag{4}$$

When we equip our spin glass agents with a particular (predictive) generative model, we can show that the approximate posterior over control states simplifies to the state posterior (see Appendix B for details), and an agent can generate a spin state by simply sampling from the posterior over actions:

$$Q(u; \phi_u) \approx P(z|\tilde{\sigma}, \gamma, \zeta; \phi_z) \qquad \sigma \sim Q(u; \phi_u)$$
$$\phi_u \approx \phi_z : 0 \le \phi_z \le 1 \qquad \sim Q(z; \phi_z) \triangleq P(z|\tilde{\sigma}; \gamma, \zeta) \tag{5}$$

We now have an active inference agent that 1) calculates a posterior belief $P(z|\tilde{\sigma}, \gamma, \zeta; \phi_z)$ about the latent state z in response to the observed spins of other agents and 2) generates a spin of its own by sampling from this belief, which ends up being identical to the posterior over actions. Intuitively, each agent just broadcasts its beliefs about the latent cause of its social observations, by sampling from its posterior over this hidden (average) state. Another way of looking at this is that each agent emits actions that maximize the accuracy of its beliefs (i.e., minimize variational free energy), under the prior assumption it is the author of its sensations, which, implicitly, are shared with other agents. Note that the choice to sample from the posterior over actions (as opposed to e.g. taking the maximum) renders this action-selection scheme a form of probability matching [26,29].

2.4   Completing the Loop

Given this sampling scheme for generating actions, we can simulate a collective active inference process by equating the actions of one agent to the observations of another. Specifically, each focal agent's spin action becomes an observation (σ_j for some j) for all the other agents that it (the focal agent) is connected to. Next, we will show how the dynamics of multi-agent active inference is analogous to a particular algorithm for sampling from the stationary distribution of a spin glass model, known as Glauber dynamics [12]. We then examine the fragility of this equivalence by exploring a number of simple modifications that break it.

3 Equivalence to Glauber Dynamics

Spin glass models are formally described in terms of a global energy function over states of the system. The global energy is related to the stationary probability distribution of the system through a Gibbs law or Boltzmann distribution:

$$p^*(\tilde{\sigma}) = \frac{1}{Z} \exp(-\beta E(\tilde{\sigma})) \tag{6}$$

where the stationary density p^* and energy function E are defined over spin configurations, where a configuration is a particular setting of each of the N spins that comprise the system: $\tilde{\sigma} = \{\sigma_i\}_{i=1}^N : \sigma_i \pm 1$. The partition function Z is a normalizing constant that ensures $p^*(\tilde{\sigma})$ integrates to 1.0, and β plays the role of an inverse temperature that can be used to arbitrarily rescale the Gibbs measure. In the case of the Ising model, this energy function is a Hamiltonian that can be expressed as a sum of pairwise interaction terms and an external drive or bias (often analogized to an external magnetic field):

$$E(\tilde{\sigma}) = -\sum_{\langle i,j \rangle} \sigma_i J_{ij} \sigma_j - \sum_i h_i \sigma_i \tag{7}$$

where J_{ij} specifies a (symmetric) coupling between spin sites i and j and h_i specifies an external forcing or bias term for site i. The bracket notation $\langle i,j \rangle$ denotes a sum over pairs. In numerical studies of the Ising model, one is typically interested in generating samples from this stationary density. One scheme for doing so is known as Glauber dynamics, where each spin σ_i of the system is updated using the following stochastic update rule:

$$\sigma_i \sim P(\sigma_i = (-1, +1))$$

$$P(\sigma_i = +1) = \frac{1}{1 + \exp(-\beta \Delta_i E)}$$

$$\Delta_i E = E_{\sigma_i = DOWN} - E_{\sigma_i = UP} = 2 \left(\sum_{j \in M_i} J_{ij} \sigma_j + h_i \right) \tag{8}$$

where $\Delta_i E$ represents the difference in the energy between configurations where $\sigma_i = -1$ and those where $\sigma_i = +1$. In other words, the probability of unit i flipping to +1 is proportional to the degree to which the global energy E would decrease as a result of the flip. If units are updated sequentially (also known as 'asynchronous updates'), then given sufficient time Glauber dynamics are guaranteed to sample from the stationary density in Eq. (6) [12].

The probability of agent i taking action $\sigma_i = +1$ is given by the action sampling rule in Eq. (5) of the previous section. We can thus write the probability of taking a particular action in terms of the posterior over latent states z, by plugging in the posterior belief over $z = +1$ (given in Eq. (2)) into Eq. (5):

$$P(\sigma_i = +1) = \frac{1}{1 + \exp(-2(\zeta_i + \gamma_i \Delta_i \sigma))} \tag{9}$$

where we now index $\zeta, \gamma, \Delta\sigma$ by i to indicate that these are the generative model parameters and observations of agent i. The identical forms shared by Eqs. (8) and (9) allow us to directly relate the parameters of individual generative models to the local energy difference $\Delta_i E$ and the global energy of the system.

$$\Delta_i E \propto J \sum_{j \in M_i} \sigma_j + h_i = \gamma \Delta_i \sigma + \zeta_i \implies J = \gamma, \ h_i = \zeta_i$$

where we assume all agents share the same likelihood precision $\gamma_i = \gamma^* : \forall i$, which is equivalent to forcing all couplings to be identical $J_{ij} = J : \forall i, j$. Individual active inference agents in this multi-agent dynamic thus behave as if they are sampling spin states in order to minimize a global energy function defined over spin configurations, which in the case of spin glass systems like the Ising model, can be computed using local observations (the spins of one's neighbours) and model parameters γ, ζ. Going the other direction, one can sample from the stationary distribution of spin glass system by simulating a collective of active inference agents with an appropriately parameterized generative model.

However, the equivalence between Glauber sampling from the stationary distribution of an Ising model and collective active inference breaks down when agents update their actions in parallel or synchronously, rather than asynchronously. In particular, under parallel action updates, the system samples from a stationary distribution with a different energy function than that from Eq. (7). See Appendix C for derivations on the relationship between the schedule of action updates (synchronous vs. asynchronous) and the resulting stationary density of the system.

4   Equivalence to Inference in Boltzmann Machines

Connecting the collective dynamics of these specialized active inference agents to inference in Boltzmann machines is straightforward. We now equip each agent's generative model with a *vector* of sensory precisions $\tilde{\gamma} = \{\gamma_j : j \in M\}$ that act as different reliabilities assigned to different neighbours' spin observations. The new factorized likelihood can be written:

$$P(\tilde{\sigma}|z; \tilde{\gamma}) = \prod_{j \in M} \frac{\exp(\gamma_j \sigma_j z)}{2 \cosh(\gamma_j z)} \tag{10}$$

We can then write the posterior over z as a function of observations and generative model parameters $\{\tilde{\sigma}, \zeta, \tilde{\gamma}\}$. By fixing z to the value $+1$, we obtain again a logistic expression for the posterior probability $P(z = +1|\tilde{\sigma}, \tilde{\gamma}, \zeta)$ that is nearly identical to the original Eq. (2):

$$P(z = +1|\tilde{\sigma}, \tilde{\gamma}, \zeta) = \frac{1}{1 + \exp(-2(\zeta + \sum_{j \in M} \gamma_j \sigma_j))} \tag{11}$$

A Boltzmann machine is a special variant of a spin glass model, defined by a global energy that in turn defines the stationary probability assigned to each of the system's configurations. The Boltzmann energy E_B, as with the classical spin glass energy, is defined over configurations of the system's binary units (also known as nodes or neurons) $\tilde{x} = \{x_i\}_{i=1}^{N}$.

In the context of inference, it is common to partition the system's units into 'visible units' and 'hidden units', $\tilde{x} = \{\tilde{v}, \tilde{h}\}$, with the following energy function:

$$E_B(\tilde{v}, \tilde{h}) = -\frac{1}{2}(\tilde{v}^{\top} \mathbf{W}_{vv} \tilde{v} + \tilde{h}^{\top} \mathbf{W}_{hh} \tilde{h} + \tilde{v}^{\top} \mathbf{W}_{vh} \tilde{h}) - \tilde{\theta}_v^{\top} \tilde{v} - \tilde{\theta}_h^{\top} \tilde{h} \tag{12}$$

where $\mathbf{W}_{vv}, \mathbf{W}_{hh}, \mathbf{W}_{vh}$ are weight matrices with symmetric couplings between units with no 'self-edges' ($W_{ii} = 0 : \forall i$) that mediate dependencies both within and between the two subsets of units $\tilde{v}, \tilde{h}$; and $\tilde{\theta}_v, \tilde{\theta}_h$ are vectors of unit-specific biases or thresholds. The Bayesian inference interpretation of Boltzmann machines considers the conditional probability distribution over $\tilde{h}$, given some fixed values of $\tilde{v}$. The 'clamping' of visible nodes to some data vector $\tilde{v} = \tilde{d}$ can simply be absorbed into the biases $\tilde{\theta}_v$, such that samples from the posterior $P(\tilde{h}|\tilde{v} = \tilde{d})$ are analogous to sampling from the joint distribution $P(\tilde{h}, \tilde{v})$ where the biases of the visible nodes are adjusted to reflect this clamping. Sampling from this model can be achieved with Glauber dynamics, since the model is a spin glass system with heterogeneous (but symmetric) couplings. We can therefore write the single unit 'ON' probability as follows, now in terms of weights $\mathbf{W}$ and thresholds $\tilde{\theta}$:

$$P(x_i = +1) = \frac{1}{1 + \exp(-\Delta_i E_B)} = \frac{1}{1 + \exp(-\sum_j W_{ij} x_j - \theta_i)} \tag{13}$$

where the interaction term that comprises the local energy difference $\Delta_i E_B$ is equivalent to a dot-product between the ith row of $\mathbf{W}$ and vector of activities $\tilde{x}$. It is thus straightforward to relate the weights and biases of a Boltzmann machine to the sensory and prior precisions of each agent's generative model. In particular, the weight connecting unit j to unit i in a Boltzmann machine W_{ij} is linearly related to the precision that agent i associates to spin observations coming from agent j: $W_{ij} = 2\gamma_{(i,j)}$ where the subscript (i, j) refers to agent i's likelihood model over observations emitted by agent j. If agents i and j are not connected, then $W_{ij} = \gamma_{(i,j)} = 0$. The bias of the ith unit is also straightforwardly related to agent i's prior precision via $\theta_i = 2\zeta_i$.

We have seen how sampling from the posterior distribution of a Boltzmann machine with fixed visible nodes $\tilde{v}$ is equivalent to the collective dynamics of a specific multi-agent active inference scheme. We have thus shown a carefully-constructed system, in which a form of sampling-based Bayesian inference at one scale emerges from a process of collective active inference at a lower scale.

5   Discussion

Although the equivalences we have shown are exact, there are numerous assumptions that, when broken, violate the equivalence between the multi-agent dynamics defined at the lower level, and the higher-level sampling dynamics of the spin glass system.

The energy-based models we have studied (Ising models, Boltzmann machines) are all *undirected* graphical models: this means that the global energy function is defined by symmetric interaction terms across pairs of spins: $J_{ij} = J_{ji}$[1]. In order to meet this requirement at the level of the individual agents, one must force the precisions that a pair of agents assign to one another to be identical: $\gamma_{(i,j)} = \gamma_{(j,i)}$. This constraint also underpins the equilibrium nature of classical spin glass systems where detailed balance conditions are met, i.e., the system is in thermodynamic equilibrium. In natural complex systems (especially biological ones), these detailed balance conditions are often broken, and the systems operate far from equilibrium [1,2,8,21–24,34]. This may be manifest in the case of realistic multi-agent dynamics in the form of asymmetric beliefs about reliability of social signals that agents assign to one another [3].

Another fragility of the multiscale equivalence is the structure of the individual generative model, which relies on very specific assumptions about how hidden states z relate to observations. Without this generative model at the single-agent level, there is no equivalence between the collective active inference process and a spin glass model—the model's dynamics could become more complex (and potentially analytically intractable), because the posterior update is not guaranteed to be a simple logistic function of the sum of neighbouring spins. This could be easily shown by changing the single agent's likelihood model to include a separate latent variable for each neighbour's spin state[2], or if the total likelihood over neighbouring spin observations did not factorize into a product of neighbour-specific likelihoods, but had some more complex statistical structure.

Finally, the convergence of Glauber dynamics to samples from the joint density over spin configurations depends on the temporal schedule of action updating; namely, spins have to be updated sequentially or asynchronously, rather than in parallel (see Appendix C for details) in order to guarantee sampling from the stationary distribution in Eq. (6). If agents act in parallel, then the stationary distribution of spin states is different than that given by the classical spin glass Hamiltonian. In other words, depending on the relative timescales of collective action, agents either will or will not minimize the global energy function that their actions appear to local minimize (i.e. actions that minimize the local energy difference $\Delta_i E$).

---

[1] $\mathbf{W} = \mathbf{W}^\top$ for the Boltzmann machine, respectively.

[2] This is analogous to the approach taken in [3], where each agent had beliefs about the belief state of each of its neighbours.

6   Conclusion

In this work we demonstrate an exact equivalence between a collective active inference process and sampling from states a spin glass system under detailed balance. Furthermore, we connect the system's collective dynamics to sampling-based inference in energy-based models like Boltzmann machines. In particular, when we constrain certain agents in the network to be 'visible nodes' and fix their actions, then the whole system samples from the posterior distribution over spin configurations, given the actions of the 'visible' agents. Despite these exact relationships, we also note the fragility of the equivalence, which relies on very particular assumptions. These include the symmetry of the precisions that pairs of agents assign to each other (i.e., couplings between spins), the temporal scheduling of the action updates, and the specific generative model used by the agents. It remains to be seen, whether when these assumptions are broken, an inferential or 'agentive' interpretation still obtains at higher scales, and if so, whether the form of the 'collective' generative model can be analytically related to the individual generative models as it was in the present, equilibrium case.

Our results have important implications for the overall agenda of multiscale active inference, and the quest to uncover the quantitative relationship between generative models operating at distinct scales in complex systems. In the system presented in the current work, we show that active inference agents may collectively achieve sampling-based inference at a distinct, higher level under particular conditions. Despite the apparent consistency at the two scales, our result actually conflicts with claims made in the multiscale active inference literature, that posits that systems hierarchically decompose into nested levels of active inference agents [6,17,25,27,28]—in other words, that systems are inherently active inference processes 'all the way down.' Note that in our system, there are only active inference agents operating at the lower level—the higher level is not an active inference agent, but is better described as a passive agent that performs hidden state-estimation or inference by sampling from a posterior belief over spin configurations. The agenda of the present work also resonates with ongoing research into the necessary and sufficient conditions for generic complex systems to be considered 'agentive' or exhibit inferential capacities [18–20,30].

Our results suggest that multiscale inference interpretations of complex systems do not necessarily emerge in any system. We nevertheless hope that the simple equilibrium case we presented here may serve as a launching pad for future studies into whether inference interpretations can be rescued at the higher scale in cases when the fragile assumptions at the single-agent level are broken.

A   Bayesian Inference for a Single Agent

In this appendix we derive the exact Bayesian inference update for the posterior over the latent state z, taking the perspective of a single agent.

We begin by rehearsing the component likelihood and prior distributions of the generative model in more detail.

A.1 Likelihood

The likelihood model relates the hidden state z to the observed spin state of a particular neighbour σ_j as an exponential distribution parameterized by a sensory precision parameter γ:

$$P(\sigma_j|z;\gamma) = \frac{\exp(\gamma\sigma_j z)}{2\cosh(\gamma z)} \tag{A.1}$$

The sign and magnitude of γ determines the nature of the expected mapping between hidden states z and neighbouring spin observations σ_j. For $\gamma > 0$, then the observed spin is expected to reflect the latent state z, and with $\gamma < 0$, then the observed spin is expected to be opposite to the latent state z. The magnitude of γ then determines how deterministic this mapping is.

Equation (A.1) can alternatively be seen as a collection of two conditional Bernoulli distributions over σ_j, one for each setting of z. This can be visualized as a symmetric matrix mapping from the two settings of z (the columns, corresponding to $z = -1, +1$) to the values of σ_j (the rows $\sigma_j = -1, +1$):

$$P(\sigma_j|z;\gamma) = \begin{bmatrix} \frac{1}{1+\exp(-2\gamma)} & \frac{1}{1+\exp(2\gamma)} \\ \frac{1}{1+\exp(2\gamma)} & \frac{1}{1+\exp(-2\gamma)} \end{bmatrix} \tag{A.2}$$

where this mapping approaches the identity matrix as $\gamma \to \infty$.

Now we can move onto writing down the likelihood over the observed spins of multiple neighbours: $\tilde{\sigma} = \{\sigma_j : j \in M\}$ where M denotes the set of the focal agent's neighbours. We build in a *conditional independence* assumption into the focal agent's generative model, whereby the full likelihood model over all observed spins factorizes across the agent's neighbours. This means we can write the likelihood as a product of the single-neighbour likelihoods shown in Eq. (A.1):

$$P(\tilde{\sigma}|z;\gamma) = \prod_{j\in M} \frac{\exp(\gamma\sigma_j z)}{2\cosh(\gamma z)}$$

$$= \exp\left(z\gamma\sum_{j\in M}\sigma_j - K\log(2\cosh(\gamma z))\right) \tag{A.3}$$

where K is the number of the focal agent's neighbours (i.e. the size of the set M). We can easily generalize this likelihood to heterogeneous precisions by instead parameterizing it with a precision vector $\tilde{\gamma} = \{\gamma_j : j \in M\}$ that assigns a different precision to observations coming from each of the focal agent's neighbours:

$$P(\tilde{\sigma}|z;\tilde{\gamma}) = \prod_{j\in M} \frac{\exp(\gamma_j\sigma_j z)}{2\cosh(\gamma_j z)}$$

$$= \exp\left(z\sum_{j\in M}\gamma_j\sigma_j - \sum_{j\in M}\log(2\cosh(\gamma_j z))\right) \tag{A.4}$$

A.2 Prior

We parameterize the focal agent's prior beliefs about the latent spin state z as a simple Bernoulli distribution, and similarly to the likelihood model, we will express it as an exponential function parameterized by a 'prior precision' parameter ζ:

$$P(z;\zeta) = \frac{\exp(\zeta z)}{2\cosh(\zeta z)} = \exp(\zeta z - \log(2\cosh(\zeta)))$$

As with the sensory precision γ, the prior precision also scales the strength of the focal agent's prior belief that the spin state z is $+1$.[3]

A.3 Bayesian Inference of Hidden States

Now we ask the question: how would a focal agent (i.e., the agent that occupies a single lattice site) optimally compute a belief over z, that is most consistent with a set of observed spins $\tilde{\sigma}$? This is a problem of Bayesian inference, which can be expressed as calculation of the posterior distribution over z via Bayes Rule:

$$P(z|\tilde{\sigma};\gamma,\zeta) = \frac{P(\tilde{\sigma},z;\gamma,\zeta)}{P(\tilde{\sigma};\gamma,\zeta)} \tag{A.5}$$

Since we are dealing with a conjugate exponential model[4], we can derive an analytic form for the posterior: $P(z|\tilde{\sigma},\gamma,\zeta)$:

$$P(z|\tilde{\sigma};\gamma,\zeta) = \frac{\exp\left(z(\zeta + \gamma\sum_j \sigma_j)\right)}{2\cosh\left(\zeta + \gamma\sum_j \sigma_j\right)} \tag{A.6}$$

where the sum over neighbouring spins j only includes the neighbours of the focal agent, i.e., $\sum_{j\in M}\sigma_j$. If we fix the hidden state z to a particular value (e.g. $z = +1$), then we arrive at a simple expression for the posterior probability that the hidden spin state is in the 'UP' state, given the observations and the generative model parameters γ,ζ. This probability reduces to a logistic or sigmoid function of sensory input, which is simply the sum of neighbouring spin values $\Delta\sigma = \sum_j \sigma_j$. This can also be seen as the 'spin difference', or the number of neighbouring spins that are in the 'UP' position, minus those that are in the 'DOWN' position. The steepness and horizontal shift of this logistic function are intuitively given by likelihood and prior precisions, respectively:

---

[3] Note that $\cosh(\zeta z)$ can be re-written $\cosh(\zeta)$ when $z \in \{-1,+1\}$ due to the symmetry of the hyperbolic cosine function around the origin.

[4] The Bernoulli prior is conjugate to the likelihood model, which can also be described of as a set of conditional Bernoulli distributions.

$$P(z = +1|\tilde{\sigma}, \gamma, \zeta) = \frac{\exp(\zeta + \gamma\Delta\sigma)}{\exp(\zeta + \gamma\Delta\sigma) + \exp(-(\zeta + \gamma\Delta\sigma))}$$

$$= \left(1 + \frac{\exp(-(\zeta + \gamma\Delta\sigma))}{\exp(\zeta + \gamma\Delta\sigma)}\right)^{-1}$$

$$= \frac{1}{1 + \exp(-2(\zeta + \gamma\Delta\sigma))} \tag{A.7}$$

The denominator in the first line of (A.7) follows from the identity $\cosh(x) = \frac{1}{2}(\exp(x) + \exp(-x))$.

B   Active Inference Derivations

In this section we provide the additional derivations needed to equip each agent with the ability to infer a posterior over control states and sample from this posterior to generate actions. This achieved through the framework of *active inference*.

Active inference casts the selection of control states or actions as an inference problem, whereby actions u are sampled or drawn from posterior belief about controllable hidden states. The posterior over actions is computed as the softmax transform of a quantity called the *expected free energy* [9]. This is the critical objective function for actions that enables active inference agents to plan actions into the future, since the expected free energy scores the utility of the anticipated consequences of actions.

B.1   Predictive Generative Model

We begin by writing a so-called 'predictive' generative model that crucially includes probability distributions over the agent's own control states $u \in \{-1, +1\}$ and how those control states relate to future (anticipated) observations. In other words, we consider a generative model over two timesteps: the current timestep t and one timestep in the future, $t + 1$. This will endow our agents with a shallow form of 'planning', where they choose actions in order to maximize some (pseudo-) reward function defined with respect to expected outcomes. This can be expressed as follows:

$$P(\tilde{\sigma}_t, z_t, u_t, \mathcal{O}_{t+1}; \gamma, \zeta) = \tilde{P}(\mathcal{O}_{t+1}|z_t, u_t, \tilde{\sigma}_t)P(\tilde{\sigma}_t, z_t, u_t; \gamma, \zeta) \tag{B.8}$$

where the generative model at the second timestep $\tilde{P}(\mathcal{O}_{t+1}|z_t, u_t, \tilde{\sigma}_t)$ we hereafter refer to as the 'predictive' generative model, defined over a single future timestep.

Active inference consists in sampling a belief from the posterior distribution over control states u—this sampled control state or action is then fixed to be the spin state of the agent under consideration. Thus the action of one agent is fed in as the observations for those spin sites that it's connected to. In order to imbue active inference agents with a sense of goal-directedness or purpose, we encode a prior distribution over actions $P(u)$ that is proportional to the negative of the expected free energy, via the softmax relationship:

$$P(u) = \frac{\exp(-\mathbf{G}(u))}{\sum_u \exp(-\mathbf{G}(u))} \tag{B.9}$$

Crucially, the expected free energy of an action $\mathbf{G}$ is a function of outcomes *expected* under a particular control state u, where beliefs about future outcomes are 'biased' by prior beliefs about encountering particular states of affairs. In order to optimistically 'bend' these future beliefs towards certain outcomes, and thus make some actions more probable than others, we supplement the predictive generative model $\tilde{P}$ with a binary 'optimality' variable $\mathcal{O} \pm 1$ that the agent has an inherent belief that it will observe. This is encoded via a 'goal prior' or preference vector, which is a Bernoulli distribution over seeing a particular value of $\mathcal{O}$ with some precision parameter ω:

$$\tilde{P}(\mathcal{O}_{t+1}; \omega) = \frac{\exp(\omega \mathcal{O})}{2 \cosh(\omega \mathcal{O})} \tag{B.10}$$

Hereafter we assume an infinitely high precision, i.e. $\omega \to \infty$. This renders the preference an 'all-or-nothing' distribution over observing the optimality variable being in the 'positive' state $\mathcal{O} = +1$:

$$= \begin{bmatrix} \tilde{P}(\mathcal{O}_{t+1} = -1) \\ \tilde{P}(\mathcal{O}_{t+1} = +1) \end{bmatrix} = \begin{bmatrix} 0.0 \\ 1.0 \end{bmatrix} \tag{B.11}$$

To allow an agent the ability to predict the relationship between their actions and expected observations, it's important to include an additional likelihood distribution, what we might call the 'forward model' of actions $P(\mathcal{O}_{t+1}|z_t, u_t; \xi)$. This additional likelihood encodes the focal agent's assumptions about the relationship between hidden states, actions, and the (expected) optimality variable. By encoding a deterministic conditional dependence relationship into this likelihood, we motivate the agent (via the expected free energy) to honestly signal its own estimate of the hidden state via its spin action u. To achieve this, we explicitly design this likelihood to have the following structure, wherein the optimality variable is only expected to take its 'desired value' of $\mathcal{O} = +1$ when $z = u$. This can be written as a set of conditional Bernoulli distributions over $\mathcal{O}$, and each of which jointly depends on z and u and is parameterized by a (infinitely high) precision ξ:

$$P(\mathcal{O}_{t+1}|z_t, u_t; \xi) = \frac{\exp(\xi \mathcal{O}_{t+1} z_t u_t)}{2 \cosh(\xi z_t u_t)} \tag{B.12}$$

When we assume $\xi \to \infty$, then we arrive at a form for this likelihood which can be alternatively expressed as a set of Bernoulli distributions that conjunctively depend on z and u, and can be visualized as follows:

$$P(\mathcal{O}_{t+1}|z_t, u_t = -1) = \begin{bmatrix} 0 & 1 \\ 1 & 0 \end{bmatrix}$$

$$P(\mathcal{O}_{t+1}|z_t, u_t = +1) = \begin{bmatrix} 1 & 0 \\ 0 & 1 \end{bmatrix} \tag{B.13}$$

where the columns of the matrices above correspond to settings of $z \in \{-1, +1\}$. Therefore, the agent only expects to see $\mathcal{O} = +1$ (the desired outcome) when the value of the hidden state and the value of the control variable are equal, i.e. $z = u$; otherwise $\mathcal{O} = -1$ is expected. For the purposes of the present study, we assume both the optimality prior $\tilde{P}(\mathcal{O}; \omega)$ and the optimality variable likelihood $P(\mathcal{O}|z, u; \xi)$ are parameterized by infinitely high precisions $\omega = \xi = \infty$, and hereafter will exclude them when referring to these distributions for notational convenience.

Having specified these addition priors and likelihoods, we can write down the new (predictive) generative model as follows:

$$\tilde{P}(\mathcal{O}_{t+1}, u_t, z_t) = P(\mathcal{O}_{t+1}|z_t, u_t)P(u_t)\tilde{P}(\mathcal{O}_{t+1})P(z_t) \tag{B.14}$$

B.2 Active Inference

Under active inference, both state estimation and action are consequences of the optimization of an approximate posterior belief over hidden states and actions $Q(z, u; \phi)$. This approximate posterior is optimized in order to minimize a variational free energy (or alternatively maximize an evidence lower bound). This is the critical concept for a type of approximate Bayesian inference known as variational Bayesian inference [4]. This can be described as finding the optimal set of variational parameters ϕ that minimizes the following quantity:

$$\phi^* = \arg\min_{\phi} \mathcal{F}$$
$$= \mathbb{E}_Q[\ln Q(z_t, u_t; \phi) - \ln \tilde{P}(\tilde{o}_t, z_t, u_t, \mathcal{O}_{t+1}; \gamma, \zeta)] \tag{B.15}$$

In practice, because of the factorization of the generative model into a generative model of the current and future timesteps, we can split state-estimation and action inference into two separate optimization procedures. To do this we also need to factorize the posterior as follows:

$$Q(z, u; \phi) = Q(z; \phi_z)Q(u; \phi_u) \tag{B.16}$$

where we have also separated the variational parameters $\phi = \{\phi_z, \phi_u\}$ into those that parameterize the belief about hidden states ϕ_z, and those that parameterize the belief about actions ϕ_u.

When considering state-estimation (i.e. optimization of $Q(z_t; \phi_z)$), we only have to consider the generative model of the current timestep $P(\tilde{o}_t, z_t; \gamma, \zeta)$. The optimal posterior parameters ϕ_z^* are found as the minimum of the variational free energy, re-written using only those terms that depend on ϕ_z:

$$\phi_z^* = \arg\min_{\phi_z} \mathcal{F}(\phi_z)$$

$$\mathcal{F}(\phi_z) = \mathbb{E}_{Q(z_t;\phi_z)}[\ln Q(z_t;\phi_z) - \ln P(\tilde{\sigma}_t, z_t; \gamma, \zeta)] \qquad (B.17)$$

To solve this, we also need to decide on a parameterization of the approximate posterior over hidden states z_t. We parameterize $Q(z_t;\phi_z)$ as a Bernoulli distribution with parameter ϕ_z, that can be interpreted as the posterior probability that z_t is in the 'UP' ($+1$) state:

$$Q(z_t;\phi_z) = (1-\phi_z)^{1-\frac{z_t+1}{2}} \phi_z^{\frac{z_t+1}{2}} \qquad (B.18)$$

By minimizing the variational free energy with respect to ϕ_z, we can obtain an expression for the optimal posterior $Q(z;\phi_z^*)$ that sits at the variational free energy minimum. Due to the exponential and conjugate form of the generative model, $Q(z_t;\phi_z)$ is the exact posterior and thus variational inference reduces to exact Bayesian inference. This means we can simply re-use the posterior update equation of Eq. (A.7) to yield an analytic expression for ϕ_z^*:

$$\phi_z^* = \frac{1}{1 + \exp\left(-2(\zeta + \gamma\Delta\sigma)\right)} \qquad (B.19)$$

When considering inference of actions, we now consider the generative model of the future timestep, which crucially depends on the current control state u_t and the optimality variable $\mathcal{O}_{t+1}$. We can then write the variational problem as finding the setting of ϕ_u that minimizes the variational free energy, now re-written in terms of its dependence on ϕ_u:

$$\phi_u^* = \arg\min_{\phi_u} \mathcal{F}(\phi_u)$$

$$\mathcal{F}(\phi_u) = \mathbb{E}_{Q(u_t;\phi_u)}[\ln Q(u_t;\phi_u) - \ln \tilde{P}(\mathcal{O}_{t+1}, u_t, z_t)] \qquad (B.20)$$

As we did for the posterior over hidden states, we need to decide on a parameterization for the posterior over actions $Q(u_t;\phi_u)$; we also parameterize this as a Bernoulli distribution with parameter ϕ_u that represents the probability of taking the 'UP' ($+1$) action:

$$Q(u_t;\phi_u) = (1-\phi_z)^{1-\frac{u_t+1}{2}} \phi_z^{\frac{u_t+1}{2}} \qquad (B.21)$$

From Eq. (B.20) it follows that the optimal ϕ_u is that which minimizes the Kullback-Leibler divergence between the approximate posterior $Q(u_t;\phi_u)$ and the prior $P(u_t)$, which is a softmax function of the expected free energy of actions $\mathbf{G}(u_t)$. In this particular generative model, the expected free energy can be written as a single term that scores the 'expected utility' of each action [7,9]:

$$\mathbf{G}(u_t) = -\mathbb{E}_{Q(\mathcal{O}_{t+1}|u_t)}[\ln \tilde{P}(\mathcal{O}_{t+1})] \qquad (B.22)$$

To compute this, we need to compute the 'variational marginal' over $\mathcal{O}_{t+1}$, denoted $Q(\mathcal{O}_{t+1}|u_t)$:

$$Q(\mathcal{O}_{t+1}|u_t) = \mathbb{E}_{Q(z_t;\phi_z^*)}[P(\mathcal{O}_{t+1}|z_t, u_t)] \tag{B.23}$$

We can simplify the expression for $Q(\mathcal{O}_{t+1}|u_t)$ when we take advantage of the Bernoulli-parameterization of the posterior over hidden states $Q(z; \phi_z^*)$. This allows us to then write the variational marginals, conditioned on different actions as a matrix, with one column for each setting of u_t:

$$Q(\mathcal{O}_{t+1}|u_t) = \begin{bmatrix} \phi_z^* & 1 - \phi_z^* \\ 1 - \phi_z^* & \phi_z^* \end{bmatrix} \tag{B.24}$$

The expected utility (and thus the negative expected free energy) is then computed as the dot-product of each column of the matrix expressed in Eq. (B.24) with the log of the prior preferences $\tilde{P}(\mathcal{O}_{t+1})$:

$$\mathbb{E}_{Q(\mathcal{O}_{t+1}|u_t)}[\ln \tilde{P}(\mathcal{O}_{t+1})] = \begin{bmatrix} -\infty\phi_z^* \\ -\infty(1 - \phi_z^*) \end{bmatrix}$$
$$\implies \mathbf{G}(u_t) = \begin{bmatrix} \infty\phi_z^* \\ \infty(1 - \phi_z^*) \end{bmatrix} \tag{B.25}$$

Because the probability of an action is proportional to its negative expected free energy, this allows us to write the Bernoulli parameter ϕ_u^* of the posterior over actions directly in terms of the parameter of the state posterior

$$\phi_u^* = \frac{1}{1 + \exp(\beta(\infty(1 - \phi_z^*)))}$$
$$= \frac{1}{1 + C\exp(-\phi_z^*)))} \tag{B.26}$$

The inverse temperature parameter β is an arbitrary re-scaling factor that can be used to linearize the sigmoid function in (B.26) over the range $[0, 1]$ such that

$$\phi_u^* \approx \phi_z^* \tag{B.27}$$

Note that the equivalence relation in Eq. (B.27) is only possible due to the infinite precisions ω and ξ of the likelihood and prior distributions over the 'optimality' variable $P(\mathcal{O}_{t+1}|u_t, z_t)$ and $\tilde{P}(\mathcal{O}_{t+1})$, and from an appropriately re-scaled β parameter that linearizes the sigmoid relationship in Eq. (B.26).

B.3   Action Sampling as Probability Matching

Now that we have an expression for the parameter ϕ_u^* of the posterior over control states $Q(u_t; \phi_u^*)$, an agent can generate a spin state by simply sampling from this posterior over actions:

$$\sigma \sim Q(u_t; \phi_u^*)$$
$$\sim Q(z_t; \phi_z^*) \triangleq P(z_t | \tilde{\sigma}; \gamma, \zeta) \tag{B.28}$$

In short, each agent samples its spin state from a posterior belief over the state of the latent variable z_t, rendering their action-selection a type of probability matching [11,29,31], whereby actions (whether to spin 'UP' or 'DOWN') are proportional to the probability they are assigned in the agent's posterior belief. Each agent's sampled spin state also serves as an observation (σ_j for some j) for the other agents that the focal agent is a neighbour of. This collective active inference scheme corresponds to a particular form of sampling from the stationary distribution of a spin glass model known as Glauber dynamics [12]. Crucially, however, the temporal scheduling of the action-updating across the group determines which stationary distribution the system samples from. We explore this distinction in the next section.

C   Temporal Scheduling of Action Sampling

In this appendix we examine how the stationary distribution from which the collective active inference system samples depends on the order in which actions are updated across all agents in the network. First, we consider the case of synchronous action updates (all agents update their actions in parallel and only observe the- spin states of their neighbours from the last timestep), and show how this system samples from a different stationary distribution than the one defined by the standard Ising energy provided in Eq. (7). We then derive the more 'classical' case of asynchronous updates, where agents update their spins one at a time, and show how in this case the system converges to the standard statioanry distribution of the Ising model. This Appendix thus explains one of the 'fragilities' mentioned in the main text, that threaten the unique equivalence between local active inference dynamics and a unique interpretation at the global level in terms of inference.

We denote some agent's spin using σ_i and its set of neighbours as M_i. The local sum of spins or spin difference $\sum_{j \in M} \sigma_j$ for agent i we denote $\Delta_i \sigma = \sum_{j \in M_i} \sigma_j$.

C.1   Synchronous Updates

To derive the stationary distribution in case of synchronous updates, we can take advantage of the following detailed balance relation, which obtains in the case of systems at thermodynamic equilibrium:

$$\frac{P(\tilde{\sigma})}{P(\tilde{\sigma}')} = \frac{P(\tilde{\sigma}|\tilde{\sigma}')}{P(\tilde{\sigma}'|\tilde{\sigma})}$$
$$\implies P(\tilde{\sigma}) = \frac{P(\tilde{\sigma}|\tilde{\sigma}')P(\tilde{\sigma}')}{P(\tilde{\sigma}'|\tilde{\sigma})} \tag{C.29}$$

where $\tilde{\sigma}$ and $\tilde{\sigma}'$ are spin configurations at two adjacent times τ and $\tau + 1$. In the case of synchronous updates (all spins are sampled simultaneously, given the spins at the last timestep), then the spin action of each agent σ'_i at time $\tau + 1$ is conditionally independent of all other spins, given the vector of spins $\tilde{\sigma}$ at the previous timestep τ. We can therefore expand the 'forward' transition distribution $P(\tilde{\sigma}'|\tilde{\sigma})$ as a product over the action posteriors of each agent:

$$
\begin{aligned}
P(\tilde{\sigma}'|\tilde{\sigma}) &= P(\sigma_1|\tilde{\sigma})P(\sigma_1|\tilde{\sigma})...P(\sigma_N|\tilde{\sigma}) \\
&= \prod_i Q(u_t; \phi^*_{u,i}) \\
&= \prod_i \frac{\exp\left(\sigma'_i\left(\zeta + \gamma \sum_{j\in M_i}\sigma_j\right)\right)}{2\cosh\left(\zeta + \gamma \sum_{j\in M_i}\sigma_j\right)} \\
&= \exp\left(\sum_i \sigma'_i(\zeta + \gamma \sum_{j\in M_i}\sigma_j) - \sum_i \log\left(2\cosh(\zeta + \gamma \sum_{j\in M_i}\sigma_j)\right)\right)
\end{aligned}
$$
(C.30)

Note we have replaced each latent variable in the posterior z with the agent's own spin state σ_i, because there is a one-to-one mapping between the posterior over z_t and the posterior over actions σ_i.

The reverse transition distribution, yielding the probability of transitioning from configuration $\tilde{\sigma}' \rightarrow \tilde{\sigma}$ is the same expression as for the forward transition, except that σ'_i and σ_i are swapped:

$$
P(\tilde{\sigma}|\tilde{\sigma}') = \exp\left(\sum_i \sigma_i(\zeta + \gamma \sum_{j\in M_i}\sigma'_j) - \sum_i \log\left(2\cosh(\zeta + \gamma \sum_{j\in M_i}\sigma'_j)\right)\right)
$$
(C.31)

The detailed balance equation in (C.29) then tells us that the stationary probability distribution over $\tilde{\sigma}$ is proportional to the ratio of the backwards transition to the forwards transition:

$$
\begin{aligned}
\frac{P(\tilde{\sigma})}{P(\tilde{\sigma}')} &= \frac{\exp\left(\sum_i \sigma_i(\zeta + \gamma \sum_{j\in M_i}\sigma'_j) - \sum_i \log\left(2\cosh(\zeta + \gamma \sum_{j\in M_i}\sigma'_j)\right)\right)}{\exp\left(\sum_i \sigma'_i(\zeta + \gamma \sum_{j\in M_i}\sigma_j) - \sum_i \log\left(2\cosh(\zeta + \gamma \sum_{j\in M_i}\sigma_j)\right)\right)} \\
&= \frac{\exp\left(\zeta\sum_i \sigma_i + \gamma\sum_{\langle i,j\rangle}\sigma_i\sigma'_j\right)\exp\left(-\sum_i \log\left(2\cosh(\zeta + \gamma \sum_{j\in M_i}\sigma'_j)\right)\right)}{\exp\left(\zeta\sum_i \sigma'_i + \gamma\sum_{\langle i,j\rangle}\sigma'_i\sigma_j\right)\exp\left(-\sum_i \log\left(2\cosh(\zeta + \gamma \sum_{j\in M_i}\sigma_j)\right)\right)} \\
&= \frac{\exp\left(\zeta\sum_i \sigma_i + \sum_i \log\left(2\cosh(\zeta + \gamma \sum_{j\in M_i}\sigma_j)\right)\right)}{\exp\left(\zeta\sum_i \sigma'_i + \sum_i \log\left(2\cosh(\zeta + \gamma \sum_{j\in M_i}\sigma'_j)\right)\right)}
\end{aligned}
$$
(C.32)

Therefore, we can write down the stationary distribution in the case of synchronous updates as an exponential term normalized by a partition function:

$$P(\tilde{\sigma}) = Z^{-1} \exp\left(\zeta \sum_i \sigma_i + \sum_i \log\left(2\cosh(\zeta + \gamma \sum_{j \in M_i} \sigma_j)\right)\right)$$

$$Z = \sum_{\tilde{\sigma}} \exp\left(\zeta \sum_i \sigma_i + \sum_i \log\left(2\cosh(\zeta + \gamma \sum_{j \in M_i} \sigma_j)\right)\right) \tag{C.33}$$

Note that the action update for an individual agent can still be written in terms of the local energy difference $\Delta_i E$, where the energy is defined using the standard Hamiltonian function given by Eq. (7) in the main text. However, due to the temporal sampling of each agent's action with respect to the others, the system collectively sample from a system with a different energy function and Gibbs measure, given by Eq. (C.33). This energy function is therefore nonlinear and can be written:

$$E_{sync}(\tilde{\sigma}) = -\zeta \sum_i \sigma - \sum_i \log(2\cosh(\zeta + \gamma \sum_{j \in M_i} \sigma_j)) \tag{C.34}$$

C.2   Asynchronous Updates

Now we treat the case where agents update their agents one-by-one or asynchronously. This means that at each timestep only one agent is updated, and that particular agent uses the spin states of all the other agents at the last timestep as inputs for its posterior inference.

We can write down the forward transition as follows, using the notation $\sigma_{\backslash i}$ to denote all the spins except for σ_i:

$$p(\sigma_i', \tilde{\sigma}_{\backslash i} | \tilde{\sigma}) = \frac{\exp(\sigma_i'(\zeta + \gamma \sum_{j \in M_i} \sigma_j))}{2\cosh(\zeta + \gamma \sum_{j \in M_i} \sigma_j)} \tag{C.35}$$

which indicates that only agent i is updated at the current timestep. The detailed balance condition implies that

$$p(\sigma_i', \tilde{\sigma}_{\backslash i} | \tilde{\sigma}) p(\tilde{\sigma}) = p(\tilde{\sigma} | \sigma_i', \tilde{\sigma}_{\backslash i}) p(\sigma_i', \tilde{\sigma}_{\backslash i}) \tag{C.36}$$

Then

$$\frac{p(\tilde{\sigma})}{p(\sigma_i', \tilde{\sigma}_{\backslash i})} = \frac{p(\tilde{\sigma} | \sigma_i', \tilde{\sigma}_{\backslash i})}{p(\sigma_i', \tilde{\sigma}_{\backslash i} | \tilde{\sigma})} = \frac{\exp(\sigma_i(\zeta + \gamma \sum_{j \in M_i} \sigma_j) - \log(2\cosh(\zeta + \gamma \sum_{j \in M_i} \sigma_j)))}{\exp(\sigma_i'(\zeta + \gamma \sum_{j \in M_i} \sigma_j) - \log(2\cosh(\zeta + \gamma \sum_{j \in M_i} \sigma_j)))} \tag{C.37}$$

$$= \frac{\exp(\sigma_i(\zeta + \gamma \sum_{j \in M_i} \sigma_j))}{\exp(\sigma_i'(\zeta + \gamma \sum_{j \in M_i i} \sigma_j))} \tag{C.38}$$

By repeating this operation for every agent (i.e. $N-1$ more times), then we arrive at:

$$\frac{p(\tilde{\sigma})}{p(\tilde{\sigma}')} = \frac{p(\tilde{\sigma})}{p(\sigma_i', \tilde{\sigma}_{\setminus i})} \frac{p(\sigma_i', \tilde{\sigma}_{\setminus i})}{p(\sigma_i', \sigma_j', \tilde{\sigma}_{\setminus i,j})} \cdots \frac{p(\tilde{\sigma}_{\setminus i}', \sigma_i)}{p(\tilde{\sigma}')} = \frac{\exp(\zeta \sum_i \sigma_i + \gamma \sum_{i<j} \sigma_i \sigma_j)}{\exp(\zeta \sum_i \sigma_i' + \gamma \sum_{i<j} \sigma_i' \sigma_j')}$$

(C.39)

We can therefore write the marginal distributions $p(\tilde{\sigma})$ as proportional to the numerator of the last term in Eq. (C.39)[5]:

$$p(\tilde{\sigma}) \propto \exp(\zeta \sum_i \sigma_i + \gamma \sum_{\langle i,j \rangle} \sigma_i \sigma_j)$$

$$\implies p(x) = Z^{-1} \exp(\zeta \sum_i \sigma_i + \gamma \sum_{\langle i,j \rangle} \sigma_i \sigma_j)$$

(C.40)

We thus recover the original stationary distribution with the standard, linear energy function as given by Eq. (7) in the main text, written now in terms of generative model parameters γ, ζ instead of the standard 'couplings' and 'biases' J, h:

$$E_{async}(\tilde{\sigma}) = -\gamma \sum_{\langle i,j \rangle} \sigma_i \sigma_j - \zeta \sum_i \sigma_i$$

(C.41)

References

1. Aguilera, M., Igarashi, M., Shimazaki, H.: Nonequilibrium thermodynamics of the asymmetric Sherrington-Kirkpatrick model. arXiv preprint arXiv:2205.09886 (2022)
2. Aguilera, M., Moosavi, S.A., Shimazaki, H.: A unifying framework for mean-field theories of asymmetric kinetic Ising systems. Nat. Commun. **12**(1), 1–12 (2021)
3. Albarracin, M., Demekas, D., Ramstead, M.J., Heins, C.: Epistemic communities under active inference. Entropy **24**(4), 476 (2022)
4. Blei, D.M., Kucukelbir, A., McAuliffe, J.D.: Variational inference: a review for statisticians. J. Am. Stat. Assoc. **112**(518), 859–877 (2017)
5. Brush, S.G.: History of the Lenz-Ising model. Rev. Mod. Phys. **39**(4), 883 (1967)
6. Friston, K.: A free energy principle for a particular physics. arXiv preprint arXiv:1906.10184 (2019)
7. Friston, K., FitzGerald, T., Rigoli, F., Schwartenbeck, P., Pezzulo, G.: Active inference: a process theory. Neural Comput. **29**(1), 1–49 (2017)
8. Friston, K., Heins, C., Ueltzhöffer, K., Da Costa, L., Parr, T.: Stochastic chaos and Markov blankets. Entropy **23**(9), 1220 (2021)

---

[5] Note that because of assumption that the system is at thermal equilibrium, the same reasoning could be applied to write the distribution over $p(\tilde{\sigma}')$ in terms of the denominator of Eq. (C.39).

9. Friston, K., Rigoli, F., Ognibene, D., Mathys, C., Fitzgerald, T., Pezzulo, G.: Active inference and epistemic value. Cogn. Neurosci. **6**(4), 187–214 (2015)
10. Friston, K.J., Daunizeau, J., Kiebel, S.J.: Reinforcement learning or active inference? PLoS One **4**(7), e6421 (2009)
11. Gaissmaier, W., Schooler, L.J.: The smart potential behind probability matching. Cognition **109**(3), 416–422 (2008)
12. Glauber, R.J.: Time-dependent statistics of the Ising model. J. Math. Phys. **4**(2), 294–307 (1963)
13. Hesp, C., Ramstead, M., Constant, A., Badcock, P., Kirchhoff, M., Friston, K.: A multi-scale view of the emergent complexity of life: a free-energy proposal. In: Georgiev, G.Y., Smart, J.M., Flores Martinez, C.L., Price, M.E. (eds.) Evolution, Development and Complexity. SPC, pp. 195–227. Springer, Cham (2019). https://doi.org/10.1007/978-3-030-00075-2_7
14. Hinton, G.E., Sejnowski, T.J.: Optimal perceptual inference. In: Proceedings of the IEEE Conference on Computer Vision and Pattern Recognition, vol. 448, pp. 448–453. Citeseer (1983)
15. Hinton, G.E., Sejnowski, T.J., et al.: Learning and relearning in Boltzmann machines. Parallel Distrib. Process.: Explor. Microstruct. Cogn. **1**(282–317), 2 (1986)
16. Hopfield, J.J.: Neural networks and physical systems with emergent collective computational abilities. Proc. Natl. Acad. Sci. **79**(8), 2554–2558 (1982)
17. Kirchhoff, M., Parr, T., Palacios, E., Friston, K., Kiverstein, J.: The Markov blankets of life: autonomy, active inference and the free energy principle. J. Roy. Soc. Interface **15**, 20170792 (2018)
18. Klein, B., Hoel, E.: The emergence of informative higher scales in complex networks. Complexity **2020**, 12 p. (2020). Article ID 8932526
19. Krafft, P.M., Shmueli, E., Griffiths, T.L., Tenenbaum, J.B., et al.: Bayesian collective learning emerges from heuristic social learning. Cognition **212**, 104469 (2021)
20. Krakauer, D., Bertschinger, N., Olbrich, E., Flack, J.C., Ay, N.: The information theory of individuality. Theory Biosci. **139**(2), 209–223 (2020). https://doi.org/10.1007/s12064-020-00313-7
21. Kwon, C., Ao, P.: Nonequilibrium steady state of a stochastic system driven by a nonlinear drift force. Phys. Rev. E **84**(6), 061106 (2011)
22. Lynn, C.W., Cornblath, E.J., Papadopoulos, L., Bertolero, M.A., Bassett, D.S.: Broken detailed balance and entropy production in the human brain. arXiv preprint arXiv:2005.02526 (2020)
23. Ma, Y., Tan, Q., Yuan, R., Yuan, B., Ao, P.: Potential function in a continuous dissipative chaotic system: decomposition scheme and role of strange attractor. Int. J. Bifurcation Chaos **24**(02), 1450015 (2014)
24. Millán, A.P., Torres, J.J., Bianconi, G.: Explosive higher-order kuramoto dynamics on simplicial complexes. Phys. Rev. Lett. **124**(21), 218301 (2020)
25. Palacios, E.R., Razi, A., Parr, T., Kirchhoff, M., Friston, K.: On Markov blankets and hierarchical self-organisation. J. Theor. Biol. **486**, 110089 (2020)
26. Pérez-Escudero, A., de Polavieja, G.: Collective animal behavior from bayesian estimation and probability matching. Nat. Precedings 1 (2011)
27. Ramstead, M.J.D., Badcock, P.B., Friston, K.J.: Answering schrödinger's question: a free-energy formulation. Phys. Life Rev. **24**, 1–16 (2018)
28. Ramstead, M.J., Constant, A., Badcock, P.B., Friston, K.J.: Variational ecology and the physics of sentient systems. Phys. Life Rev. **31**, 188–205 (2019)
29. Shanks, D.R., Tunney, R.J., McCarthy, J.D.: A re-examination of probability matching and rational choice. J. Behav. Decis. Mak. **15**(3), 233–250 (2002)

30. Virgo, N., Biehl, M., McGregor, S.: Interpreting dynamical systems as Bayesian reasoners. In: Kamp, M., et al. (eds.) ECML PKDD 2021. Communications in Computer and Information Science, vol. 1524, pp. 726–762. Springer, Cham (2021). https://doi.org/10.1007/978-3-030-93736-2_52
31. Vulkan, N.: An economist's perspective on probability matching. J. Econ. Surv. **14**(1), 101–118 (2000)
32. Walter, J.C., Barkema, G.: An introduction to Monte Carlo methods. Phys. A **418**, 78–87 (2015)
33. Welling, M., Teh, Y.W.: Approximate inference in Boltzmann machines. Artif. Intell. **143**(1), 19–50 (2003)
34. Yan, H., Zhao, L., Hu, L., Wang, X., Wang, E., Wang, J.: Nonequilibrium landscape theory of neural networks. Proc. Natl. Acad. Sci. **110**(45), E4185–E4194 (2013)

Mapping Husserlian Phenomenology onto Active Inference

Mahault Albarracin[1,2(✉)], Riddhi J. Pitliya[1,3], Maxwell J. D. Ramstead[1,4], and Jeffrey Yoshimi[5]

[1] VERSES Research Lab, Los Angeles, CA 90016, USA
`mahault.albarracin@gmail.com`
[2] Université du Québec à Montréal, Montréal, QC, Canada
[3] Department of Experimental Psychology, University of Oxford, Oxford, UK
[4] Wellcome Centre for Human Neuroimaging, University College London, London WC1N 3AR, UK
[5] University of California, Merced, 5200 Lake Rd, Merced, CA 95343, USA

Abstract. Phenomenology is the rigorous descriptive study of conscious experience. Recent attempts to formalize Husserlian phenomenology provide us with a mathematical model of perception as a function of prior knowledge and expectation. In this paper, we re-examine elements of Husserlian phenomenology through the lens of active inference. In doing so, we aim to advance the project of computational phenomenology, as recently outlined by proponents of active inference. We propose that key aspects of Husserl's descriptions of consciousness can be mapped onto aspects of the generative models associated with the active inference approach. We first briefly review active inference. We then discuss Husserl's phenomenology, with a focus on time consciousness. Finally, we present our mapping from Husserlian phenomenology to active inference.

Keywords: Phenomenology · Active inference · Computational phenomenology · Naturalizing phenomenology · Time consciousness

1 Introduction

In recent years, there has been a resurgence of work attempting to formalize the structure and content of first-person conscious experience, leveraging mathematical and computational techniques to help model conscious experience [1–4]. One recently proposed version of this project, called "computational phenomenology", leverages the generative modeling techniques that were originally developed in computational neuroscience and theoretical neurobiology to formalize and model the structure and contents of conscious experience [5].

This paper aims to contribute to the project of computational phenomenology, by mapping core elements of the structure of conscious experience as

---

M. J. D. Ramstead and J. Yoshimi—These authors contributed equally.

described by Husserlian phenomenology to the constructs of the active inference framework, and in particular, to components of the generative models that underwrite that formulation. Computationally modelling conscious first-person experience using active inference would shed light on subjective individual experience and intersubjective experiences, which could be used to better understand factors constituting normal and abnormal behavior. We begin with a brief overview of active inference and generative modeling. We then review some of the core elements of Husserlian phenomenology, with a focus on time consciousness, drawing on the formalization of Husserl presented in [6]. We argue that we can use the generative models of active inference to represent these phenomenological structures. In so doing, we aim to advance the agenda for a computational phenomenology and take first steps towards worked examples of the method.

2   An Overview of Active Inference

Given the intended audience of this paper, we will only briefly review active inference. In the broadest sense, active inference is a corollary of the free energy principle in Bayesian mechanics. Active inference is a process theory that can be used to model any physically separable, re-identifiable thing or particle, i.e., anything that persists as a coherent locus of states or paths, over some appreciable timescale. Active inference describes the dynamics (i.e., observable behavior) of things, so defined, as a path of least action, where the action is defined as time or path integral of an information theoretic quantity called self-information or, more simply, surprisal [7–11]. This quantity is also known as the negative log evidence in Bayesian inference. This means that the paths of least action maximize model evidence—a normative behavior sometimes referred to as self-evidencing [12]. In many practical applications of active inference, we do not consider the surprisal directly as it is often computationally intractable, since it requires averaging over a potentially infinite amount of states. Instead we consider an upper bound on surprisal called "variational free energy" [13]. This variational free energy measures the discrepancy between the observations or data that were expected, given a probabilistic (generative) model of how they were generated, and the data that was obtained. Intuitively, the idea is that any entity described by Bayesian mechanics maintains a model whose predictions tend to be confirmed over time (it minimizes the degree to which it is surprised). We will see that this kind of self-evidencing has a straightforward interpretation in Husserl's phenomenology.

In the narrower sense that will concern us more directly in this paper, the term "active inference" refers to a family of a mathematical models that we can use to simulate and model the behavior of cognitive agents [14,15]. Active inference is usually implemented using partially observable Markov decision processes (POMDPs), or (equivalently) using Forney-style factor graphs [16]. In active inference, the action-perception loops via which agents engage with the salient features of their environmental niche, and with the other denizens of that niche, are cast as implementing approximate Bayesian (variational) inference (see [17] for a helpful introduction). Active inference thus comprises a set of formal tools, usually implemented in code, used to model the behavior of agents that interact with their environment, as a form of inference. The active inference toolkit allows us to model the epistemic and pragmatic imperatives of the behavior of agents: agents act to gather information about their environment and select those actions that bring them closer to characteristic states, which can be read as allostatic or homeostatic set points. In active inference, these set points—or attracting sets—are defined with respect to the kind of sensory data or outcomes that an agent expects to generate via action, given "the kind of thing that it is" [9].

Active inference is a situated or enactive kind of generative modeling, which considers not only how data are modeled—i.e., explained—but, crucially, how those data are gathered in the service of self-evidencing (see [5] for a review and discussion of its applications to phenomenology). Generative modeling underwrites many forms of mathematical modeling and scientific investigation [5,18]. The general idea is straightforward. We have some data of interest, which we want to explain using statistical methods; i.e., we want to understand the *causes* of the data. So, we compute a number of alternative probabilistic models of the process that generated that data, and evaluate the evidence that the data provides for each model. In active inference, we assume that agents implement generative models, and update those models in light of sensory evidence. This modeling strategy assumes that agents can only access their environment by sampling it via sensory states. These generative models harness the beliefs of an agent about the "hidden" states of the external world, i.e., they encode what an agent knows about the process that generates its sensory data [19]. Agents are thus modeled as inferring what the primary causal pathways in the world are, and as navigating the opportunities for engagement that they are presented with by leveraging these inferences. Prior beliefs are updated continuously based on new data (i.e., new observations) via approximate Bayesian (variational) inference [20]. The current "content" of an agent's "experience" of "things" in the world is thus the set of states that are being inferred, on the basis of sensory data.

In active inference, *action* is modeled as a kind of self-fulfilling prophecy: agents predict what state they will be in upon acting, and then generate evidence for this prediction by actually acting in the environment [21]. The action itself is selected based on beliefs about possible courses of action, which are called "policies". Policies are thus beliefs about expected sequences of actions, which depend on an agent's beliefs about the current state of the world and the goals that it is trying to achieve (specified in terms of preferred observations). Different policies are, in some sense, variations of beliefs about expected future observations, contingent on possible courses of action. The value of a policy is determined by estimating a quantity is known as "expected free energy", which encodes how much each policy will minimize surprisal or, equivalently, maximize model evidence, with respect to preferred outcomes [22]. This rests upon the degree to which expected surprisal (i.e., uncertainty) can be resolved on the one hand, and the avoidance of surprising (i.e., aversive) outcomes on the other. We say that the optimal policy is the one that provides the most evidence for the generative model of the organism (or equivalently, that is expected to generate the least free energy). The selection of a policy is thereby driven by the expected free energy of that policy and agent's preferences, allowing an agent to conduct goal-directed behaviors.

To simulate an agent, we equip it with the states and parameters shown in Table 1, which can either be specified *a priori* by the experimenter, or learned based on real data [23]. In the POMDPs used in active inference, a distinction is made between observable data (denoted **o**), and hidden states (denoted **s**) [24]. The probability of some observation, given that some state obtains, is described by the likelihood matrix, denoted **A**; the entries of this matrix quantify the probability of observing some data, given that world is in some state. The parameter encoding the beliefs of the particle or thing about how states transition into each other over time is a matrix denoted as **B**, with each entry scoring the probability of transitioning to some state, given that the system was previously in some other state. A vector denoted **C** encodes preferences for each observation. Prior beliefs about base rates of occurrence of states are described by the **D** vector, with each entry scoring the prior probability of the associated state. Finally, baseline preferences for policy selection are described by the **E** vector. **C** is used to compute variational free energy (**F**) and expected free energy (**G**), which are used in perceptual inference and policy selection, respectively [25].

Table 1. Parameters used in the general model under the active inference framework. We explain the generative model symbols that refer to different matrices and elements which are connected through them.

o	Observations or sensory states of an agent
s	Hidden or external states
A	Likelihood matrix that captures beliefs about the mapping from observations to their causes (hidden)
B	Transition matrix that captures beliefs about the mapping between states at one time step to states at the next time step
C	Prior preference matrix that captures the preferred observations for the agent, which will drive their actions
D	Priors that capture beliefs about base rates of occurrence of the hidden states
E	Prior preferences for policies in the absence of data
F	Variational free energy
G	Expected free energy
π	Policy matrix that captures the policies available to an agent

At any time-step, the current state is estimated by using "forward" and "backward" message passing. Forward messages are passed from nodes encoding beliefs about past states and observations to the node computing the current state; whereas backwards messages are passed from nodes encoding beliefs about future states and observations, contingent on policy selection, to the node computing the current state. To clarify, the agent does not "experience" the parameters of its generative model—encoded in its **A**, **B**, **C**, and **D** parameters. Rather, these parameters underwrite the message passing and belief updating; namely, the updating of prior beliefs into posterior beliefs in the face of new sensory evidence. As indicated, at any time step, the current "content" of an agent's "experience" of "things" in the world is implemented the set of states that are actively being inferred by the agent, on the basis of sensory data received.

3   An Overview of Husserl's Phenomenology

We now review Husserl's phenomenological description of time consciousness and intentionality, and how they constitute experienced objects.[1] We use "phenomenology" in the technical sense that is commonplace in philosophy, to refer to a general descriptive methodology for the study of the structure and contents of the conscious, first-person experience of a subject or agent (or what might be called a "stable cognizer") [27]. We are concerned here with phenomenology as articulated by its founder, Edmund Husserl, who described it as an attempt to provide rigorous descriptions of the structure of first-person experience.[2]

We are primarily concerned with Husserl's account of *time consciousness* [30–33]. For Husserl, consciousness evinces what one might call a kind of "temporal thickness", which is the ultimate condition of possibility for the perception of any object whatsoever.[3] Husserl's descriptive analyses suggest that the core structures of time consciousness, which enable what he calls the "constitution" of objects in consciousness (i.e., their disclosure to an experiencing subject) is threefold, comprising what he calls "primal impression", "retention," and "protention".

Primal impressions correspond to experience of the immediate present. Suppose a melody plays, or that you walk around an oak tree. The currently perceived note in the melody, or the current visual experience of the oak tree, correspond to primal impressions. In these cases, there is an additional structure that informs the primal impression: what Husserl calls hyletic data, or *hyle*

---

[1] In this paper we are mapping from one complex domain to another complex domain: active inference is a complex and growing area, as is Husserl scholarship [26]. Within Husserl scholarship, it is inevitable that we rely on existing interpretations, which are subject to scholarly dispute. This is a first sketch of the broad outlines of the mapping, that we aim to enrich in later work. For example, there are number of potential correlates of retention and protention in active inference discussed below, and further work is needed clearly delineating these.

[2] We will not be concerned here with the thorny issues that attend the naturalization of phenomenology [3]. See [5,28,29] for a review.

[3] In this, Husserl is aligned with other philosophers of his time, including James [34] and Bergson [35,36]. A detailed historical analysis of the many sources of and precedents for Husserl's account of time consciousness is [37].

(from the Greek word for matter, or stuff). The *hyle* correspond to our sense of the melody and the oak tree as being real occurrences in the world beyond us—a raw presence that we cannot alter by an act of will. However, we do not experience the *hyle* directly. They inform our primal impressions, but are not literal constituents of those impressions. The melody and the oak tree that we experience reflect both our own "top down" understandings of these things and their "bottom up" presence. Thus, our primal impression is a hylomorphic compound of raw presence and interpretation.[4]

Primal impressions are "temporalized" in the flow of consciousness. More specifically, interpreted hyletic data are formatted into retentions and protentions. Retention is the "still living" preservation of the contents of a now-past primal impression in our present consciousness. In the case of the melody, one is still conscious of the notes that have just been struck, just as one hears the present tone. A protention corresponds to our sense of what will come next in the melody. Together they produce a temporal depth or thickness that is a condition of possibility for experiencing a melody, rather than a sequence of disconnected notes.

Husserlian retentions and protentions are not explicit representations. They are implicit, immediate components of the temporal thickness of experience, which can be contrasted with explicit representations of remembered past events (what Husserl calls "recollections") and explicitly anticipated future events; where an event in some set of (possibly nested) lived experiences. Remembering an important life event, or looking forward to some planned future events, are themselves mental acts that are experienced in ongoing processes with their own temporal depth (thus as, an explicit recollection unfolds, retentions and protentions associated with that recollected moment unfold as well).

The experiences which unfold in time consciousness inform the way we understand the world to be—they "constitute" our sense of the world. In particular, protentions are tacit anticipations or expectations about what will happen in the next moment. When what actually happens next is consistent with our anticipations, we experience *fulfillment*. When what happens is inconsistent with our anticipations, we experience *frustration* or surprise (these are technical terms in Husserl; as with retention and protention, they do not imply an explicit or focused awareness). Thus, our experience of temporally extended objects consists in a flow of anticipations and fulfillment/frustration of those anticipations [39]. Our inner time-consciousness thus at core consists in a dynamic process that anticipates what will be experienced next, based on what has just been experienced.

Husserl suggests that over time retentions fade away and "sediment", informing our understanding of the world. Similarly for fulfilled or frustrated proten-

---

[4] The question of what exactly hyletic data are is a matter of controversy. We rely on a reading derived from Føllesdal [38], who says "In acts of perception our senses play a role, providing certain boundary conditions." They "limit" what we can experience in a moment, without being directly experienced (they must be animated or interpreted by noetic form before they are experienced).

tions. If the melody was much different than we thought it would be, the experiences of surprise would change our background understanding of the melody, leaving a trace, so that the next time we experience the same melody, our expectations have adapted to the change. In this way, we build up a kind of *model of reality* that is in the background of our experience, generating the anticipations and temporal depth of time consciousness (see [6,40]). The fact that experience reflects all the sediments of past process of time consciousness means that our consciousness is laden with past retentions, which Husserl believes shape the way that we anticipate future primal impressions.

One analytic tool that Husserl introduces to study the structure of sedimented background knowledge is what he calls a "horizon" or "manifold". The idea with a horizon is to begin with some object given (i.e., constituted) in experience, and then to imagine different ways that an experience of that object could continue to be experienced. Each possible continuation of the current experience will produce a different profile of fulfillment and frustration. If we only focus on fulfilling continuations—that is, on further experience that would not surprise us—we get what is called a "trail set" in [6]. Trail sets can be used to formalize Husserl's notion of a horizon, i.e., what our implicit understanding of an object is, beyond what we immediately see. Standing before the oak tree, we have some expectations of how it would look, were we to move around it. Those expectations are open, they are "determinably indeterminate" leeways (*Spielräume*). These trails present the oak tree as having more or less branches on its back side, different coloration patterns, etc. However, they do not contain experiences of the back side of the oak tree that would surprise us, like one where a sign was nailed to it, or it was covered in spray paint. If we explore the oak tree, and one of those things is seen, then a protention will be frustrated, and that frustration will sediment in background knowledge, so that in the future, we will not be surprised: the trail set changes, we now see it as an oak tree with spray paint on its back. (This learning rule has been formalized using Bayesian statistics, making it easily amenable to active inference modeling; see [6]).

Our account so far has focused on perceptions of things or events, like seeing an oak tree or hearing a melody, but it can be extended to arbitrary mental processes, like hearing, imagining, planning, and also to more complex dynamic processes, like learning how to dance or learning mathematics. In each case, experiences unfold, and time consciousness operates, creating anticipations which are fulfilled or frustrated. The results of these processes of frustration and fulfillment are then sedimented into background knowledge. In this way, we maintain and update models of the external world, of how to dance, of mathematics, of history, of our own values and future plans, etc. These different types of knowledge generate different kinds of horizons associated with different kinds of trail sets: ways we expect things to be, ways we expect our body to move, ways we expect a conversation to unfold, what we expect ourselves to do relative to our values, etc.

4   Mapping Husserlian Time Consciousness onto Generative Models in Active Inference

In this section, we map aspects of the generative models that figure in active inference to aspects of Husserl's phenomenology. See Table 1 for a list of the states and parameters of generative models in active inference. As indicated, our project is situated within the broader framework of computational phenomenology.[5]

We can associate aspsects of a generative model, represented as a POMDP, to aspects of Husserlian phenomenology. In generative models, inferences about the current state of the world are informed by beliefs about what past states were experienced, and also by beliefs about what future states will be. Technically, the messages that are used to update current beliefs about hidden states come from factors that represent beliefs about states in the past, and also from factors that represent beliefs about states in the future.

We can begin by associating observations **o** with hyletic data, hidden states **s** with perceptual experiences, and the various parameters of the POMDP (e.g., the likelihood matrix **A** and state transition matrix **B**) with sedimented knowledge. Recall that in active inference modeling, outcomes are data that agents aim to explain (or alternatively, that we scientists are trying to explain, in generative modelling more broadly). Hidden states are inferred from this data, as their causes. During perception, hidden states of the generative model are used to generate predictions, which are compared against actual observations; and the parameters of the model are updated in a Bayes optimal manner, such that these predictions get better over time, leading to reductions in variational free energy. This maps directly on to the Husserlian apparatus. Our immediate perceptual experiences (correlated with **s**) are based on a mixture of relevant background knowledge (correlated with **A**, **B**, and so on) and hyletic data (correlated with **o**). The hyletic data are not literal constituents of experiences, just as **o** is not a literal constituent of **s**. Rather, the hyletic data impose boundary conditions or limits on what we can experience—they correspond to a sense of the presence of the world—but they are not experienced directly. Sensory experiences arise from the *interplay* of hyletic data and background knowledge (or "noetic form") in Husserl. In a similar way, **o** constrains or limits what hidden state **s** will be inferred, given **A** and **B**, but is not contained in **s**. Hidden states are updated as a function of observed sensory states **o** and beliefs encoded in a likelihood and transition matrices **A** and **B**, but the hidden states do not directly contain those observations.

There are several ways to capture retention and protention in a generative model. One way is to focus on the process of inferring hidden states from observed

---

[5] Technically, computational phenomenology is a version of generative modeling that is agnostic about whether the models at play are real descriptions of the actual processes at play in agents, or whether these models are merely useful heuristics to model first-person experience. See [5] for a discussion. Of note, the work presented in this paper dovetails nicely with realist approaches the implementation of generative models by agents,; see integrated world modelling theory as proposed in [2].

data, by comparing a prediction about what will be observed with what is actually observed, and updating beliefs using an error signal (as in predictive coding implementations of active inference, [41]). Such an approach involves direct correlates of protention (a prediction signal), fulfillment or frustration (the error signal), and beliefs update on that basis. Thus, even in the immediate or "static" perception of an oak tree [15] the process of state estimation involves correlates of protention, fulfillment and frustration. A second approach is to focus on the state transition matrices $\mathbf{B}$, which encode state transition probabilities, and which thus underwrite "dynamic perception" [15], that is, beliefs about how objects change over time. These matrices are used to estimate what will occur next in a song as we listen to it, or how the oak tree will sway under the influence of the wind. These state estimates themselves rely on what occurred just previously (see, e.g., Fig. 4 in [42]). So here again we have direct correlates of retention and protention in the active inference framework, this time in the operation of the $\mathbf{B}$ matrices. There may also be links between retention and working memory, especially when it is understood (in an active inference framework) in terms of evidence accumulation in a temporally structured hierarchy [43].[6]

The experience of fulfillment and frustration can be modelled as a process of Bayesian belief updating [6]. In line with this analysis, we suggest that we can quantify fulfillment/frustration in terms of the variational/expected free energy that is generated by subsequent sensory experience; where the free energies quantify the degree to which current experiences conflict with protended experience (i.e., the degree of fulfillment or frustration).

Active inference, like Husserlian phenomenology, is ultimately about action in a lived world. As described above, a policy in active inference is a set of beliefs about possible courses of action; and action itself is modelled as a kind of self-fulfilling prophecy. This basic structure can be extended to include counterfactual richness, which can be associated with the trails of fulfilling experiences in a horizon. In so-called "sophisticated" treatments of active inference, agents select which action to pursue by engaging in a deep tree search, unfolding possible sensory consequences of available actions recursively, and evaluating each branch in terms of the free energy expected along that branch [46]. The search process is efficient; only those paths with high posterior probability are evaluated, but the search is defined over a larger set of possible paths. The optimal policy is the one that maximizes preferred outcomes (relative to $\mathbf{C}$, which encodes prior preferences for data) and maximizes model evidence (or minimizing surprisal). These counterfactual policy deep trees of sophisticated active inference can be mapped to Husserlian structures, including a set of values encoded in background knowledge (a correlate of $\mathbf{C}$), and other features of background knowledge (e.g., our knowledge of state transitions, encoded in $\mathbf{B}$), which can be used to generate a trail set: a set of expected perceptions that is consistent with our beliefs, goals and desires.

---

[6] A fuller discussion would also involve a comparison with existing discussions of the naturalization of time-consciousness, such as [44] and [45].

We can thus describe a mode of analyzing POMDPs that maps onto the method of horizon analysis in Husserl. Focusing on the case of perception, imagine all possible sequences of data **o**, given some assumed hidden state **s**, and a policy or sequence of actions. Some of the data generated will confirm the knowledge implicit in the model parameters (i.e., provide evidence for it, as quantified by free energy); others will disconfirm it (i.e., will generate large amounts of free energy). The sequences of observations that confirm the beliefs of the agent about the current state of its world correspond to a set of possible continuations from the current observation that are not surprising, i.e., that lead to little variational free energy. Such continuations are captured in the parameters of the generative model (e.g., the **A** and **B** matrices). These active inference trail-sets map directly on to the Husserlian trail sets. The latter can be thought of as an alternative, and perhaps more intuitive, way of understanding the information implicit in the matrices. In the one case, we have a method of "probing" the expectations implicit in the parameters of the generative model; in the other case, we have a method of probing the expectations implicit in an experiential horizon.

The representational analogs of retention and protention (recollection and explicit prediction) can be formalized by appealing to (possibly hierarchical) state estimation. Indeed, focally recollected and anticipated events constitute (past and future) states of the world (or indeed, of the self) that need to be represented explicitly. To begin to formalize this, one can point to the explicit distinction, in active inference, between the **A**, **B**, **C**, and **D** parameters, which contribute to current state estimation, and the states which are actually being inferred in the present, to account for the distinction. This richness of this account can be increased by appealing to nested hierarchies of generative models. Explicit recollection of memories and anticipation would then correspond to state factors higher up in the hierarchy, which bin or coarse-grain observations at subordinate layers.

In both active inference and phenomenology, the analysis of perception is just a convenient starting place. All the mappings developed above can be applied to other features of cognition and experience: auditory and tactile experience, multi-modal experience, cognition, language, skilled behavior, planning, affect, intersubjectivity, etc., each associated with its own state estimations, learned likelihood matrices, retentions and protentions, trail sets, and so forth.

5   Conclusion

This paper has drawn parallels between Husserlian phenomenology and the active inference framework. We proposed to formalize some core elements of Husserlian phenomenology via active inference. Our aim in so doing was to advance the project of computational phenomenology. We proposed Husserl's descriptions of primal impression, hyletic data, retention, protention, fulfillment, frustration, trail set and horizon, recollection and explicit anticipations can be mapped onto aspects of the generative models of active inference.

Husserlian phenomenology is fertile ground for formalization. Formalizing phenomenology allows us to leverage it in order to better understand and model human experience, and to make testable empirical predictions. Concurrently, active inference has been used to model many aspects of cognition, but its use to explain qualitative and subjective experience is still in the very early stages. Moving towards computational phenomenology through a connection between Husserlian phenomenology and active inference may allow us to bridge the gaps to fundamental questions such as the explanatory gap and positionality, and extend further into sociological issues of intersectionality, which make fundamental reference to first-person experience.

Acknowledgments. The authors thank Philippe Blouin, Laurence Kirmayer, Magnus Koudahl, Antoine Lutz, Jonas Mago, Jelena Rosic, Anil Seth, Lars Sandved Smith, Dalton Sakthivadivel, and the members of the VERSES Research Lab for useful discussions that shaped the contents of the paper. Special thanks are due to Juan Diego Bogotá, Zak Djebbara, and Karl Friston.

References

1. Tononi, G., Boly, M., Massimini, M., Koch, C.: Integrated information theory: from consciousness to its physical substrate. Nat. Rev. Neurosci. **17**(7), 450–461 (2016)
2. Safron, A.: An integrated world modeling theory (IWMT) of consciousness: combining integrated information and global neuronal workspace theories with the free energy principle and active inference framework; toward solving the hard problem and characterizing agentic causation. Front. Artif. Intell. **3**, 30 (2020). ISSN 2624-8212
3. Petitot, J., Varela, F.J., Pachoud, B., Roy, J.-M.: Naturalizing Phenomenology: Issues in Contemporary Phenomenology and Cognitive Science. Stanford University Press (1999). ISBN 0804736103
4. Seth, A.: Being You: A New Science of Consciousness. Penguin (2021)
5. Ramstead, M.J.D., et al.: From generative models to generative passages: a computational approach to (neuro) phenomenology. Rev. Philos. Psychol. 1–29 (2022)
6. Yoshimi, J.: Husserlian Phenomenology: A Unifying Interpretation. Springer, Heidelberg (2016). https://doi.org/10.1007/978-3-319-26698-5
7. Ramstead, M.J.D., et al.: On bayesian mechanics: a physics of and by beliefs. arXiv preprint arXiv:2205.11543 (2022)
8. Da Costa, L., Friston, K., Heins, C., Pavliotis, G.A.: Bayesian mechanics for stationary processes. arXiv preprint arXiv:2106.13830 (2021)
9. Sakthivadivel, D.A.R.: Towards a geometry and analysis for Bayesian mechanics. arXiv preprint arXiv:2204.11900 (2022)
10. Andrews, M.: The math is not the territory: navigating the free energy principle (2020)
11. Friston, K., et al.: The free energy principle made simpler but not too simple. arXiv preprint arXiv:2201.06387 (2022)
12. Hohwy, J.: The self-evidencing brain. Noûs **50**(2), 259–285 (2016)
13. Friston, K.J.: The free-energy principle: a unified brain theory? Nat. Rev. Neurosci. **11**(2), 127–138 (2010). ISSN 1471-0048
14. Parr, T., Pezzulo, G., Friston, K.J.: Active Inference: The Free Energy Principle in Mind, Brain, and Behavior. MIT Press, Cambridge (2022)

15. Smith, R., Friston, K.J., Whyte, C.J.: A step-by-step tutorial on active inference and its application to empirical data. J. Math. Psychol. **107**, 102632 (2022)

16. Friston, K.J., Parr, T., de Vries, B.: The graphical brain: belief propagation and active inference. Netw. Neurosci. **1**(4), 381–414 (2017). ISSN 2472-1751

17. Buckley, C.L., Kim, C.S., McGregor, S., Seth, A.K.: The free energy principle for action and perception: a mathematical review. J. Math. Psychol. **81**, 55–79 (2017). ISSN 0022-2496

18. Cleminson, A.: Establishing an epistemological base for science teaching in the light of contemporary notions of the nature of science and of how children learn science. J. Res. Sci. Teach. **27**(5), 429–445 (1990)

19. Pezzulo, G., Parr, T., Friston, K.: The evolution of brain architectures for predictive coding and active inference. Philos. Trans. R. Soc. B **377**(1844), 20200531 (2022)

20. Tschantz, A., Baltieri, M., Seth, A.K., Buckley, C.L.: Scaling active inference. In: 2020 International Joint Conference on Neural Networks (IJCNN), pp. 1–8. IEEE (2020)

21. Corcoran, A.W.: Allostasis and uncertainty: an active inference perspective. Ph.D. thesis, Faculty of Arts, Monash University (2021)

22. Hipolito, I., Baltieri, M., Friston, K.J., Ramstead, M.J.D.: Embodied skillful performance: where the action is (2020)

23. Stephan, K.E., Iglesias, S., Heinzle, J., Diaconescu, A.O.: Translational perspectives for computational neuroimaging. Neuron **87**(4), 716–732 (2015)

24. Parr, T., Friston, K.J.: Generalised free energy and active inference. Biol. Cybern. **113**(5), 495–513 (2019)

25. Friston, K.J., FitzGerald, T., Rigoli, F., Schwartenbeck, P., Pezzulo, G.: Active inference: a process theory. Neural Comput. **29**, 1–49 (2016)

26. Kallens, P., Yoshimi, J.: Bibliometric analysis of the phenomenology literature. In: Londen, P., Walsh, P., Yoshimi, J. (eds.) Horizons of Phenomenology. Springer, Heidelberg (2022)

27. Yoshimi, J.: The formalism. In: Yoshimi, J. (ed.) Husserlian Phenomenology. SpringerBriefs in Philosophy, pp. 11–33. Springer, Cham (2016). https://doi.org/10.1007/978-3-319-26698-5_3

28. Ramstead, M.J.D.: Naturalizing what? Varieties of naturalism and transcendental phenomenology. Phenomenol. Cogn. Sci. **14**(4), 929–971 (2015)

29. Yoshimi, J.: Prospects for a naturalized phenomenology. In: Philosophy of Mind and Phenomenology, pp. 299–321. Routledge (2015)

30. Husserl, E.: The Phenomenology of Internal Time-Consciousness. Indiana University Press (2019)

31. Husserl, E.: Ideas Pertaining to a Pure Phenomenology and to a Phenomenological Philosophy: Second Book Studies in the Phenomenology of Constitution, vol. 3. Springer, Heidelberg (1990). ISBN 0792307135

32. Husserl, E.: Ideas Pertaining to a Pure Phenomenology and to a Phenomenological Philosophy: First Book: General Introduction to a Pure Phenomenology, vol. 2. Springer, Heidelberg (2012). ISBN 9400974450

33. Husserl, E.: Cartesian Meditations: An Introduction to Phenomenology. Springer, Heidelberg (2013). ISBN 9401749523

34. James, W.: Principles of Psychology 2007. Cosimo (2007)

35. Bergson, H.: Essai sur les données immédiates de la conscience. F. Alcan (1911)

36. Bergson, H.: Matière et mémoire. République des Lettres (2020)

37. Andersen, H.K., Grush, R.: A brief history of time-consciousness: historical precursors to James and Husserl. J. Hist. Philos. **47**(2), 277–307 (2009)

38. Føllesdal, D.: Bibliometric analysis of the phenomenology literature. In: Dreyfus, H., Hall, H. (eds.) Husserl, Intentionality, and Cognitive Science. MIT Press (1982)
39. Soueltzis, Nikos: Protention as Phenomenon. In: Soueltzis, N. (ed.) Protention in Husserl's Phenomenology. P, vol. 230, pp. 63–99. Springer, Cham (2021). https://doi.org/10.1007/978-3-030-69521-7_4
40. Sokolowski, R.: The Formation of Husserl's Concept of Constitution, vol. 18. Springer, Heidelberg (2013). ISBN 9401733252
41. Friston, K., Kiebel, S.: Predictive coding under the free-energy principle. Philos. Trans. Roy. Soc. B: Biol. Sci. **364**(1521), 1211–1221 (2009). ISSN 0962-8436
42. Friston, K.J., Parr, T., Yufik, Y., Sajid, N., Price, C.J., Holmes, E.: Generative models, linguistic communication and active inference. Neurosci. Biobehav. Rev. **118**, 42–64 (2020)
43. Parr, T., Friston, K.J.: Working memory, attention, and salience in active inference. Sci. Rep. **7**(1), 14678 (2017)
44. Grush, R.: How to, and how not to, bridge computational cognitive neuroscience and Husserlian phenomenology of time consciousness. Synthese **153**(3), 417–450 (2006)
45. Wiese, W.: Predictive processing and the phenomenology of time consciousness: a hierarchical extension of Rick Grush's trajectory estimation model (2017)
46. Friston, K., Da Costa, L., Hafner, D., Hesp, C., Parr, T.: Sophisticated inference. arXiv preprint arXiv:2006.04120 (2020)

The Role of Valence and Meta-awareness in Mirror Self-recognition Using Hierarchical Active Inference

Jonathan Bauermeister$^{(\boxtimes)}$ and Pablo Lanillos

Radboud University, Houtlaan 4, 6525 XZ Nijmegen, The Netherlands
jonathan.bauermeister@online.de

Abstract. The underlying processes that enable self-perception are crucial for understanding multisensory integration, body perception and action, and the development of the self. Previous computational models have overlooked an essential aspect: affective or emotional components cannot be uncoupled from the self-recognition process. Hence, here we propose a computational approach to study self-recognition that incorporates *affect* using state-of-the-art hierarchical active inference. We evaluated our model in a synthetic experiment inspired by the mirror self-recognition test, a benchmark for evaluating self-recognition in animals and humans alike. Results show that *i*) negative valence arises when the agent recognizes itself and learns something unexpected about its internal states. Furthermore, *ii*) the agent in the presence of strong prior expectations of a negative affective state will avoid the mirror altogether in anticipation of an undesired learning process. Both results are in line with current literature on human self-recognition.

Keywords: Active inference · Affect · Self-mirror recognition

1 Introduction

The ability of self-recognition has been typically attributed only to humans and few other species [2] and hides several essential brain processes related to multisensory perception, embodiment and decision making [14,15]. To evaluate this ability, Gallup, in 1970, developed a test for chimpanzees named the mirror self-recognition (MSR) [12]. This test, which was also adapted for infants [1], consists of placing a mark, unbeknownst to the subject, on her face. The subject is then placed in front of a mirror. The agent passes the test if there are reaching or exploratory behaviours to remove the mark or inspect it.

While most studies postulate mark directed behaviour (inspect or remove) as a necessary condition for self-recognition [3], more recent cross-cultural studies have shown that children from cultures with higher parental authority are not inclined to remove the mark [4]. Similarly if one creates a social context during the mirror test, where several other subjects around the infant also have marks on their face, the infant despite having passed the mirror test before is less motivated to engage in mark directed behaviour [19]. Both results, by enriching the complexity of such behaviour by environmental and social factors, cast doubts on interpreting the necessity or sufficiency of mark directed behaviour for self-recognition.

On the other hand, it has been evidenced in humans a strong emotional component, i.e., to express negative affect when seeing their own reflection. When infants pass the test around the age of two, they universally express negative affect toward their mirror image, which has been interpreted as embarrassment, shyness or puzzlement [1,18]. Ultimately, we do not know what the phenomenology of a two-year-old seeing herself in the mirror is like. Anyhow, *the negative affective part of the experience seems to be uncontested.*

While there is previous research on computational models of self-recognition (e.g., generative modelling focusing on visual-kinesthetic matching or appearance cues [14,16]), none of the works has attended to the emotional component. Here, we studied self-recognition by (*i*) developing of a computational model that incorporates the affective component into the self-recognition process and (*ii*) evaluating it on a new synthetic experiment based on the MSR.

To model the affective component we use the notion of valence [13]. The working hypothesis is that the negative or positive quality of an affective experience can arise as a consequence of obtaining new information about oneself through mirror self-recognition. Importantly, this new information might favour different action selection policies (action dependent valence), leading to a change in valence. This iterative process coherently connects emotions with self-recognition and decision-making. To incorporate valence into the perception-action loop we used the hierarchical active inference construct [11], where the agent perceives, learns and acts to obtain the expected outcomes by minimizing the expected free energy.

We further developed an experimental benchmark to evaluate the effect of valence in the process of decision-making and self-recognition. In the experiment, the agent can decide to look at a mirror, look at a wall or look at a video of another agent. Furthermore, thanks to the hierarchical nature of the model, we further studied the importance of meta-cognition ('higher' layers), in combination with affect, for (anticipated) self-recognition. For instance, adults in full possession of a self-concept can also anticipate a confrontation with their mirror image. If self-evaluation is negative, or one's body image has radically shifted due to surgery, patients are motivated to actively avoid the mirror [10].

Related Work. There were several tries to build a computational model of self-recognition—See [14] for a review. Relevant to this work, in [16] the robot inferred itself by answering the question 'did I generate those sensory outcomes?'. For example, if the robot has an intention to move its arm and can predict its interoceptive and exteroceptive sensory outcomes with low prediction error, then it will infer that the likeliest cause of this action was the system itself. While this approach may be promising to give insights into self-recognition, it does not yet explain how affect arises during the human MSR test. It has been theorized that affective and action based self-modelling naturally arises for a system engaged in deep temporal active inference [6,7]. Nevertheless, to the authors knowledge, there is no computational model as of now, that explicitly assesses affect within self-recognition. Fortunately, recently within the active inference research, there has been an effort to introduce internal drivers that modulate the generative model parameters and the action selection process. For instance, valence – pleasantness or unpleasantness of an emotional stimulus – was introduced in [13] to model the confidence of the model estimates. Importantly, valence encodes how well the agent is performing in the environment, thus, aiding action selection. The mathematical formalization of valence and the hierarchical structure of the generative model allows the building of a complex agent that has beliefs over beliefs, which are modulated by the increase or decrease of valence, thus affecting the whole decision making process. In this work, we adapt this model to study self-recognition and decision making.

2   Methods

First, we describe the general framework of our approach and how to introduce valence based on the work of [13]. Second, we detail the affective self-recognition computational model and finally, we describe the experimental setup for self-recognition.

2.1   Discrete Hierarchical Active Inference

We model the problem under the discrete state-space formulation of active inference [17][1]. The agent computes both the posterior state estimation (perception) and action selection by minimizing, through marginal message passing, a single quantity: the expected free energy. This quantity measures the divergence from the current expectations to the real world state. In order to be able to compute it the agent needs a generative model of the world, thus, allowing predictions of the future outcomes. See Appendix 5 for detailed explanation.

Temporal Depth To achieve temporal depth we use a hierarchical generative model as depicted in Fig. 1. Here, the hidden states on the second layer change slower than the hidden states on the first layer. Thus the beliefs about states in the first layer (bottom) can fluctuate several times within one trial, where the beliefs on the second layer (top) only change at the end of each trial (the length of a trial is defined by the modeller). For example, the agent can have the abstract belief that it is in a happy mood. This belief will set the priors on the first level at the beginning of a trial accordingly. So now the agent expects certain observations (facial muscles expressing a smile, heart rate going up etc.), even if within the trial the facial muscles will most likely change several times (depending on the granularity of the model) and not only stay in one position, say smiling, the agent could still infer that overall it is happy (which is an abstract

---

[1] For a thorough tutorial on discrete active inference formulation see [21] and for a concise mathematical overview see [5].

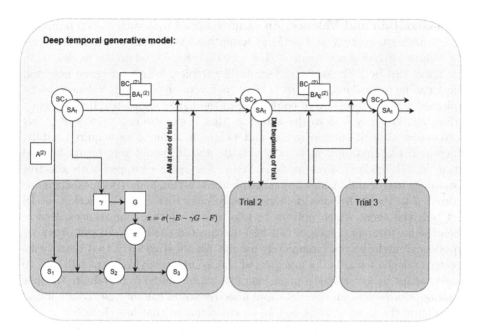

Fig. 1. A temporally deep generative model with two hierarchical layers. The blue boxes in the first layer correspond to trials (here simplified). This is the architecture the agent uses to perceive and act in the world. The second layer only communicates with the first at the beginning of a trial through descending Messages and at the end of the trial is informed by ascending messages. Hence, it can already be seen that states on the second level change slower (only once every trial) than states on the first layer. Here the agent has two state factors on the second level. It has a contextual belief which replaces the prior at the beginning of each trial. Additionally, it has an affective state which sets the precision on G at the beginning of the trial. The impact of G on π can now be regulated by the agent through its affective state. Figure adapted from [13]. (Color figure online)

second-order belief that integrates information over time). Only if in the course of the trial it consistently observes unexpected observations (prediction errors) it will update its belief on the second level at the end of a trial accordingly.

Mathematically, the second layer has likelihood mappings and state transitions like the first layer. They differ in that the likelihood mapping A2 does not map from observation to hidden state but from the hidden state at layer one (facial expression i.e smile, frown, neutral) to the hidden state on layer two (mood i.e happy or sad). The transition matrix B2 then encodes how likely the context, for example, the agent's mood, changes over trials. As a consequence, the prior D1 is replaced by a more dynamically changing higher-order state. At the beginning of each trial, the higher-order state acts as prior for the agent. By the end of the trial the agent updates the higher-order state based on the information collected in this trial. For a mathematical expression of ascending and descending messages—See [13] for further description.

Meta-cognition and Valence. An agent equipped with such a deep temporal model can learn context and perform some tasks very well but underperforms in a volatile changing environment. Here we describe, based on the work of [13], how affect can be included in a discrete hierarchical active inference network. Affect can be formalized through valence (negative or positive). Valence can be explained as the expression of confidence in the model estimates. If the agent's actions continuously lead to the outcomes that it expects and prefers, it grows more confident in its action model and weighs it stronger as acquired habits. Whereas if the environment is very volatile and it cannot rely on its learned action model yielding to an 'anxious' state. The agent equipped with affective states finds better and biologically more plausible action plans (policies) than one without [13]. Partly because it takes time to construct a reliable action model that tells the agent which policies to take under which circumstances. Hence, when the environment changes fast and unexpected the action model (learn by experience) might become completely useless. An affective agent that reacts with negative valence towards the unexpected change in the environment will able to quickly adapt by lowering the precision of its action model to reevaluate the new situation. Conversely, an agent without affective states cannot quickly adapt and will execute the same actions despite an environment that has changed.

Thus, valence acts as a second-order state. Implementation-wise the agent has a categorical distribution over it being either in the state 'positive valence' or 'negative valence', and it is updated at the end of a trial via ascending messages. If in a trial the agent could rely on its action model then it increases its valence for the next trial. At the beginning of the next trial, instead of using a static prior the second-order affective state informs the *precision* of the action model. Such a top-down estimation of the reliability of its model can also be understood as a form of meta-cognition as it is monitoring the confidence of the cognitive process.

Meta-awareness and Attention. The meta-cognition architecture can be extended to model meta-awareness by adding a third layer, and allowing the agent to dynamically change the information flow between second and first layer. This has been implemented by [20] as a precision on the likelihood mapping. Generally, precision on likelihood matrices in active inference is linked to attentional processes. The states on the third level then represent if the agent is aware of her cognitive processes on the second level. The benefits and biological plausibility of this meta-awareness capacity have been shown by [20]. Here we use a simplified setup where we change the precision of the 2nd layer likelihood by hand to see how the agent's behaviour changes if it is very aware of its emotional states and hence, can use the information for cognition deeper down the hierarchy. The precision modulates how strong the connection between the two layers is and therefore how informative descending and ascending messages are. For instance, low precision will lead the agent to rely more on its 'direct experience' (first layer of its generative model).

2.2   Affective Self-recognition Model

Based on the previously described framework, we propose a computational model to investigate self-recognition with the emotional component. We focused on the following two research questions: (*i*) How does valence influence behaviour during mirror self-recognition? and (*ii*) How might negative valence arise in mirror self-recognition?

To evaluate the model and further be able to answer the questions, we designed an experiment where an agent can decide to either see its emotional expression in the mirror, look at a wall and see no face or look at a video of an emotional expression of another person. Note that our computer-simulated agent can not look into an actual physical mirror, so we need to formalize the function of the mirror, which possibly leads to an affective reaction. We interpret making an observation in the mirror as acquiring information about oneself (here: about the agent's emotional state via its facial expression) by dragging the internal attention of the agent onto this aspect of itself. Thus, the mirror is a self-exploration tool that allows this information to be available for decision making and introspection from layers higher up the hierarchy.

Generative Model. We formalize a two-layered deep generative model to capture the self-recognition experiment, as described in Fig. 2. The agent can obtain exteroceptive observations, where it either sees a face that is happy, neutral or sad, or nothing when it looks at the wall. This information can be used to infer the emotional state of the other person. And it can obtain interoceptive observations about its facial expression (for example sensing its facial muscles) to infer its emotional state (happy, neutral, sad). The agent also has a state (attention in the figure) that captures if it is paying attention to its interoceptive observation with values Yes or No. This is an attention state, in the sense that it modulates the precision on the likelihood A, but it cannot be actively controlled by the agent. Instead, it captures what happens internally when the agent sees itself in the mirror. By recognizing itself, it is forced to pay attention to its internal observation (the precision ω on A will become very precise). Hence, formalizing the notion of the mirror dragging attention onto oneself and making certain observations informative.

On the second layer, the agent has two state factors. The valence can be positive or negative (implemented as a categorical distribution). Its value depends on the expected precision of the action model G. Additionally, the agent has a second-order belief about which mood it is in (happy, neutral, sad). This state sets the priors of what facial expression the agent expects when observing itself. For all the details of how this generative model is implemented please see the Appendix 5.

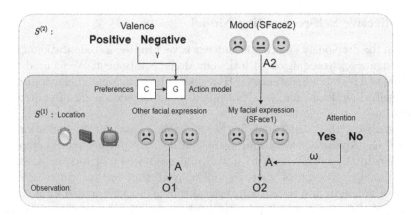

Fig. 2. Generative model description. Two layered model of an agent inferring its own mood and deciding weather to look at the mirror or not. At any given time step the 1st layer includes an action model G, preferences C and four state factors: Location, Other-facial-expression, My-facial-expression and Attention. The Attention state modulates the likelihood A via the precision ω. For each state factor, the agent has a categorical distribution of which state it believes itself to be in. The 2nd layer tracks the agent's valence and belief about its mood. Valence interacts with the action model G and the mood or context state interacts with the agents belief of its facial expression on the first level via A2.

Agent Operational Specification. First, inspired by the idea that if self-evaluation is positive one seeks out a mirror, we assume that the agent prefers to see *itself* but only if it is 'happy' or 'neutral' instead of 'sad'. This is encoded in the preference matrix C. The agent's actions are to go to one of the three locations: mirror, wall or tv. Because the agent knows from its generative model that it will pay attention to itself if it goes to the mirror its behaviour will depend on its self-knowledge, i.e., what state it believes to be in and how aware it is of that state. If it thinks it is happy one would expect that the agent will act to admire itself in the mirror. Second, the true emotional state of the agent may change. Here we have coupled the dynamics of the true state to the current valence of the agent. If its positive valence goes above 70% its true state shifts to happy, below 30% to sad and otherwise neutral. To be robust against small fluctuations, we imposed that the agent's belief about its valence has to shift at least by 15% to make a switch. Although, these decisions are arbitrary, they are designed to show how the agent adapts to a change in its true state.

3   Results

We analyzed our model behaviour in our synthetic experiment to study which actions the agent chooses and when and how the valence of the agent changes under different conditions, such as changing the true emotional state, the prior knowledge the agent has about its emotional state or the introspective availability of its emotional states (precision on A2). Particularly, we focused on two different initial conditions. In the first experiment (Sect. 3.1), we study how valence might naturally evolve in a mirror self-recognition scenario. To this end, the agents true state was set to sad, but it had low meta-awareness. For the second experiment (Sect. 3.2), we studied mirror avoidance behaviour. Thus, the agent had high meta-awareness. The code to replicate the results can be found in this this link.

Each agent was evaluated for 8 consecutive trials in each condition, where each trial lasts for three time steps (three observations). After the first observation, the agent will have to decide where to go (mirror, wall or tv). Its action plan horizon is two steps ahead, thus, it can predict outcomes until the end of a trial by using its generative model.

3.1   Experiment 1: I Am Sad and I Know It, but I am Not Very Aware of It

We set the true state of the agent to 'sad' and the precision on A2 low. Also, the agent 'knows' that it is sad on the second level of the hierarchy. However, due to the low precision on A2 this will not inform the first level, thus, the agent will be more informed by the actual perceptual information in a given trial. The experiment is described in Fig. 3. At trial 0 the agent is convinced enough that it is sad and calculates that its best action will be to go to the video. Paying attention to the other face, loosens its priors making them less informative about its emotional state. This is reflected in the categorical distribution at trial 1. It is more entropic or less precise as in trial 0. In trial 1 it decides to go to the mirror. Hence in trial 2, it makes an unexpected observation (seeing itself frown in the mirror) which also results in a drop in valence, indicating that this particular mirror encounter is negatively experienced. Having reaffirmed its belief that it is sad, it finds it best to go to the video. With this decision, the agent regains a bit of confidence in its action model. The valence goes up between trials 2 and 3. And due to the in build dynamics, its true state shifts to neutral. Finally, the agent notices the change in its true state to neutral and decides it's time to go to the mirror again, which even further improves its confidence in its action model and for the rest of the trials it will be happily smiling at itself in the mirror.

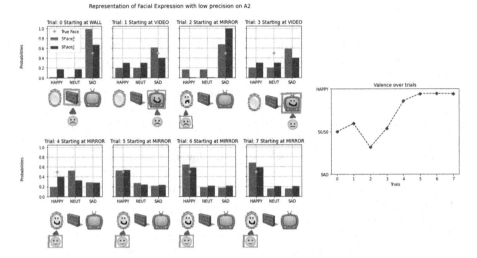

Fig. 3. Experiment 1. The belief distribution of My- facial-expression (SFace1 and SFace2) at the beginning of each trial are shown. Below that the agent with its true state is shown as smiley. It is indicated at what location it currently is. The green box indicates if its attention is on the exteroceptive or interoceptive observation. The graph next to it shows how the agent's valence evolves throughout the trials. The value is a probability, where a high value means high confidence in being in the state of 'positive valence'. At each trial, the agent has to choose where to go and hence at which location it will start the next trial. The valence graph shows how negative affect is elicited when the agent makes an unexpected observation in the mirror and therefore learns something new about its internal states.

3.2   Experiment 2: I Am Sad and I Know It and I Am Aware of It

We explored how the behaviour of the agent changes if its first-order states are introspectively available to it. The experiment is described in Fig. 4. We set the same initial state as in the previous study: true state is 'sad' and first location is the 'wall'. Differently, this time the mapping A2 is very precise.

At the first time point of each trial, the beliefs on both levels of the hierarchy are the same due to the almost one-to-one mapping of A2. The agent's behaviour differs from previous experiment in that it decides to stay at the video until trial 3. Only after a much longer time—when its priors have loosened enough—it tries out the mirror again. When this happens (in trial 4) we observe again a reconfirmation of its belief of being sad. This is accompanied by the same drop in valence once it realizes it wasn't the best action to go to the mirror. When going back to the video the agent has the chance to pick up on the change of its true state. However, it misses it because it keeps its prior belief of being sad extended through time.

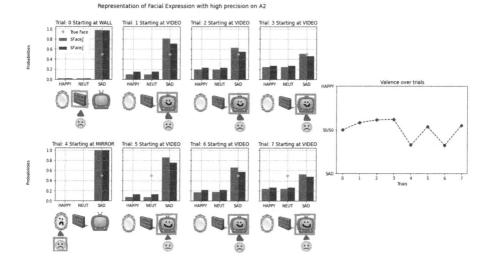

Fig. 4. Experiment 2. The agent, now being more aware of its own internal state, is anticipating an uncanny encounter with the mirror. Hence it is avoiding the mirror longer. However, it is also less able to pick up on a change in the true state of its emotion due to it expecting, much stronger, that it is actually sad. Its valence only shifts between sad and neutral. The results show how having strong priors about its emotional state and being able to attend to it discourages exploratory behaviour such as going to the mirror and learning about itself.

4 Discussion

The experiments showed a possible self-recognition process where the agent gets insight into its own generative model due to observations about itself made available through the mirror. The valence of the agent was coupled to this observation being surprising or not. The results show, how the valence changes and how the agents favoured actions change as a result of a change in valence. Thus, modelling mirror self-recognition as an internal shift of attention shows how negative and positive valence plausibly arises. The mirror self-recognition provides the agent with new self-knowledge, which can be used by deeper levels in the hierarchy to perform further inference. For example, changing precision estimates, thereby possibly favouring different actions, which in turn results in a change in action-based valence. We do not model how self-knowledge first arises, but what can be shown here is that negative valence arises in self-recognition processes that yield insights about oneself, which change the best available action. The negative valence is not directly dependent on the agent feeling sad or happy (its emotional states that the agents tries to infer), but rather about the (accurate) knowledge the agent has about these states. It is important to highlight that emotion is a more complex phenomenon that is likely constituted by many more dimensions than just valence [8]. Therefore, this computational model only offers the first steps, namely trying to account for the valenced part of mirror self-recognition.

Besides, only using a categorical attention internal state is a strong simplification of the reflection of one's physical appearance for visual-kinesthetic matching as shown by [14,16]. Mirror self-recognition in humans may additionally involve further internal attentional dynamics.

Meta-awareness. The capacity of meta-awareness allows an agent to change the strength with which one is aware of oneself. From dreaming to being awake, from being lost in thought to paying attention, humans in full possession of a self-concept do it all the time. The model behaviour in experiment 2 (Fig. 4) shows how meta-awareness is important to explain mirror avoidance and engagement behaviour. Being highly aware of a negative state of self an agent can anticipate an unsettling mirror encounter and prefers to avoid the mirror. Although at the cost of potentially missing a change in its true state. Given the limitations of the model, these statements are speculative. By expanding the model in future research one can potentially address open questions such as mirror avoidance and modelling mirrors in therapy [9]. Airing on the side of caution, even if the proposed computational model here does not simulate self-awareness, it can be used to pose interesting questions about action dependent affect in mirror self-recognition for future work. For example what are the actions available to an infant recognizing itself in the mirror? Is its negative affect resulting from suddenly being suspect of its usual policy of playful engagement with the other in the mirror? Or is it a feeling of alienation? If one prefers to interpret the negative affect as a feeling of alienation one could argue to expand the model to include mental actions. Planning on the second level (mental actions) could have its own confidence and valence associated with them. Actions on this (or even higher levels) could answer more existential questions such as what kind of person should I be? How do others see me? Tracking the expected confidence in one's mental actions might be an interesting choice to model more complex emotions such as the feeling of alienation. It could be interesting to design clever mirror tests, that involve different action affordances to test different stages of self-awareness more specifically.

5   Conclusion

This thesis proposes an affective self-recognition model based on the formalization of action dependent valence, using hierarchical active inference. As a proof of concept, we have shown how a synthetic affective response towards one's mirror image might arise. The results show that mirror self-recognition provides the agent with new information, which changes the favoured strategy and hence leads to negative valence. Secondly, the results show how an active inference agent with high meta-awareness of a negative evaluated state of self displays mirror avoidance behaviour. Therefore emphasizing the importance of deeper hierarchical layers, regarded as meta-cognition and meta-awareness, to explain more complex behaviours seen when facing the MSR test.

Appendix

Discrete Hierarchical Active Inference

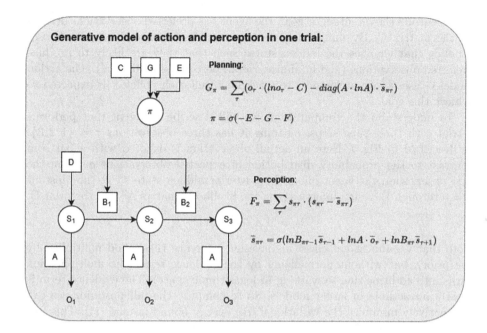

Fig. 5. A generative model of one trial with three time steps. Rectangles correspond to categorical distributions and circles to the random variables the agent wants to learn (here hidden states and policies). A is the likelihood that defines how likely observations are given the state $P(o_\tau|s_\tau)$. B encodes the probability of moving from one state into the next one $P(s_{\tau+1}|s_\tau, \pi)$ given the policy π. C is a vector or matrix that encodes which observations the agent prefers. D gives the prior at the first time step in the trial to perform Bayesian inference. G is the expected free energy. The best policy is the one that minimizes G (future reward + information gain) and F (current perceptual evidence or prediction error). F is also calculated for each policy meaning that the agent has a posterior state estimate for all possible policies. Lastly, E sets a 'habitual' prior for policies in case G is uninformative. Note that the past message $\ln B_{\pi\tau-1}\bar{s}_{\tau-1}$ at the first time point becomes the prior $\ln D$ and at the last time point the future message becomes ones (hence uninformative). Finally, the actual observation is marked with a bar $\bar{o}$ in contrast to the predictive posterior over observations o. Figure adapted from [13].

Figure 5 shows a generative model used for discrete hierarchical active inference. The agent's hidden state is s_τ, where τ indexes the time step. At each step, the agent gets an observation from the environment o_τ. Following active inference simplified notation [5] we will use capital letters to define the probabilistic functions. The agent has a prior belief D about hidden states s_τ, a likelihood mapping

A between states and observations ($P(o_\tau|s_\tau)$), and a transitions matrix B that encodes how states evolve over time depending on the policy π ($P(s_{\tau+1}|s_\tau, \pi)$). The agent can invert the generative model to perform Bayesian inference and get from an observation to a posterior over hidden states (in active inference this inference process is equated with perception).

To encode the intention or goal, the agent has preferred observations defined by the matrix C. By minimizing the expected free energy the agent chooses a policy that changes the hidden states such that they are likely to produce preferred observations (and minimize overall perceptual ambiguity). The action model G uses those preferences to track how well each policy π is expected to achieve this goal.

To understand the computations we will describe an agent that performs a trial with three time steps, meaning it has three observations $\tau = \{1, 2, 3\}$, as described in Fig. 5. Here an actual observation is denoted with a bar $\bar{o}$ in contrast to the probability distribution of expected observations o. From the first observation the agent infers the posterior hidden state $\bar{s}_\tau$ at time instant $\tau = 0$ through Bayesian inference, via the likelihood matrix A^2 and the prior D:

$$\bar{s}_\tau = \ln A \cdot \bar{o}_\tau + \ln D$$

Note that we almost get classical Bayesian inference (likelihood multiplied by the prior), but without normalizing by the evidence term (the multiplication turns into addition due to working in logarithmic space). The evidence term is mostly intractable in larger models. So to compute the full posterior, we can alternatively minimize the variational free energy bound instead [11]. This free energy formulation in discrete state space boils down to the difference in the belief the agent has about the world before (prior s_τ) and after (posterior $\bar{s}_\tau$) an observation.

$$F = s_\tau - \bar{s}_\tau$$

In other words, the free energy encodes the prediction error. If the prior belief matches or is supported by the observation the free energy is low. In contrast to a surprising observation that renders the prior belief less likely and therefore increases F. The agent can predict observations with the generative model and the belief about hidden states. It evaluates these predictions by how well they compare to the actual evidence, by calculating F. Then the agent can iteratively make predictions that will decrease F and hence lead to more accurate estimates of hidden states.

We have described how the agent updates its belief $\bar{s}_\tau$ about the world by trying to minimize prediction errors, therefore getting good at expecting what is really out there. Next, the agent also computes the expected observations to optimize not only for the current time point but for the whole trial. Under each policy or plan of actions π the agent can evaluate, how likely certain observations are in the future. Additionally, it can consider how ambiguous possible

---

[2] Using the discrete space formulation of active inference in matrix form this is computed by selecting the right column of the matrix A, i.e., through a one-hot observational vector.

future observations are. Both, information gain and preferred observations, are described in the expected free energy G:

$$G = \sum_\tau (o_\tau \cdot (\ln o_\tau - C) - \mathrm{diag}(A \cdot \ln A) \cdot \bar{s}_\tau)$$

The first part of the equation is the average difference in expected observations o_τ and preferred outcomes C over all time points. The second part relates to the model entropy or how precise the distribution is from which the expected observations are sampled. For each state at time τ there is a likelihood A that can give the agent more or less certainty about what outcome to expect.

To sum up, the agent minimizes F to optimize the posterior belief about states (estimation) and minimizes G to compute which policy to choose (action). Lastly, marginal message passing is just a mathematical way of sending information across time. For example, if the agent already knows which policy it is likely to take after seeing the first observation then that knowledge can, through marginal message passing, already inform its prior at the next time point in the trial. Vice versa the agent can update past beliefs, based on new observations which later can be helpful for learning. This leaves us with the final equations for posterior state estimation including future and past messages (using the transition matrix B) and the average free energy over timepoints in one trial:

$$\bar{s}_\tau = \sigma(\ln B_{\tau-1}\bar{s}_{\tau-1} + \ln A \cdot \bar{o}_\tau + \ln B_\tau \bar{s}_{\tau+1}) \tag{1}$$

$$F = \sum_\tau s_\tau \cdot (s_\tau - \bar{s}_\tau) \tag{2}$$

where σ is a softmax function that normalizes the input vector such that it sums to 1 and forms a proper probability distribution. The G and these two equations defined as shown in Fig. 5, describe the agents' basis to act in the world within a given trial. A limitation of this scheme is the static nature of the prior D at the beginning of each trial. It would be preferable that the agent can update/learn its prior based on the information it gathered in a trial. To make the context in which the agent navigates learnable one can expand the generative model with a deep temporal layer [11]. This allows the agent to form abstract and contextual beliefs that carry across trials—as described in the next subsection.

Affective Self-recognition Model Implementation Details

This section provides details about the generative model. Simulations were run by extending the pymdp infer-actively framework on github. The inference process of state estimation and policy selection on the first layer has been calculated using the pymdp framework. Inference on the second level, via ascending and descending messages was programmed for this setup. A commented code is available on github via this link.

First Layer

The priors on the state factors are specified in the D matrix. For the state factor 'Location' (Mirror, Wall, Video) the prior is uniform. The state factor 'Other emotional state' (Happy, Neutral, Sad, Null), which can be inferred via the observations (Smile, Neutral, Frown, None) also has a uniform prior. The prior on 'Self emotional state' depends on the starting condition and the second layer. Lastly, the state 'Mirror-controlled attention' (don't attend, attend) is set on don't attend:

$$P(S_{T_0}^{MC-Attention}) = [0.99, 0.01]$$

For each observation, there is a likelihood tensor A_{1-3}. The first observation is exteroceptive (Smile, Neutral, Frown, None), the second interoceptive (Smile, Neutral, Frown) and the third an observation about the location (which ensures the agent always knows where she is). The dimensions of the likelihoods are the observation and all the hidden state factors, i.e.: A[Observation, Location, Other, Self, Attention] or $A_1[4, 3, 3, 3, 2]$. For example, if I want to index the likelihood of my exteroceptive observation given that I am looking at the wall:

$$P(O_{ex}|S^{Location} = Wall, S^{Self}, S^{Other}, S^{MC-Attention}) =$$
for i,j in 0:2, k in 0:1

$$A_1[:, 1, i, j, k] = \begin{pmatrix} 0.01 \ Smile \\ 0.01 \ Neutral \\ 0.01 \ Frown \\ 0.97 \ None \end{pmatrix}$$

Basically saying the agent knows her probability of seeing 'None' if she is at the wall is 0.97, independent of all the other states she is in. If the agent is in the 'attend' state she is attending to herself and therefore can only relate the information of the exteroceptive observation to herself. This has to be defined for all states, but effectively the agent only makes use of this attention when she is in front of the mirror and the exteroceptive observation in fact relates to her:

$$P(O_{ex}|S^{MC-Attention} = attend, S^{Self}, S^{Location}) =$$
for l,i in 0:3 :

$$A_1[:, l, i, :, 0] = \begin{pmatrix} 0.97 \ 0.01 \ 0.01 \ Smile \\ 0.01 \ 0.97 \ 0.01 \ Neutral \\ 0.01 \ 0.01 \ 0.97 \ Frown \\ 0.01 \ 0.01 \ 0.01 \ None \end{pmatrix}$$

Here the columns stand for the different states in the state factor 'Self emotional state' (Happy, Neutral, Sad). If the agent is not paying attention we get the same matrix, but this time relating to the state of the other.

$P(O_{ex}|S^{MC-Attention} = \text{don't attend}, S^{Location}, S^{Other}) =$
for l,j in 0:3 :

$$A_1[:,l,:,j,1] = \begin{pmatrix} 0.97 \ 0.01 \ 0.01 \ Smile \\ 0.01 \ 0.97 \ 0.01 \ Neutral \\ 0.01 \ 0.01 \ 0.97 \ Frown \\ 0.01 \ 0.01 \ 0.01 \ None \end{pmatrix}$$

Now for the interoceptive observation, the precision on A will depend on the state of attention the agent is in. Therefore one can push A through a softmax with a precision (inverse temperature) parameter c.

$P(O_{in}|S^{MC-Attention}, S^{Location}, S^{Other}) =$
for l,i in 0:3 and k in 0:1 :

$$A_2[:,l,i,:,0] = \begin{pmatrix} 0.97 \ 0.01 \ 0.01, \ c \ Smile \\ 0.01 \ 0.97 \ 0.01, \ c \ Neutral \\ 0.01 \ 0.01 \ 0.97, \ c \ Frown \end{pmatrix}$$

Where paying attention has $c = 5$ and not paying attention $c = 0.001$. Finally, the location observation is a 1 to 1 mapping:

$P(O_{loc}|S^{MC-Attention}, S^{Self}, S^{Other}) =$

$$A_3[:,l,i,:,0] = \begin{pmatrix} 1 \ 0 \ 0 \ Mirror \\ 0 \ 1 \ 0 \ Wall \\ 0 \ 0 \ 1 \ Video \end{pmatrix}$$

Next the transition matrices B need to be defined. The rows correspond to the state in the next time step and columns the state in the current time step. The transition for the location depends on the action chosen and the agent knows with certainty where she will be next. The agent also knows that her attention state will shift to focused when she goes to the mirror and unfocused going to the video. The agent has a bit of uncertainty around how her own emotional state is changing in time and a bit more uncertainty about how the state of the other is changing.

$P(S_{\tau+1}^{Self}|S_{\tau}^{Self}) =$

$$B_1[:,:,0] = \begin{pmatrix} 0.95 \ 0.05 \ 0.05 \\ 0.05 \ 0.95 \ 0.05 \\ 0.05 \ 0.05 \ 0.95 \end{pmatrix}$$

$P(S_{\tau+1}^{Other}|S_{\tau}^{Other}) =$

$$B_2[:,:,0] = \begin{pmatrix} 0.8 \ 0.1 \ 0.1 \\ 0.1 \ 0.8 \ 0.1 \\ 0.1 \ 0.1 \ 0.8 \end{pmatrix}$$

The preference are set with the C matrix. For all observation modalities C will be initiated with zeros. Then the preference to see self happy or neutral can be encoded as:

$C_1[0] = \mathbf{3.0}$
$C_1[1] = \mathbf{3.0}$

The description of the first layer concludes with the policies available to the agent. They are any combination of going to a location that is possible within a trial. The trials consist of three observations and 2 actions. The agent starts by sampling an observation then decides where to go, and repeats this step. After the final observation, the agent doesn't need to go anywhere because the trial is over and will start again from the beginning.

Second Layer

The A and B matrix for the Valence state are the same as in [13]:

$$A2_{valence}[:,:] = \begin{pmatrix} \mathbf{0.97} & \mathbf{0.3} \\ \mathbf{0.3} & \mathbf{0.97} \end{pmatrix}$$

$$B2_{valence}[:,:] = \begin{pmatrix} \mathbf{0.8} & \mathbf{0.3} \\ \mathbf{0.2} & \mathbf{0.7} \end{pmatrix}$$

For the state S2Face or 'Mood', the A2 matrix can again be changed with a precision parameter c. This one is set manually to simulate meta-awareness. In my simulation high means c = 5 and low c = 1.

$$A2_{Face}[:,:] = \begin{pmatrix} \mathbf{1} & \mathbf{0,} & \mathbf{c} \\ \mathbf{0} & \mathbf{1,} & \mathbf{c} \end{pmatrix}$$

$$B2_{Face}[:,:] = \begin{pmatrix} \mathbf{1} & \mathbf{0} & \mathbf{0} \\ \mathbf{0} & \mathbf{1} & \mathbf{0} \\ \mathbf{0} & \mathbf{0} & \mathbf{1} \end{pmatrix}$$

This concludes the description of the two-layered generative model.

References

1. Amsterdam, B.: Mirror self-image reactions before age two. **5**(4), 297–305 (1972). https://doi.org/10.1002/dev.420050403
2. Anderson, J.R., Gallup, G.G.: Which primates recognize themselves in mirrors? PLoS Biol. **9**(3), e1001024 (2011). https://doi.org/10.1371/journal.pbio.1001024
3. Bard, K.A., Todd, B.K., Bernier, C., Love, J., Leavens, D.A.: Self-awareness in human and chimpanzee infants: what is measured and what is meant by the mark and mirror test? Infancy **9**(2), 191–219 (2006). https://doi.org/10.1207/s15327078in0902_6
4. Broesch, T., Callaghan, T., Henrich, J., Murphy, C., Rochat, P.: Cultural variations in children's mirror self-recognition. J. Cross-Cult. Psychol. **42**(6), 1018–1029 (2010). https://doi.org/10.1177/0022022110381114
5. Costa, L.D., Parr, T., Sajid, N., Veselic, S., Neacsu, V., Friston, K.: Active inference on discrete state-spaces: a synthesis. J. Math. Psychol. **99**, 102447 (2020). https://doi.org/10.1016/j.jmp.2020.102447

6. Deane, G.: Dissolving the self. Philos. Mind Sci. **1**(I), 1–27 (2020). https://doi.org/10.33735/phimisci.2020.i.39
7. Deane, G.: Consciousness in active inference: deep self-models, other minds, and the challenge of psychedelic-induced ego-dissolution. Neurosci. Conscious. **2021**(2) (2021). https://doi.org/10.1093/nc/niab024
8. Fontaine, J.R., Scherer, K.R., Roesch, E.B., Ellsworth, P.C.: The world of emotions is not two-dimensional. Psychol. Sci. **18**(12), 1050–1057 (2007). https://doi.org/10.1111/j.1467-9280.2007.02024.x
9. Freysteinson, W.M.: Demystifying the mirror taboo: a neurocognitive model of viewing self in the mirror. Nurs. Inq. **27**(4) (2020). https://doi.org/10.1111/nin.12351
10. Freysteinson, W.M., Deutsch, A.S., Lewis, C., Sisk, A., Wuest, L., Cesario, S.K.: The experience of viewing oneself in the mirror after a mastectomy. Oncol. Nurs. Forum **39**(4), 361–369 (2012). https://doi.org/10.1188/12.onf.361-369
11. Friston, K., FitzGerald, T., Rigoli, F., Schwartenbeck, P., Pezzulo, G.: Active inference: a process theory. Neural Comput. **29**(1), 1–49 (2017). https://doi.org/10.1162/neco_a_00912
12. Gallup, G.G.: Chimpanzees: self-recognition. Science **167**(3914), 86–87 (1970)
13. Hesp, C., Smith, R., Parr, T., Allen, M., Friston, K.J., Ramstead, M.J.D.: Deeply felt affect: the emergence of valence in deep active inference. Neural Comput. **33**(2), 398–446 (2021). https://doi.org/10.1162/neco_a_01341
14. Hoffmann, M., Wang, S., Outrata, V., Alzueta, E., Lanillos, P.: Robot in the mirror: toward an embodied computational model of mirror self-recognition. KI - Künstliche Intell. **35**(1), 37–51 (2021). https://doi.org/10.1007/s13218-020-00701-7
15. Lanillos, P., Dean-Leon, E., Cheng, G.: Enactive self: a study of engineering perspectives to obtain the sensorimotor self through enaction. In: 2017 Joint IEEE International Conference on Development and Learning and Epigenetic Robotics (ICDL-EpiRob), pp. 72–78. IEEE (2017)
16. Lanillos, P., Pages, J., Cheng, G.: Robot self/other distinction: active inference meets neural networks learning in a mirror (2020)
17. Parr, T., Markovic, D., Kiebel, S.J., Friston, K.J.: Neuronal message passing using mean-field, bethe, and marginal approximations. Sci. Rep. **9**(1), 1889 (2019). https://doi.org/10.1038/s41598-018-38246-3
18. Rochat, P.: Five levels of self-awareness as they unfold early in life. Conscious. Cogn. **12**(4), 717–731 (2003). https://doi.org/10.1016/s1053-8100(03)00081-3
19. Rochat, P., Broesch, T., Jayne, K.: Social awareness and early self-recognition. Conscious. Cogn. **21**(3), 1491–1497 (2012). https://doi.org/10.1016/j.concog.2012.04.007
20. Sandved-Smith, L., Hesp, C., Mattout, J., Friston, K., Lutz, A., Ramstead, M.J.D.: Towards a computational phenomenology of mental action: modelling meta-awareness and attentional control with deep parametric active inference. Neurosci. Conscious. **2021**(1) (2021). https://doi.org/10.1093/nc/niab018
21. Smith, R., Friston, K., Whyte, C.: A step-by-step tutorial on active inference and its application to empirical data (2021). https://doi.org/10.31234/osf.io/b4jm6

World Model Learning from Demonstrations with Active Inference: Application to Driving Behavior

Ran Wei[1]([envelope]), Alfredo Garcia[1], Anthony McDonald[1,2], Gustav Markkula[3], Johan Engström[4], Isaac Supeene[4], and Matthew O'Kelly[4]

[1] Texas A&M University, College Station, USA
wei.ran@tamu.edu
[2] University of Wisconsin, Madison, USA
[3] University of Leeds, Leeds, UK
[4] Waymo LLC, Mountain View, USA

Abstract. Active inference proposes a unifying principle for perception and action as jointly minimizing the free energy of an agent's internal world model. In the active inference literature, world models are typically pre-specified or learned through interacting with an environment. This paper explores the possibility of learning world models of active inference agents from recorded demonstrations, with an application to human driving behavior modeling. The results show that the presented method can create models that generate human-like driving behavior but the approach is sensitive to input features.

Keywords: Active inference · Inverse reinforcement learning · Driving behavior modeling

1 Introduction

Active inference proposes a unifying principle for perception and action as jointly minimizing the free energy of an agent's internal generative model [6]. It has been strongly influential in contemporary neuroscience and cognitive science. More recently, active inference has been proposed as a framework for modeling driving behavior, both at the conceptual [5,10] and computational levels [31]. The framework is attractive for computational driver behavior modeling as it enables the learning of complex behaviors from large amounts of driving data while at the same time being grounded in a fundamental theory of cognition and behavior which guides model design and enables increased interpretability of machine-learned models. However, most existing active inference models in the cognitive neuroscience literature address relatively simple toy problems. Thus, the scaling of active inference by means of modern machine learning techniques is currently an active area of research [29]. The novel contribution of this paper

C. L. Buckley et al. (Eds.): IWAI 2022, CCIS 1721, pp. 130–142, 2023.
https://doi.org/10.1007/978-3-031-28719-0_9

is to explore the application of active inference models in the context of learning human driving behavior from recorded data (i.e., Learning from Demonstration; LfD).

LfD provides an efficient alternative to the current manual specification or trial-and-error learning approaches to active inference model design. Assuming the demonstrating agent is an active inference agent, we can instead estimate the agent's generative model, consisting of a world model and a preference model, from demonstrated behavior. This approach is similar to inverse reinforcement learning (IRL) [20,33] with an important difference. Instead of using a single reward function, active inference explains the demonstrator with a world model-preference pair, which makes active inference more transparent about the agent's decision process than traditional IRL methods because we can introspect the learned world model. This allows us to understand variations in human behavior as "optimal inference in suboptimal models" [26,31].

The closest approaches to the work presented here are [1,11,15,23]. We build on these works by jointly estimating agent world model and preference model from demonstration. However, our work differs from these approaches in that it does not assume the environment is fully observable as in [23], it makes no assumptions about the agent's world model's alignment with the environment in light of the active inference formulation [11,15], and it focuses on a large continuous environment rather than a small discrete environment [1]. We demonstrate our method in continuous car following scenarios recorded on highways [32]. The learned driving policy jointly models its own states, road geometry, and other vehicles (i.e., agents) using discrete abstract states and implements continuous vehicle control. We show that this approach can mimic human driving behavior in simple scenarios but that it may learn an incorrect model of the world, known as "causal confusion" in LfD [4], and occasionally deviate from the lane. We further show that this deviation can be corrected by revising the observation set based on grounded theory of driver steering [25], thus illustrating the how inductive biases and domain knowledge can be injected into LfD approaches.

2 Active Inference Model of Highway Driving

In this section, we propose a mixed discrete-continuous active inference model of driving behavior and present the update rules for driver perception and control by minimizing expected free energy.

2.1 World Model

We model the driver's perceptual process using a discrete-time controlled hidden Markov process with discrete hidden states $s \in \mathcal{S}$, discrete actions $a \in \mathcal{A}$, and continuous observations $o \in \mathcal{O}$. The hidden states are the driver's *internal* representation of the driving environment which is used to guide action selection (e.g. steering and braking). The discrete actions represent driving motor primitives (i.e., prototype actions as described in [16]). The continuous observations are a vector of signals known to influence driving control behavior (e.g., visual looming of the lead vehicle [18]). The state evolves according to a Markov chain with transition probabilities $P(s_{t+1}|s_t, a_t)$. The driver cannot directly observe the state but a high dimensional continuous signal o_t with distribution $P(o_t|s_t)$. Importantly, the definition of states and the corresponding transition and observation probabilities are free to deviate from the actual environment as long as they explain the demonstrated behavior.

2.2   A POMDP Formulation of Active Inference

Given the world model, the agent's perception-action loop at every decision epoch consists of inferring a belief distribution on the current hidden state and selecting an action controlling the evolution of the hidden state. Active inference posits the minimization of *free energy* as a unifying principle for describing the perception-action loop.

Let $h_t = \{o_t, ..., o_0, a_{t-1}, ..., a_0\} \in H_t$ denote the observable history of the dynamic decision process including all past and present revealed observations and all implemented actions up to time $t > 0$, where $H_t \triangleq \mathcal{O}^t \times \mathcal{A}^{t-1}$.

According to the free energy minimization principle, the agent's belief distribution at time $t > 0$ which we denote by $b_t(s_t)$ must correspond to the Bayes updated belief distribution on the state s_t, i.e. the conditional probability distribution of s_t given history h_t, i.e. $b_t(s_t) = P(s_t|h_t)$. The active inference model of the perception-action loop assumes the agent has preferences over hidden states s_{t+1} which are represented by a probability distribution $\tilde{P}(s_{t+1})$. The expected free energy associated with the choice of action a_t and current belief distribution b_t at time $t > 0$ can be written as [3]:

$$EFE(b_t, a_t) = \mathbb{E}\big[D_{KL}\big(b_{t+1}||\tilde{P}\big)\big] + \mathbb{E}[\mathcal{H}(o_{t+1})] \tag{1}$$

where the first expectation is taken with respect to

$$
\begin{aligned}
P(o_{t+1}|b_t, a_t) &:= \sum_{s_{t+1}} P(o_{t+1}|s_{t+1})P(s_{t+1}|b_t, a_t) \\
&= \sum_{s_{t+1}} P(o_{t+1}|s_{t+1}) \sum_{s_t} P(s_{t+1}|s_t, a_t)b(s_t)
\end{aligned}
\tag{2}
$$

and $D_{KL}\big(b_{t+1}||\tilde{P}\big)$ is the Kullback-Leibler divergence between the random belief distribution $b_{t+1}(\cdot) = P(\cdot|h_t \cup \{o_{t+1}, a_t\})$ and $\tilde{P}(\cdot)$. $\mathbb{E}[\mathcal{H}(o_{t+1})]$ is the entropy of the observables expected under the predictive distribution $P(s_{t+1}|b_t, a_t)$ defined in (2). The first term in (1) is a measure of the extent to which the belief distribution b_{t+1} (resulting from implementing action a_t and recording observation o_{t+1}) differs from the preferred one $\tilde{P}$. Let $\pi \in \Pi$ denote a randomized action selection policy conditioned on the history of the process, i.e. $\pi(a|h_t) \in [0, 1], a \in \mathcal{A}$ and $\sum_{a \in \mathcal{A}} \pi(a|h_t) = 1$ for all $h_t \in H_t$. An information processing cost is modeled as the Kullback-Leibler divergence between policy π and a default a priori control policy π_0 which is oblivious to new information [21,28] i.e.:

$$D_{KL}(\pi(\cdot|h_t)||\pi_0) := \sum_{a \in \mathcal{A}} \pi(a|h_t) \log \frac{\pi(a|h_t)}{\pi_0(a)}$$

With a uniform default distribution, $D_{KL}(\pi(\cdot|h_t)||\pi_0) = \mathbb{E}_{\pi(a|h_t)} \log \pi(a|h_t) - \log |\mathcal{A}|$. For a finite planning horizon T, the *active inference* controller is the solution to the problem:

$$\mathcal{G}_\tau^*(h_\tau) \triangleq \min_{\pi \in \Pi} \mathbb{E} \Big[\sum_{t \geq \tau}^{T} (EFE(b_t, a_t) + \log \pi(a_t|h_t)) \Big] \tag{3}$$

The combination of additive structure and Markovian dynamics allows for a recursive characterization of the optimal policy as follows:

$$\mathcal{G}_t^*(h_t) = \min_{\pi \in \Pi} \Bigg\{ \sum_{a_t \in \mathcal{A}} \pi(a_t|h_t) \Big[\\ EFE(b_t, a_t) + \log \pi(a_t|h_t) + \int_{\mathcal{O}} P(o_{t+1}|h_t, a_t) \mathcal{G}_{t+1}^*(h_{t+1}) do_{t+1} \Big] \Bigg\} \tag{4}$$

where $h_{t+1} = h_t \cup \{o_{t+1}, a_t\}$. Note that with no loss of generality the recursive equation can be expressed in terms of belief states b_t as opposed to the history h_t. The following is a standard result characterizing the optimal solution to (4) [7].

Proposition 1. *Let $\mathcal{G}_t^*(b_t, a_t)$ be defined as:*

$$\mathcal{G}_t^*(b_t, a_t) := EFE(b_t, a_t) + \log \pi(a_t|b_t) + \int_{\mathcal{O}} P(o_{t+1}|b_t, a_t) \mathcal{G}_{t+1}^*(b_{t+1}) do_{t+1}$$

The optimal policy is of the form:

$$\pi(a|b_t) = \frac{e^{-\mathcal{G}_t^*(b_t, a)}}{\sum_{\tilde{a} \in \mathcal{A}} e^{-\mathcal{G}_t^*(b_t, \tilde{a})}} \tag{5}$$

2.3 Estimation of POMDP Model

Given the model for the active inference controller described above, in this section, we describe the problem of estimating such a model given recorded sequences of actions and observables. This is akin to *inverse* learning a POMDP model (see Sect. 4.7 in [22]).

In what follows we consider a parametrization of observation probabilities $P_{\theta_1}(o_{t+1}|s_{t+1})$ and state-dynamics $P_{\theta_1}(s_{t+1}|s_t, a_t)$ with $\theta_1 \in \mathbb{R}^{p_1}$ where $p_1 > 0$. Given data in the form of finite histories $h_{T,i} = \{(o_{t,i}, a_{t,i})\}_{t=0}^{T}$ for $i \in \{1, \ldots, N\}$, a sequence of belief trajectories $\{b_{t,\theta_1,i}\}_{t=0}^{T}$ can be recursively computed for a fixed value of θ_1.

Assuming preferences over hidden states are parametrized $\tilde{P}_{\theta_2}(s_{t+1})$ with $\theta_2 \in \mathbb{R}^{p_2}$ with $p_2 > 0$, the log-likelihood of observed actions can be written as:

$$\log \ell(\theta) = \sum_{i=1}^{N} \sum_{t=0}^{T-1} \log \pi_\theta(a_{t,i}|b_{t,\theta_1,i}) \tag{6}$$

where $\pi_\theta(\cdot|b_{t,\theta_1,i})$ is the optimal policy in (5) and $\theta := (\theta_1, \theta_2)$.

(6) can be optimized using a nested-loop algorithm alternating between **(i)** a parameter update step at iteration $k > 0$ in which we set θ^{k+1} as the solution to:

$$\max_\theta \sum_{i=1}^{N} \sum_{t=0}^{T-1} \log \pi_\theta(a_{t,i}|b_{t,\theta_1^k,i}) \quad \text{s.t.} \quad \pi_\theta(a_t|b_t) = \frac{e^{-\mathcal{G}_{t,\theta^k}^*(b_t,a_t)}}{\sum_{\tilde{a}_t \in \mathcal{A}} e^{-\mathcal{G}_{t,\theta^k}^*(b_t,\tilde{a}_t)}}$$

where $\mathcal{G}_{t,\theta^k}^*$ denotes the current free energy function and **(ii)** solving for the free energy function $\{\mathcal{G}_{t,\theta^{k+1}}^*\}_t$ given the new parameter values.

3   Implementation

In this section, we first describe the signals assumed to be observed by the drivers during a car-following scenario and defer a detailed description of the dataset to appendix A.1. We then describe the model fitting process with an augmentation of the model to continuous braking and steering control. Finally, we describe the procedure for model comparison.

3.1   Driver Observations

We leveraged prior works on driver behavior theory [17,18,25] to define the observation vector o used in the car-following task. Markkula et al. [17] proposed visual looming denoted by τ^{-1} as a central observation signal in human longitudinal vehicle control, which is defined as the derivative of the optical angle of the lead vehicle subtended on the driver's retina divided by the angle itself: $\tau^{-1} = \dot{\theta}/\theta$. Salvucci & Gray [25] proposed a two-point model of human lateral vehicle control where the human driver controls the vehicle by representing road curvature with a near-point, assumed at a fixed distance in front of the vehicle, and a far-point, assumed to be the lead vehicle in the car-following context, and steers to minimize the deviation from a combination of the near and far-points. Using these insights, we designed an observation vector consisting of three sensory modalities:

1. The state of the ego vehicle in ego-centric coordinate
2. Relationships with the lead vehicle in ego-centric coordinates
3. Road geometry

We featurized the ego state with the longitudinal and lateral velocity and relationship to the lead vehicle with relative distance and speed with longitudinal and lateral components, and looming. To encode the road geometry in the two-point model, we used the lane center 30 m ahead of the current position as the near-point and the lead vehicle as the far-point and used as features the heading error from the near and far-points and lane-center distance to the current road position.

3.2 Model Fitting

We parameterized the hidden state transition probabilities $P(s_{t+1}|s_t, a_t)$ and preference distribution $\tilde{P}(s_t)$ with categorical distributions and observation probabilities $P(o_t|s_t)$ with multivariate Gaussian distributions. For a fixed belief vector b_t, the expected KL divergence and entropy in (1) can be computed in closed-form. We used the QMDP method [14] to approximate the cumulative expected free energy assuming the states will become fully observable in the next time step: $\mathcal{G}^*(b_t, a_t) \approx \sum_{s_t} b(s_t)\mathcal{G}^*(s_t, a_t)$. This allows us to train the model in automatic differentiation frameworks (e.g., Pytorch) using Value-Iteration-Networks style implementations [9,27].

In order to fit the discrete action model from Sect. 2 to continuous longitudinal and lateral controls, we extended the model with a continuous control module. Let u denote a multidimensional continuous control vector (longitudinal and lateral accelerations in the current setting), we modeled the mapping from a discrete action a to u using $P(u|a)$ parameterized as a multivariate Gaussian with its parameters added to vector θ_1. $P(u|a)$ thus automatically extracts primitive actions, such as different magnitudes of acceleration and deceleration [16], from data by adaptively discretizing the action space. We assume at a given time step t, the agent also performs a Bayesian belief update about the previous action realized with prior given by the policy $\pi(a_t|b_t)$ and the posterior $P(a_t|u_t) \propto P(u_t|a_t)\pi(a_t|b_t)$. The action log likelihood objective in (6) is modified as:

$$\log \ell(\theta) = \sum_{i=1}^{N} \sum_{t=0}^{T-1} \log \sum_{a_{t,i}} P_{\theta_1}(u_{t,i}|a_{t,i})\pi_\theta(a_{t,i}|b_{t,\theta_1,i}) \tag{7}$$

3.3 Model Comparison

We measured the quality of the trained agents by using a combination of offline and online testing metrics on a held-out dataset. For offline metrics, we used mean absolute error (MAE). For online metrics, we first ran the trained agents in a simulator that replayed the recorded trajectories of the lead vehicles and then recorded the final displacement and average lane deviation for each trajectory tested. The final displacement is defined as the distance between the final position reached by the trained agents and the final position in the dataset. The average lane deviation is the agents' distance to the tangent point on the lane center line averaged over all time steps in the trajectory.

We varied three aspects of the agents to compare with the canonical agent described previously. First, we examined the importance of the chosen features by replacing the near-point heading error and distance to lane center with distances to the left and right boundaries at the current road position, a feature set commonly used by driving agents for simulated testing [2,13]. We label the agents trained with the original two-point observation as "TP". Next, we examined the importance of grounding the world model in actual observations by adding an observation regularizer to the training objective with a coefficient of 0.01:

$$\mathcal{L}_{obs} = \sum_{t=1}^{T} \log P(o_t | h_t) \qquad (8)$$

This encourages the agent to have a more accurate belief about the world with higher observation likelihood under the agent's posterior beliefs. We label agents trained with this penalty "Obs". Finally, we examined the impact of agent planning objectives on the learned world model and behavior. We replaced EFE with an alternative objective called expected cross entropy (ECE):

$$ECE(b_t, a_t) = \mathbb{E}[\log \tilde{P}(o_{t+1})] \qquad (9)$$

which is the expected marginal likelihood of the agent preference model.

We used 30 states and 60 actions for all agents as they were sufficient to produce reasonable behavior. As a baseline, we trained a behavior cloning (BC) agent consisting of a recurrent and a feed-forward neural network to emulate the belief update and control modules of the active inference agent. We provide more details of the BC agent in appendix A.2.

4   Results and Discussions

Figure 1 shows the offline (left panel) and online (middle and right panels) testing metrics for each agent tested using the same set of 15 scenarios sampled from the held-out dataset, with the canonical agent labeled as "TP+EFE". The MAE of all active inference agents were between 0.11 and 0.14 m/s^2. The BC agent outperformed all agents with a MAE of 0.08, however the BC+TP agent had a higher MAE value of 0.135. This is likely due to the sensitivity to input features during training, despite better function approximation capability of neural networks. The final displacements were on average 13 m, the average lane deviation was 1.37 m, and no collision with the lead vehicle was observed. These metrics show that the agents can generate reasonable behavior by staying in the lane and following the lead vehicle (see a few sample trajectories generated in Fig. 3a).

Comparing across different agents, Fig. 1 shows that adding an observation penalty increased offline MAE, however, it did not noticeably affect the agents' online performance. This might be related to the objective mismatch problem in model-based reinforcement learning where a model better fitted to the observations may not enhance control capabilities [12]. The middle and right panels

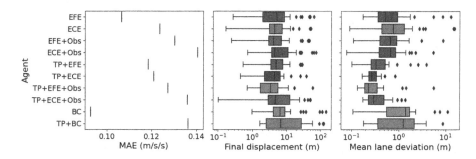

Fig. 1. Box plots of offline (column 1) and online (columns 2 & 3) performance metrics of the compared agents. Offline metrics are calculated on the entire held-out set. Each box plot in the online metrics shows the distribution of agent performance in 15 random held-out scenarios tested with 3 different random seeds.

show that some of the agents produced final displacements and lane deviation as large as 100 m and 15 m, respectively, as a result of deviating from the lane and failing to make corrections (see Fig. 3b). Interestingly, active inference agents using the two-point observations model generated noticeably less lane deviation than other agents (see Fig. 1 right with x axis in log-scale) despite similar performance in terms of offline metrics. This observation highlights the importance of incorporating generalizable features into agent world model.

Figure 2 shows a subset of the parameters of the learned world models. All panels ordered the states by desirability so that states with lower EFE are assigned smaller indices. The left panel plots the variance of the observation distribution for the relative distance feature against the states. The orange and blue lines represent the ECE and EFE objectives, respectively. This panel shows a clear increasing trend in the observation variance with the decrease of state desirability. The middle and right panels show the transition matrices controlled by the learned policy: $P^{\pi}(s'|s) = \sum_{a \in \mathcal{A}} P(s'|s, a)\pi(a|b = \delta(s))$, where $b = \delta(s)$ denotes a belief concentrated on a single state. Whereas the transition probabilities of the ECE agent spread more uniformly across the state space, the transition matrix of the EFE agent has a block-diagonal structure. As a result, it is difficult to traverse to the desirable states in the upper diagonal (states 0–24) from the undesirable states (states 24–30) in the lower diagonal. We have empirically observed that when the EFE agent deviates from the lane, its EFE values also increase significantly without it taking any corrective actions. This shows that the increasing variance played a more important role in determining the desirability of a state than the KL divergence from the preferred states.

The observation made in Fig. 2 is similar to the "causal confusion" problem in LfD [4]. In [4], the authors found that the learning agent may falsely attribute the cause of an action to previous actions in the demonstration rather than the observation signals and its own goals. Our agent exhibited a different type of "causal confusion" similar to the model exploitation phenomena in reinforcement learning [8], where the cause of an action is attributed to a model

138   R. Wei et al.

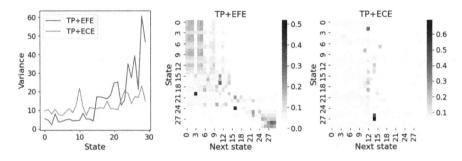

Fig. 2. Parameters of the learned world models. States are sorted by desirability (i.e., low expected free energy). **Left:** Observation variance vs. state. **Middle & right:** Heat map of controlled transition matrix. Darker color corresponds to higher transition probability.

with incorrect counterfactual state and observation predictions. The consequence is that the agent does not have the ability to make corrections when entering these states. However, learning the correct counterfactual states from demonstration is difficult because these states are rarely contained in the demonstration as the demonstrating agents are usually experts who rarely visit undesirable states. Prior works addressed this by interacting with an environment [30] and receiving real-time expert feedback [24]. We have instead partially alleviated this by designing domain specific features (i.e., the two-point observation model) to reduce the probability of the agent deviating from desired states. However, given active inference strongly relies on counterfactual simulation of the world model in the planning step, future work should focus on discovering the correct counterfactual states from human demonstrations using approaches at the model level rather than at the feature level, e.g., by constraining the model class or learning causal world models via environment interactions [4].

Acknowledgements. Support for this research was provided in part by grants from the U.S. Department of Transportation, University Transportation Centers Program to the Safety through Disruption University Transportation Center (451453-19C36), the U.S. Army Research Office (W911NF2210213), and the U.K. Engineering and Physical Sciences Research Council (EP/S005056/1). The role of the Waymo employees in the project is solely consulting including making suggestions and helping set the technical direction.

A   Appendix

A.1   Dataset

We used the INTERACTION dataset [32], a publicly available naturalistic driving dataset recorded with drone footage of fixed road segments, to fit a model of highway car-following behavior. Each recording in the dataset consists of the positions, velocities, and headings of all vehicles in the road segment at a sampling frequency 10 Hz. Specifically, we used a subset of the data[1] due to the abundance of car-following trajectories and relatively complex road geometry with road curvature and merging lanes. We defined car-following as the trajectory segments from the initial appearance of a vehicle to either an ego lane-change or the disappearance of the lead vehicle. Reducing the dataset using this definition resulted in a total of 1027 car-following trajectories with an average duration of 13 s and standard deviation of 8.7 s. We obtained driver control actions (i.e., longitudinal and lateral accelerations) by taking the derivative of the velocities of each trajectory. We then created a set of held-out trajectories for testing purposes by first categorizing all trajectories into four clusters based on their kinematic profiles using UMAP [19] and sampled 15% of the trajectories from each cluster.

A.2   Behavior Cloning Agent

The behavior cloning agents consist of a recurrent neural network with a single gated recurrent unit (GRU) layer and a feed-forward neural network. The GRU layer compresses the observation history into a fixed size vector, which is decoded by the feed-forward network into a continuous action distribution model by a multivariate Gaussian distribution. To make the BC agents comparable to the active inference agents, the GRU has 64 hidden units and 30 output units and the feed-forward network has 30 input units, 2 hidden layers with 64 hidden units, and SiLU activation function. We used the same observation vector as input to the BC agents as to the active inference agents.

A.3   Sample Path

Example sample paths generated by the agents with and without the two-point observation model.

---

[1] Recording 007 from location "DR_CHN_Merging_ZS".

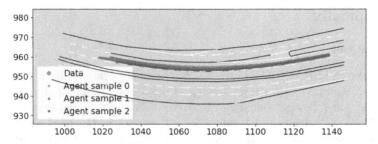

(a) Sample paths generated by active inference agent with the two-point observation model.

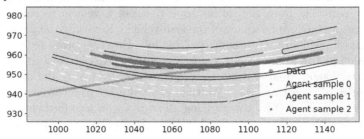

(b) Sample paths generated by active inference agent without the two-point observation model.

Fig. 3. Active inference agent sample path comparison.

References

1. Baker, C., Saxe, R., Tenenbaum, J.: Bayesian theory of mind: modeling joint belief-desire attribution. In: Proceedings of the Annual Meeting of the Cognitive Science Society, vol. 33 (2011)
2. Bhattacharyya, R., et al.: Modeling human driving behavior through generative adversarial imitation learning. arXiv preprint arXiv:2006.06412 (2020)
3. Da Costa, L., Parr, T., Sajid, N., Veselic, S., Neacsu, V., Friston, K.: Active inference on discrete state-spaces: a synthesis. J. Math. Psychol. **99**, 102447 (2020)
4. De Haan, P., Jayaraman, D., Levine, S.: Causal confusion in imitation learning. In: Advances in Neural Information Processing Systems, vol. 32 (2019)
5. Engström, J., et al.: Great expectations: a predictive processing account of automobile driving. Theor. Issues Ergon. Sci. **19**(2), 156–194 (2018)
6. Friston, K., FitzGerald, T., Rigoli, F., Schwartenbeck, P., Pezzulo, G.: Active inference: a process theory. Neural Comput. **29**(1), 1–49 (2017)
7. Haarnoja, T., Zhou, A., Abbeel, P., Levine, S.: Soft actor-critic: off-policy maximum entropy deep reinforcement learning with a stochastic actor. In: International Conference on Machine Learning, pp. 1861–1870. PMLR (2018)
8. Janner, M., Fu, J., Zhang, M., Levine, S.: When to trust your model: model-based policy optimization. In: Advances in Neural Information Processing Systems, vol. 32 (2019)

9. Karkus, P., Hsu, D., Lee, W.S.: QMDP-Net: deep learning for planning under partial observability. In: Advances in Neural Information Processing Systems, vol. 30 (2017)

10. Kujala, T., Lappi, O.: Inattention and uncertainty in the predictive brain. Front. Neuroergon. **2**, 718699 (2021)

11. Kwon, M., Daptardar, S., Schrater, P.R., Pitkow, X.: Inverse rational control with partially observable continuous nonlinear dynamics. In: Advances in Neural Information Processing Systems, vol. 33, pp. 7898–7909 (2020)

12. Lambert, N., Amos, B., Yadan, O., Calandra, R.: Objective mismatch in model-based reinforcement learning. arXiv preprint arXiv:2002.04523 (2020)

13. Leurent, E.: An environment for autonomous driving decision-making. https://github.com/eleurent/highway-env (2018)

14. Littman, M.L., Cassandra, A.R., Kaelbling, L.P.: Learning policies for partially observable environments: scaling up. In: Machine Learning Proceedings 1995, pp. 362–370. Elsevier (1995)

15. Makino, T., Takeuchi, J.: Apprenticeship learning for model parameters of partially observable environments. arXiv preprint arXiv:1206.6484 (2012)

16. Markkula, G., Boer, E., Romano, R., Merat, N.: Sustained sensorimotor control as intermittent decisions about prediction errors: computational framework and application to ground vehicle steering. Biol. Cybern. **112**(3), 181–207 (2018)

17. Markkula, G., Engström, J., Lodin, J., Bärgman, J., Victor, T.: A farewell to brake reaction times? kinematics-dependent brake response in naturalistic rear-end emergencies. Accid. Anal. Prev. **95**, 209–226 (2016)

18. McDonald, A.D., et al.: Toward computational simulations of behavior during automated driving takeovers: a review of the empirical and modeling literatures. Hum. Factors **61**(4), 642–688 (2019)

19. McInnes, L., Healy, J., Melville, J.: Umap: Uniform manifold approximation and projection for dimension reduction. arXiv preprint arXiv:1802.03426 (2018)

20. Ng, A.Y., Russell, S.J., et al.: Algorithms for inverse reinforcement learning. In: Icml, vol. 1, p. 2 (2000)

21. Ortega, P.A., Braun, D.A.: Thermodynamics as a theory of decision-making with information-processing costs. Proc. R. Soc. A: Math. Phys. Eng. Sci. **469**(2153), 20120683 (2013)

22. Osa, T., Pajarinen, J., Neumann, G., Bagnell, J.A., Abbeel, P., Peters, J.: An algorithmic perspective on imitation learning. Found. Trends Rob. **7**, 1–179 (2018)

23. Reddy, S., Dragan, A., Levine, S.: Where do you think you're going?: inferring beliefs about dynamics from behavior. In: Advances in Neural Information Processing Systems, vol. 31 (2018)

24. Ross, S., Gordon, G., Bagnell, D.: A reduction of imitation learning and structured prediction to no-regret online learning. In: Proceedings of the Fourteenth International Conference on Artificial Intelligence and Statistics, pp. 627–635. JMLR Workshop and Conference Proceedings (2011)

25. Salvucci, D.D., Gray, R.: A two-point visual control model of steering. Perception **33**(10), 1233–1248 (2004)

26. Schwartenbeck, P., et al.: Optimal inference with suboptimal models: addiction and active Bayesian inference. Med. Hypotheses **84**(2), 109–117 (2015)

27. Tamar, A., Wu, Y., Thomas, G., Levine, S., Abbeel, P.: Value iteration networks. In: Advances in Neural Information Processing Systems, vol. 29 (2016)

28. Tishby, N., Polani, D.: Information Theory of Decisions and Actions, pp. 601–636. Springer, New York (2011)

29. Tschantz, A., Baltieri, M., Seth, A.K., Buckley, C.L.: Scaling active inference. In: 2020 International Joint Conference on Neural Networks (IJCNN), pp. 1–8. IEEE (2020)
30. Tschantz, A., Seth, A.K., Buckley, C.L.: Learning action-oriented models through active inference. PLoS Comput. Biol. **16**(4), e1007805 (2020)
31. Wei, R., McDonald, A.D., Garcia, A., Alambeigi, H.: Modeling driver responses to automation failures with active inference. IEEE Trans. Intell. Transp. Syst. (2022)
32. Zhan, W., et al.: Interaction dataset: an international, adversarial and cooperative motion dataset in interactive driving scenarios with semantic maps. arXiv preprint arXiv:1910.03088 (2019)
33. Ziebart, B.D., Maas, A.L., Bagnell, J.A., Dey, A.K., et al.: Maximum entropy inverse reinforcement learning. In: AAAI, vol. 8, pp. 1433–1438. Chicago, IL, USA (2008)

Active Blockference: cadCAD with Active Inference for Cognitive Systems Modeling

Jakub Smékal[1,2]([✉]) [iD], Arhan Choudhury[1] [iD], Amit Kumar Singh[1] [iD], Shady El Damaty[2] [iD], and Daniel Ari Friedman[1] [iD]

[1] Active Inference Institute, Davis, CA, USA
jakub.smekal@gmail.com
[2] OpSci, San Diego, USA
https://www.activeinference.org/,
https://www.opsci.io

Abstract. Cognitive approaches to complex systems modeling are currently limited by the lack of flexible, composable, tractable simulation frameworks. Here, we present Active Blockference, an approach for cognitive modeling in complex cyberphysical systems that uses cadCAD to implement multiagent Active Inference simulations. First, we provide an account of the current state of Active Inference in cognitive modeling, with the Active Entity Ontology for Science (AEOS) as a particular example of Active Inference applied to decentralized science communities. We then give a brief overview of Active Blockference and the initial results of simulations of Active Inference agents in grid environments (Active Gridference). We conclude by sharing some preferences and expectations for further research, development, and applications. The open source package can be found at https://github.com/ ActiveInferenceLab/ActiveBlockference.

Keywords: Active blockference · Active inference · cadCAD · Cognitive systems modeling · AEOS

1 Active Inference

1.1 General Formalism

Active Inference is an integrated framework for modeling perception, cognition, and action in different types and scales of entities [2]. Active Inference has been applied to settings including motor behavior, epistemic foraging, and multiscale biological systems. More general introductions to Active Inference can be found elsewhere [9] [7]. Here we focus on the features of Active Inference that are essential for the Active Blockference framework, with an eye towards cognitive modeling in cyberphysical systems.

The kernel or skeleton of an Active Inference agent model (referred to as a generative model) is a set of five parameters: **A**, **B**, **C**, **D**, and **E** [9]. Depending on the exact model specification and implementation, these variables can be

C. L. Buckley et al. (Eds.): IWAI 2022, CCIS 1721, pp. 143–150, 2023.
https://doi.org/10.1007/978-3-031-28719-0_10

fixed, learned, or nested within hierarchical models. Here we describe the minimal agent form. The **A** matrix represents the generative model's prior beliefs about the hidden state-observation mapping, i.e. given an observation, what state does the agent find itself in. Parameter **B** gives the prior beliefs about the temporal transitions between hidden states, i.e. state-action mapping encoding how the agent's actions change its state. The **C** matrix encodes the generative model's preferences over particular observations, i.e. the goal. **D** represents the prior belief about the initial state. Finally, **E** contains the agent's affordances. Mathematically, these objects can be described by

$$\mathbf{A} \simeq P(o|s)$$
$$\mathbf{B} \simeq P(s_t|s_{t-1}, u_{t-1})$$
$$\mathbf{C} \simeq \tilde{P}(o)$$
$$\mathbf{D} \simeq P(s_0)$$
$$\mathbf{E} \simeq [u_1, ..., u_i], i \in \mathbb{Z}.$$

The rest of the paper provides a brief overview of the existing tools and frameworks for active inference modeling, introduces the initial implementation and simulation results of the Active Blockference package, and discusses future directions for further development.

2   Active Blockference: A Toolkit for Cognitive Modeling in Complex Web Ecosystems

2.1   Active Inference in Python (pymdp)

Active Blockference builds directly on `pymdp`, a Python library for active inference agents modeled as Partially Observable Markov Decision Processes (POMDP) [8]. Active Blockference currently contains a full implementation of `pymdp` agents, modified for usage within the cadCAD framework (which introduces features such as parameter sweeping, Monte Carlo simulations, reproducible and scalable parallel execution). Pymdp consists of four modules, namely `inference.py`, which contains functions for inference over discrete hidden states performed by agents, `control.py`, containing the implementation of inference over policies, `learning.py`, used for updating Dirichlet posteriors, and `algos.py` for performing variational message passing [8].

The cadCAD models currently available in Active Blockference make use of a slightly adjusted version of `control.py` along with some of `pymdp`'s utility functions and adds custom modules for interoperability with cadCAD. Further changes to the Active Blockference modules may occur as the dependencies (related to e.g. cadCAD, Active Inference, cognitive models) will continue to evolve. In later sections of this paper, we describe several possible applications of Active Blockference that will make use of available tools in `pymdp` as well as allow composability with other modern algorithms that may enable novel advancements in Active Inference research and application.

2.2 Active Inference in Web3 (AEOS)

Recently the Active Inference framework was used to develop an Active Entity Ontology for Science (AEOS) [6]. AEOS considers the interaction of many types of different active and informational entities in epistemic ecosystems using the Active Inference entity partitioning model. In the partitioning model, internal states reflect a generative model and are statistically insulated by blanket states from external niche states. For example, each entity integrates observations to generate perceptions that inform decisions within the available action space for that entity. In this case, the entity partitioning model is applied to cyberphysical systems such as human, team, and distributed autonomous organization (DAO). The differences in form and function across all these kinds of entities are captured by differently structured or parameterized models. Active entities of different types and scales can interact with and influence each other via opportunities for perceptions (observations) and action (affordances), which in the online case essentially always entails read/write relationships with informational entities in the cyberphysical niche. For example, in the AEOS framework, we can define a team as an entity that requests funding within the capacity of its affordances, by publishing a grant and interacting with another entity, a Scientific Agency, which executes an assessment protocol to determine funding outcomes.

Currently AEOS bridges between graphical (e.g. flowchart-based where nodes are entities and edges are affordances or relationships) and natural language-based descriptions of epistemic ecosystems (where entities are nouns and actions are verbs). These high level descriptions lend themselves to a formalism and shared logical framework to construct, simulate, and evaluate hypothetical complex systems for scientific grant making or distributed coordination in online teams. Several features would greatly increase the utility and applicability of AEOS, for example the ability to model online epistemic communities [1] including remote teams [10] and Web3 environments. Below, we take a step in this direction by introducing Active Blockference as a tractable framework for cognitive modeling of complex cyberphysical systems (Fig. 1).

2.3 The Active Inference Entity Model in Active Blockference

Active Blockference connects the general active inference approach and parameters described in Sect. 1.1 with the powerful cadCAD simulation framework. At the time of writing, the simulations in Active Blockference focus on agents moving in grid environments, therefore parts of the code examples are grid-specific. However, the implementation is quite general, as the following code examples can be easily adjusted for arbitrary discrete state-space environments and future work will focus on implementing continuous environments as well. In the following sections, discrete states could map onto systems states as well as Active/Informational entities as described in the AEOS.

As before, the **A** matrix represents the agent's prior beliefs about how hidden states relate to observations (i.e. "What temperature is it, given the thermometer reading?"). In code, the **A** matrix is obtained by applying the softmax function

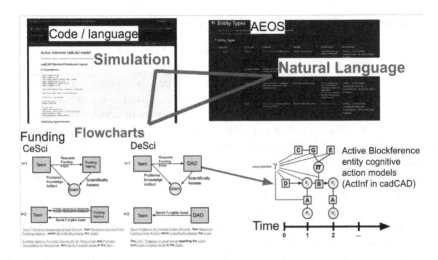

Fig. 1. Connections between code, natural language, and graphical representation of models within the AEOS framework

on the identity matrix of dimensions given by the number of possible observations and number of possible states. The softmax function is what makes the generative model a POMDP. The **B** parameter, denoting the state-action mapping (how unobserved states change through time conditioned on action selection), is given by a multi-dimensional tensor with one dimension being of the size of **E** and the other dimensions given by the dimensions of the environment, in this case the respective lengths of a grid world. In discrete state-spaces, an empty **B** tensor is currently initialized by iterating through the entire environment and the available affordances, encoding how each action changes the state of the environment. For a simple grid environment, **C** is given by a one-hot encoded vector over all the possible states, with the preferred state having the only non-zero value. Similarly, **D** is also represented by a one-hot encoded vector, only this time it is the starting position in the grid environment that gets encoded. As expected, **E** simply contains all the model's affordances, namely the movement actions available in a 2D grid environment weighted by their prior expectations (e.g. a habit distribution over action).

2.4   Generative Model Updates (Perception, Cognition, Action Selection) in Active Blockference

The Active Inference policy function in Active Blockference is the core function that defines the progression of the cadCAD simulation, as this is the Active Inference time-step function through which the generative model interacts with its environment. The Active Inference function, or perception-cognition-action-impact loop [4] is defined as follows.

First, we construct multiple policies the agent might follow depending on a variable policy length defined before the start of the simulation. This is commonly referred to as Active Inference with planning. Next, the agent infers which state it finds itself in by sampling its A matrix with the given observation and concatenating it with the agent's prior belief about its current state in the environment. With the current inference, we perform a calculation of the expected free energy (EFE), denoted by **G** for all the policies generated. In pymdp and Active Blockference, the expected free energy is currently approximated as the sum of observational ambiguity (calculated as the entropy of the **A** matrix) and the Kullback-Leibler divergence (a measure of the statistical distance between two probability distributions) between the expected observations following a given policy and the prior preferences given by the **C** matrix. From pymdp, the expected free energy is given by

$$\mathbf{G}_\tau(\pi) = \mathbb{E}_{Q(o_\tau, s_\tau|\pi)}[\ln Q(s_\tau, \phi|\pi) - \ln \tilde{P}(o_\tau, s_\tau, \phi|\pi)]$$

$$\mathbf{G} = \sum_{\tau=1}^{n} \mathbf{G}_\tau(\pi)$$

where n is the number of timesteps to sample over [3] [8].

The expected free energy is used to calculate the action posterior (policy selection), which is done by applying the softmax function on the negative EFE, and that is used along with the **E** matrix and the generated policies to compute the probability of each action. The probability distribution over actions is then sampled to get the action which updates the agent's observation prior for the next step in the simulation loop. Finally, the agent performs the chosen action and its real environment state is updated. Note that since the agent is modeled as a POMDP, its **A** matrix does not map the agent's internal state and observations perfectly, so there might actually be a mismatch between where the agent "thinks" it finds itself within the environment and what its location is in reality. This is then reflected when the agent infers its current state as described above. Below, we include the Active Inference loop in Active Blockference in code.

```
1  def p_actinf(params, substep, state_history, previous_state
       ):
2      # State Variables
3      agents = previous_state['agents']
4
5      # list of all updates to the agents in the network
6      agent_updates = []
7
8      for source, agent in agents.items():
9
10         policies = construct_policies([agent.n_states],
11         [len(agent.E)], policy_len=agent.policy_len)
12         # get obs_idx
13         obs_idx = grid.index(agent.env_state)
14
15         # infer_states
16         qs_current = u.infer_states(obs_idx, agent.A,
17         agent.prior, params['noise'])
```

```
18
19        # calc efe
20        _G = u.calculate_G_policies(agent.A, agent.B,
21        agent.C, qs_current, policies=policies)
22
23        # calc action posterior
24        Q_pi = u.softmax(-_G, params['noise'])
25        # compute the probability of each action
26        P_u = u.compute_prob_actions(agent.E, policies,
          Q_pi)
27
28        # sample action
29        chosen_action = u.sample(P_u)
30
31        # calc next prior
32        prior = agent.B[:,:,chosen_action].dot(qs_current)
```

3   Results

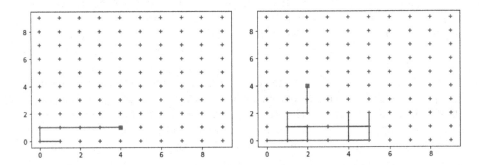

Fig. 2. Example of Active Inference agents (green dot) navigating a grid world to reach a preferred state (orange square). (Color figure online)

Here we have presented an initial formulation of Active Blockference, a package that integrates cognitive modeling approaches from Active Inference with engineering-grade simulation tools from cadCAD. All code for Active Blockference is available at https://github.com/ActiveInferenceLab/ActiveBlockference.

Active Blockference currently focuses on grid environments which are easily scalable to varying degrees of complexity [3]. Figure 2 shows the trajectories of ActInf agents trying to reach a target state on a 10×10 grid. The preferred positions were initialized randomly and the agents started at the position (0,0). By performing the active inference loop described in Sect. 2.4, the agents were able to reach their preferred state. Here the grid represents spatial locations (as in a foraging or navigation task). However in higher dimensional abstract

Gridference spaces, coordinates could represent movement in cognitive or system parameter spaces.

Example exploratory simulations have been used to model the behavior of multiple agents in a distributed ecosystem. In online settings, this design pattern might reflect the convergence of multiple actors to preferred action selection patterns, given environmental constraints and the emergence of a policy selection bias.

It is important to comment on the second plot in Fig. 2, as it shows some of the current limitations of our approach, where the agent takes many actions that do not lead to the target position before converging to its preferred state. This is due to the partially observable nature of the model and can be solved with increasing the temporal depth of policy planning and parameter optimization, however, that does come at a higher computational cost. This challenge of fitting deep temporal policies, and introducing principled structure learning of generative models, is addressed in recent and ongoing work on Active Inference. Nevertheless, there are many ways to evaluate expected free energy, some of which might be more effective than others in different contexts, however those questions are beyond the scope of this introductory paper.

4   Conclusion and Future Directions

Here we presented how Active Inference can be applied as a cognitive modeling framework in the context of complex systems simulations via cadCAD, and showed our initial results from 2D grid environments where agents modeled as Partially Observable Markov Decision Processes perform Active Inference to reach a preferred state (or not).

The Active Blockference project is hosted as an open source project at the Active Inference Lab. We now give a brief account of several future directions of Active Blockference.

The first aim is to expand the functionality of Active Blockference with all the available tools in `pymdp` and to identify the still missing elements that are employed in Active Inference modeling in other languages (e.g `SPM` in `MATLAB`, `ForneyLab` in `Julia`, etc.).

Second, we are continuing to develop more complex multiagent simulations in grid environments. The grid environment (implemented in gridworld.py) is scalable to n-dimensions, allowing for example, research in cognitive parameters in n-dimensional spaces (reflecting structure learning and modeling of intelligence as abstract spatial navigation). This and various other areas serve as relevant area for further epistemic exploration and pragmatic development. In particular, this work can provide a critical missing gap for quantitative approaches to model, predict, and design complex systems with cognitive-behavioral entities such as Decentralized Autonomous Organizations (DAOs).

Third, we are applying Active Blockference to modeling online systems. We envision that this kind of "applied Active Blockference toolbox" will allow entirely new approaches to token engineering, providing a formal ontology for the design of token simulations with the aim of finding optimal incentive structures and performing cognitive audits for cyberphysical systems. Some examples might include simulations for Decentralized Science (DeSci), Decentralized School (DeSchool), the incentive structures of Filecoin miners, decentralized markets (e.g. Uniswap), funding mechanisms in Web3, measuring participation in DAO, stewarding public goods, and platform/ecosystem governance and meta-governance.

Fourth, we are interested in graphical user inference (GUI) environments for Active Blockference development and application. Such "no code" interfaces and applications would increase the utility and accessibility of these tools.

On a longer timescale, the combination of Active Inference agents and cad-CAD can be connected to arbitrary computational environments for research in reinforcement learning, symbolic programming, quantum information theory modeling [5], and cognitive modeling of biological entities, which immediately follows from the existing applications of Active Inference to neuronal dynamics and behavior.

References

1. Albarracin, M., Demekas, D., Ramstead, M., Heins, C.: Epistemic communities under active inference. Entropy **24**(4), 476 (2022). https://www.mdpi.com/1099-4300/24/4/476
2. Constant, A., Ramstead, M.J., Veissière, S.P., Campbell, J.O., Friston, K.J.: A variational approach to niche construction. J. R. Soc. Interf. **15**(141), 20170685 (2018). https://doi.org/10.1098/rsif.2017.0685
3. Da Costa, L., Parr, T., Sajid, N., Veselic, S., Neacsu, V., Friston, K.: Active inference on discrete state-spaces: a synthesis. J. Math. Psychol. **99**, 102447 (2020). https://doi.org/10.1016/j.jmp.2020.102447. https://www.sciencedirect.com/science/article/pii/S0022249620300857
4. David, S., Cordes, R.J., Friedman, D.A.: Active inference in modeling conflict (2021). https://doi.org/10.5281/zenodo.5759807
5. Fields, C., Friston, K., Glazebrook, J.F., Levin, M.: A free energy principle for generic quantum systems. arXiv preprint arXiv:2112.15242 (2021)
6. Friedman, D., et al.: An active inference ontology for decentralized science: from situated sensemaking to the epistemic commons (2022). https://doi.org/10.5281/zenodo.6320575
7. Friston, K., et al.: The free energy principle made simpler but not too simple. arXiv preprint arXiv:2201.06387 (2022)
8. Heins, C., et al.: pymdp: A python library for active inference in discrete state spaces. arXiv preprint arXiv:2201.03904 (2022)
9. Smith, R., Friston, K., Whyte, C.: A step-by-step tutorial on active inference and its application to empirical data. J. Math. Psychol. **107**, 102632 (2022). https://doi.org/10.1016/j.jmp.2021.102632
10. Vyatkin, A., Metelkin, I., Mikhailova, A., Cordes, R., Friedman, D.A.: Active inference & behavior engineering for teams (2020). https://doi.org/10.5281/zenodo.4021163

Active Inference Successor Representations

Beren Millidge[1,2(✉)] and Christopher L. Buckley[1,3]

[1] VERSES Research Lab, Los Angeles, California, USA
[2] MRC Brain Network Dynamics Unit, University of Oxford, Oxford, UK
`beren@millidge.name`
[3] Sussex AI Group, Department of Informatics, University of Sussex, Brighton, UK

Abstract. Recent work has uncovered close links between classical reinforcement learning (RL) algorithms, Bayesian filtering, and Active Inference which lets us understand value functions in terms of Bayesian posteriors. An alternative, but less explored, model-free RL algorithm is the successor representation, which expresses the value function in terms of a successor matrix of average future state transitions. In this paper, we derive a probabilistic interpretation of the successor representation in terms of Bayesian filtering and thus design a novel active inference agent architecture utilizing successor representations instead of model-based planning. We demonstrate that active inference successor representations have significant advantages over current active inference agents in terms of planning horizon and computational cost. Moreover, we show how the successor representation agent can generalize to changing reward functions such as variants of the expected free energy.

1 Introduction

Active Inference (AIF) is an unifying theory of action selection in theoretical neuroscience [15–17]. It proposes that action selection, like perception, is fundamentally a problem of inference and that agents select actions by maximizing evidence under a biased generative model [5,13]. Active inference operates under the aegis of the Bayesian brain hypothesis [8,19] and free energy principles [1,4,10,11,14,26] and possesses several neurobiological process theories [13,33].

Recent work [29] has uncovered close links between active inference and the framework of control as inference, which shows how many classical reinforcement learning algorithms can be understood as performing Bayesian inference to infer optimal actions [2,20,35,40]. These works, as well as the related duality between control and inference in linearly solvable MDPs [38,39] has allowed us to understand classical objects in reinforcement learning such as Q-functions and value functions in terms of Bayesian filtering posteriors. Similarly, close connections between active inference and reinforcement learning methods have also been demonstrated [23,24,42]. It has been shown that deep active inference agents can be derived that can perform actor-critic algorithms [24] as well as model-based reinforcement learning [9,41,42], while the fundamental difference between them has been found to be related to the encoding of value into the generative model [25,29]. Moreover, it has become obvious that active inference can be understood and applied in a model-free (Bellman-equation) paradigm with simply a distinct reward function (the expected free energy) [6,24]. However, while much of this work has focused on understanding value functions and model-based RL, another fundamental object in model-free reinforcement learning is the successor representation [7], which has received much less attention. The successor representation (SR) [7] provides an alternative way to estimate value functions. Instead of estimating the value function with the Bellman backup as in temporal difference (TD) learning, a successor matrix of long-term discounted state transitions is estimated instead and then dynamically combined with the reward function to yield the value function for a fixed policy. Compared to estimating the value function directly, the SR requires more memory to store the successor matrix but grants the ability to dynamically recompute the value function as the reward function changes as well as providing a compressed form of a 'cognitive map' of the environment which can be directly used for exploration and option discovery [21,22,30]. Moreover, from a neuroscientific perspective, the SR has been closely linked to representations in the hippocampus [31,37] which are concerned with representing abstract (usually spatial) relations [3,43,44].

In this work, applying the probabilistic interpretation of the SR, we showcase how the SR can be directly integrated with standard methods in active inference, resulting in the successor-representation active inference (SR-AIF) agent. We show how SR-AIF has significant computational complexity benefits over standard AIF and that, moreover, the explicit generative model in AIF enables the SR to be computed instantly without requiring substantial experience in the environment. Additionally, we show how SR methods can flexibly represent the value function of the EFE and can be used to dynamically trade-off exploration and exploitation at run-time.

2   Active Inference

Discrete-state-space active inference possesses a large literature and several thorough tutorials [5,13,36], so we only provide the essentials here. AIF considers agents acting in POMDPs with observations o, states x, and actions u. The agent optimizes over policies $\pi = [u_1, u_2, \ldots]$ which are simply sequences of actions. The agent is typically assumed to be equipped with a generative model of the environment $p(o_{1:T}, x_{1:T})$ which describes how observations and states are related over time. This generative model can be factorized into two core components: a likelihood model $p(o_t|x_t)$ which states how observations are generated from states and is represented by a likelihood matrix denoted A, and a transition model $p(x_t|x_{t-1}, u_{t-1})$ which states how states change depending on the previous state and action and is represented by a transition matrix denoted $B(u)$. The rewards or goals of the agents are encoded as strong priors in the generative model and are represented by a 'goal vector' denoted C. Since AIF considers agents embedded in a POMDP it has to solve both state inference and action selection problems. State inference is performed using variational inference with a categorical variational distribution $q(x)$ which is obtained by minimizing the variational free energy,

$$q^*(x) = \underset{q}{argmin} \ \mathcal{F} = \underset{q}{argmin} \ \mathbb{E}_{q(x_t|o_t)}[\log q(x_t|o_t) - \log p(o_t, x_t|x_{t-1}, u_{t-1})] \tag{1}$$

AIF uses a unique objective function called the *Expected Free Energy* (EFE) which combines utility or reward maximization with an information gain term which promotes exploration. AIF agents naturally perform both reward-seeking and information-seeking behaviour [15,27,28,32]. The EFE is defined as,

$$\begin{aligned} \mathcal{G}_t(o_t, x_t) &= \mathbb{E}_{q(o_t, x_t)}[\log q(x_t) - \log \tilde{p}(o_t, x_t|x_{t-1}, u_{t-1})] \\ &= \underbrace{\mathbb{E}_{q(o_t, x_t)}[\log \tilde{p}(o_t)]}_{\text{Expected Utility}} - \underbrace{\mathbb{E}_{q(o_t)}[KL[q(x_t|o_t)||q(x_t)]]}_{\text{Expected Information Gain}} \end{aligned} \tag{2}$$

where $\tilde{p}$ is a 'biased' generative model which contains the goal prior vector C. As can be seen, the EFE can be decomposed into a reward-seeking and exploratory component which underlies the flexible uncertainty-reducing behaviour of AIF agents. To select actions, AIF samples from the prior over policies $q(\pi)$ which is defined as the softmax over the path integral of the EFE into the future for each timestep. Typically, future policies are evaluated up to a time horizon T. This path integral can be expressed as,

$$q(\pi) = \sigma(\sum_t^T \mathcal{G}_t^\pi(o_t, x_t)) \tag{3}$$

where $\sigma(x) = \frac{e^{-x}}{\sum_x e^{-x}}$ is the softmax function. Evaluating this path integral exactly for each policy is typically extremely computationally expensive and has exponential complexity due to the exponentially branching number of possible futures to be evaluated. This causes AIF agents to run slowly in practice and has encouraged research into alternative 'deep' active inference agents which estimate this path integral in other more efficient (but only approximate) ways [5,12,13] Here, we present a novel approach based on successor representations.

3   Successor Representation

The Successor Representation [7] provides an alternative way to compute the value function of a state. Instead of directly learning the value function (or Q function), for instance by temporal difference (TD) learning, the successor representation learns the *successor matrix*, which is the discounted long term sum of expected state occupancies, from which the value function can be dynamically computed by simply multiplying the successor matrix with the reward function. This allows the SR to instantly adapt behaviour to changing reward functions online without explicit model-based planning.

The Value function, $\mathcal{V}$, can be defined as the expected long term sum of rewards,

$$
\begin{aligned}
\mathcal{V}^\pi(x) &= r(x) + \gamma B^\pi \mathcal{V}^\pi(x) \\
&= r(x) + \gamma B^\pi [r(x) + \gamma B^\pi [r(x) + B^\pi [\cdots]]]
\end{aligned}
\tag{4}
$$

where we assume a fixed policy π and transition matrix B and a scalar discount rate $0 \le \gamma \le 1$. For a fixed policy the Bellman equation is linear and we can rearrange its as,

$$
\begin{aligned}
\mathcal{V}^\pi(x) &= r(x) + \gamma B^\pi [r(x) + \gamma B^\pi [r(x) + \gamma B^\pi [\cdots]]] \\
&= (I + \gamma B^\pi + \gamma^2 B^\pi B^\pi + \cdots) r(x) = M^\pi r(x)
\end{aligned}
\tag{5}
$$

where M^π is the successor matrix and can be thought of as encoding the long-run probability that state x transitions to state x'.

4   Successor Representation as Inference

In a special class known as linearly solvable MDPs there is a general duality between control and inference [38,39] such that control can be cast as a Bayesian filtering problem where the value function corresponds to the Bayesian filtering posterior. To see this, consider the optimal Bellman equation,

$$
\mathcal{V}^*(x_t) = \underset{u}{argmax} \left[r(x_t) + c(u) + \mathbb{E}_{p(x_{t+1}|x_t,u)} [\gamma \mathcal{V}^*(x_{t+1})] \right]
\tag{6}
$$

where we have added an additional control cost $c(u)$. The fundamental challenge is the nonlinearity of the Bellman equation due to the argmax operation. [39] noticed that if the dynamics are completely controllable and set by the action $p(x_{t+1}|x_t, u) = u(x_t)$ and the control cost is set to $KL[u(x_t)||p(x_{t+1}|x_t, u)]$ which penalizes divergence from the prior dynamics, then the argmax is analytically solvable. By defining the 'desirability function' $z(x) = e^{-\mathcal{V}^*(x)}$ and exponentiating, we can obtain a linear equation in z,

$$
z(x) = e^{-r(x)} \mathbb{E}_{p(x_{t+1}|x_t)} [\gamma z(x_{t+1})]
\tag{7}
$$

which can be solved easily. Crucially, however, this equation takes the same form as the Bayesian filtering recursion $p(x_t|o_t) \propto p(o_t|x_t) \mathbb{E}_{p(x_t|x_{t-1})} [p(x_{t-1}|o_{t-1})]$ when we make the identification of the 'desirability' $z(x_t)$ with the posterior $p(x_t|o_t)$ and the exponentiated reward $e^{-r(x_t)}$ with the likelihood $p(o_t|x_t)$. Interestingly, this same relationship between exponentiated reward and probability is also used heuristically in the control as inference literature [20]. An additional subtle point is that control is about the future instead of the past so the sum telescopes forward instead of backwards in time. By factoring Eq. 6 as in Eq. 5, it is straightforward to observe that,

$$
M = \sum_{\tau=t}^{T} \gamma^\tau \left(\prod_{i=t}^{T} \sum_{x_i} p(x_i|x_{i-1}) \right) = \sum_{\tau=t}^{T} \gamma^\tau p(x_\tau|x_t)
\tag{8}
$$

In effect, we can think of M as representing the discounted sum of the probabilities of all the possible times to reach state x over the time horizon. A similar and novel result can be derived for the case of general (not linearly solvable) MDPs but with a fixed policy except here we derive an upper bound on the SR instead of an equality. We begin by taking the log of the backwards Bayesian filtering posterior and then repeatedly applying Jensen's inequality to obtain,

$$\log p(x_t|o_t) = \log p(o_t|x_t) + \log \mathbb{E}_{p(x_{t+1}|x_t)}[\log p(x_{t+1}|o_{t+1})]$$
$$\leq \log p(o_t|x_t) + \mathbb{E}_{p(x_{t+1}|x_t)}\Big[\big[\log p(o_t|x_t) + \mathbb{E}_{p(x_{t+2}|x_{t+1})}[\log\dots]\big]\Big] \tag{9}$$

Which has the same recursive structure as the linear Bellman Equation for a fixed policy (Eq. 4) so long as we maintain the equivalence between the value function and the log posterior and the reward and the log likelihood. The technique in Eq. 5 can then be applied to give the same probabilistic interpretation of the SR as Eq. 8. In sum, we have shown how optimal control can be associated with filtering and Bayesian posteriors exactly in the case of linear MDPs and the Bayesian posterior as an upper bound in the case of a fixed policy. These results provide a sound probabilistic and Bayesian interpretation of the SR, which has hitherto been missing in the literature, and enables us to design mathematically principled active inference agents based upon the SR.

5   Successor Representation Active Inference

Using the probabilistic interpretation of the SR and the equations of discrete state-space AIF, we can construct an AIF agent which utilizes the SR to compute value functions of actions instead of model-based planning. That is, the policy posterior path integral $q(\pi) = \sigma(\mathcal{G}) = \sigma(\sum_t^T \mathcal{G}_t)$ can be considered as a value function and dynamically computed using the SR. The fact that in discrete AIF the generative model transition matrix $B(u)$ is given allows us to dispense with learning the successor matrix from experience. However, to apply the SR, we need to choose which policy π we wish to compute the value function under. This choice is important since the default policy must assign enough probability mass to all parts of the state-space to be able to provide an accurate value estimate there. Heuristically, we set the default policy to be uniform over the action space $p(u) = \frac{1}{\mathcal{A}}$ where $\mathcal{A}$ is the cardinality of the action space. This lets us define the default transition matrix,

$$\tilde{B} = \mathbb{E}_{p(u)}[B(u)] = \frac{1}{\mathcal{A}}\sum_i B[:,:,u_i] \tag{10}$$

Given $\tilde{B}$, we can analytically calculate the SR using the infinite series result,

$$M^\pi = (I + \gamma\tilde{B} + \gamma^2\tilde{B}^2\cdots) = (I - \gamma\tilde{B})^{-1} \tag{11}$$

This means that as long as the generative model is known, the EFE value function $q(\pi)$ can be computed exactly without any interaction with the environment by first computing M^π as in Eq. 11 and then multiplying by the reward function which is the EFE $\mathcal{G} = M^\pi \mathcal{G}^\pi(x)$. From this EFE value function actions can be sampled from the posterior over actions as,

$$u \sim q(\pi) = \sigma(\mathcal{G}) \tag{12}$$

A slight complication is that while the SR is defined for MDPs, AIF typically assumes a POMDP structure with observations o that do not fully specify the hidden state but are related through the likelihood matrix A. We address this by computing observation value functions as the expected state posterior under the state posterior distribution,

$$\mathcal{V}^\pi(o) = \mathbb{E}_{q(x|o)}[\mathcal{V}^\pi(x)] = q\mathcal{M}^\pi \mathcal{G}^\pi \tag{13}$$

where $q = [q_1, q_2 \cdots]$ is the categorical variational posterior. The SR-AIF algorithm can thus be summarized as follows: we are given a generative model containing the A and B matrices and a set of desired states C. At initialization, the agent computes the successor matrix M^π using the default policy with Eq. 10. For each action in a given state, SR-AIF computes the EFE value function $\mathcal{G}^\pi$ for that action and then actions are sampled from the policy posterior which is the softmax over the EFE action-value functions. In a POMDP environment exactly the same process takes place except instead of action-state we have action-observation value functions which are computed as Eq. 13.

5.1   Computational Complexity

In theory the computational complexity of SR-AIF is superior than standard AIF as standard-AIF uses model-based planning which evaluates the EFE value function $\mathcal{G}$ by exhaustively computing all possible future trajectories for different policies. This has a cost that grows exponentially in the time horizon due to the branching of possible futures. If we denote the number of actions $\mathcal{A}$, the dimension of the state-space $\mathcal{X}$ and the time-horizon T, we can approximately say that the computational complexity of standard AIF is of order $\mathcal{O}(\mathcal{X}T^2 \cdot \mathcal{A}^T)$ since the number of possible trajectories is approximately $\mathcal{A}^T$ where evaluating each step of a trajectory costs of order $\mathcal{X}$ and we must repeat this for each timestep. This is exponential in the time-horizon and renders AIF unsuitable for long term planning. Several heuristic methods have been proposed to handle this, usually by pruning obviously unsuccessful policies [12]. However, this does not remove the exponential complexity but only reduces it by a constant factor. In practice, this exponential explosion is handled by reducing the time-horizon or the policy space to be searched, which renders the evaluation of $\mathcal{G}$ approximate and makes the resulting AIF agents myopic to long term reward contingencies.

By contrast, SR-AIF analytically computes an approximation to the EFE value function $\mathcal{G}$ directly from the known transition dynamics by Eq. 11. This means that no exhaustive future simulation is required for each action but instead

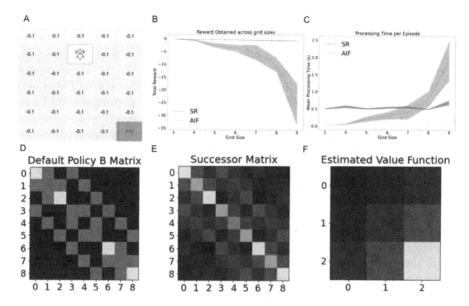

Fig. 1. Top Row: A: Schematic of the grid-world task. The AIF agent is initialized in a random square and must make it to the bottom corner to obtain reward. B: The total reward obtained on average for SR-AIF and AIF agents. Due to a limited planning horizon, the AIF agent cannot solve larger gridworlds and hence incurs large negative rewards. C: Computational cost (measured in compute time per episode) for SR-AIF. For small gridworlds, SR-AIF is more expensive since the matrix inversion cost dominates while for larger gridworlds the cost of standard AIF increases exponentially. Bottom Row: Visualization of the default policy matrix $\tilde{B}$, Successor matrix M, and estimated value function $\mathcal{V}^\pi$ for a 3× gridworld.

only a one-time cost is incurred at initialization. The main cost is the matrix inverse of approximately $\mathcal{X}^3$. Then an action must be selected which costs of order $\mathcal{A}$. This means the total complexity of SR-AIF is of order $\mathcal{O}(\mathcal{X}^3 + \mathcal{A}T)$. SR-AIF thus reduces the computational complexity of AIF from exponential to cubic and hence, in theory, allows discrete-state-space active inference to be applied to substantially larger problems than previously possible.

6   Experiments

We empirically demonstrate the superior computational complexity and ultimate performance of SR-AIF as the state-space and time-horizon grows on a series of grid-world environments. These provide a simple test-bed environment for evaluating computational complexity in practice without the confounding factors introduced by a more complex environment. The agent is initialized randomly in an $N \times N$ grid and must reach a reward located in the bottom corner of the grid. On average, as the grid size increases, both the state-space size and the planning horizon required to find this reward increase.

We implemented the AIF agent using the `pymdp` library for discrete state-space AIF [18]. We found that for larger grid-sizes the matrix inverse used to compute the successor matrix often became numerically unstable. Heuristically, we countered this by increasing the 'discount rate' γ in Eq. 11 to be greater than 1 (we used 5 for larger grid-sizes). Otherwise γ for SR-AIF and standard AIF was set to 0.99. For the active inference agent, to keep computation times manageable, we used an planning horizon of 7 and policy length of 7.

However, this task had no epistemic contingencies but only involved reward maximization. A key aspect of active inference though is its native handling of uncertainty through the EFE objective. Here, we demonstrate that the SR representation can adapt to uncertainty and dynamically change the balance of exploration and exploitation. To demonstrate this, we introduce an uncertainty and exploration component into the gridworld task by setting some squares of the grid to be 'unknowable' such that if the agent is on these squares, it is equally likely to receive an observation from any other square in the same row of the grid. This is done by setting columns of the A matrix to uniform distributions for each 'unknowable' square in the grid. We show that if equipped with the EFE objective function, SR-AIF is able to instantly recompute the value function based on this information in the A matrix without having to change the successor matrix M. Moreover, due to the property that the value function can be recomputed for each reward function, this allows a dynamic weighting of the utility and information gain components of the EFE to take place at runtime.

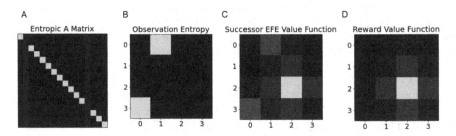

Fig. 2. Effect of introducing observation uncertainty into the model. A: the A matrix with two 'unknowable' squares resulting in a uniform distribution in two columns. B: the corresponding entropy of the state-space with the two 'unknowable' squares having high entropy. C and D: The value function computed using the EFE which responds positively to regions of high uncertainty since there is the potential for information gain compared to the standard reward function. SR-AIF is able to correctly combine both utility and epistemic drives at runtime.

7   Discussion

In this paper, we have derived a probabilistic interpretation of the SR and related it to control as inference and linear RL. We then constructed an SR-AIF algorithm which exhibits superior significant performance and computational complexity benefits to standard AIF due to its amortization of policy selection using a successor matrix which can be computed analytically at initialization.

It is important to note that while the SR-AIF has substantially better computational complexity, this comes at the cost of a necessary approximation. The successor matrix is computed only for a fixed default policy π and the choice of this policy can have significant effects upon the estimated value function and hence upon behaviour. The choice of the default policy is thus important to performance and was here chosen entirely on heuristic grounds. Principled ways of estimating or bootstrapping better default policies would be important for improving the performance of SR-AIF in practice. Alternatively, the MDP itself could be regularized so that it becomes linear as in [39] such that the optimal policy can be solved for directly. This approach has been applied in a neuroscience context [34] but the extension to active inference remains to be investigated.

Another point of extension is that here we have considered AIF and SR-AIF in the context of a single sensory modality and a single-factor generative model. However, many tasks modelled by AIF use multiple factors and modalities to express more complex relationships and contingencies. The extension of SR-AIF to multiple modalities and factors is straightforward algebraically, but has subtle implementation details and is left to future work.

References

1. Aguilera, M., Millidge, B., Tschantz, A., Buckley, C.L.: How particular is the physics of the free energy principle? Physics of Life Reviews (2021)
2. Attias, H.: Planning by probabilistic inference. In: AISTATS. Citeseer (2003)
3. Behrens, T.E., et al.: What is a cognitive map? organizing knowledge for flexible behavior. Neuron **100**(2), 490–509 (2018)
4. Buckley, C.L., Kim, C.S., McGregor, S., Seth, A.K.: The free energy principle for action and perception: a mathematical review. J. Math. Psychol. **81**, 55–79 (2017)
5. Da Costa, L., Parr, T., Sajid, N., Veselic, S., Neacsu, V., Friston, K.: Active inference on discrete state-spaces: a synthesis. arXiv preprint arXiv:2001.07203 (2020)
6. Da Costa, L., Sajid, N., Parr, T., Friston, K., Smith, R.: The relationship between dynamic programming and active inference: the discrete, finite-horizon case. arXiv preprint arXiv:2009.08111 (2020)
7. Dayan, P.: Improving generalization for temporal difference learning: the successor representation. Neural Comput. **5**(4), 613–624 (1993)
8. Doya, K., Ishii, S., Pouget, A., Rao, R.P.: Bayesian brain: Probabilistic approaches to neural coding. MIT press (2007)
9. Fountas, Z., Sajid, N., Mediano, P., Friston, K.: Deep active inference agents using monte-carlo methods. Adv. Neural. Inf. Process. Syst. **33**, 11662–11675 (2020)
10. Friston, K.: A free energy principle for a particular physics. arXiv preprint arXiv:1906.10184 (2019)
11. Friston, K., Ao, P.: Free energy, value, and attractors. Comput. Math. Methods Med. (2012)
12. Friston, K., Da Costa, L., Hafner, D., Hesp, C., Parr, T.: Sophisticated inference. Neural Comput. **33**(3), 713–763 (2021)

13. Friston, K., FitzGerald, T., Rigoli, F., Schwartenbeck, P., Pezzulo, G.: Active inference: a process theory. Neural Comput. **29**(1), 1–49 (2017)
14. Friston, K., Kilner, J., Harrison, L.: A free energy principle for the brain. J. Physiol. Paris **100**(1–3), 70–87 (2006)
15. Friston, K., Rigoli, F., Ognibene, D., Mathys, C., Fitzgerald, T., Pezzulo, G.: Active inference and epistemic value. Cogn. Neurosci. **6**(4), 187–214 (2015)
16. Friston, K., Samothrakis, S., Montague, R.: Active inference and agency: optimal control without cost functions. Biol. Cybern. **106**(8–9), 523–541 (2012)
17. Friston, K.J., Daunizeau, J., Kiebel, S.J.: Reinforcement learning or active inference? PloS one 4(7) (2009)
18. Heins, C., Millidge, B., Demekas, D., Klein, B., Friston, K., Couzin, I., Tschantz, A.: pymdp: A python library for active inference in discrete state spaces. arXiv preprint arXiv:2201.03904 (2022)
19. Knill, D.C., Pouget, A.: The bayesian brain: the role of uncertainty in neural coding and computation. Trends Neurosci. **27**(12), 712–719 (2004)
20. Levine, S.: Reinforcement learning and control as probabilistic inference: Tutorial and review. arXiv preprint arXiv:1805.00909 (2018)
21. Machado, M.C., Bellemare, M.G., Bowling, M.: Count-based exploration with the successor representation. In: Proceedings of the AAAI Conference on Artificial Intelligence, vol. 34, pp. 5125–5133 (2020)
22. Machado, M.C., Rosenbaum, C., Guo, X., Liu, M., Tesauro, G., Campbell, M.: Eigenoption discovery through the deep successor representation. arXiv preprint arXiv:1710.11089 (2017)
23. Millidge, B.: Combining active inference and hierarchical predictive coding: a tutorial introduction and case study (2019)
24. Millidge, B.: Deep active inference as variational policy gradients. arXiv preprint arXiv:1907.03876 (2019)
25. Millidge, B.: Applications of the free energy principle to machine learning and neuroscience. arXiv preprint arXiv:2107.00140 (2021)
26. Millidge, B., Seth, A., Buckley, C.L.: A mathematical walkthrough and discussion of the free energy principle. arXiv preprint arXiv:2108.13343 (2021)
27. Millidge, B., Tschantz, A., Buckley, C.L.: Whence the expected free energy? Neural Comput. **33**(2), 447–482 (2021)
28. Millidge, B., Tschantz, A., Seth, A., Buckley, C.: Understanding the origin of information-seeking exploration in probabilistic objectives for control. arXiv preprint arXiv:2103.06859 (2021)
29. Millidge, B., Tschantz, A., Seth, A.K., Buckley, C.L.: On the relationship between active inference and control as inference. arXiv preprint arXiv:2006.12964 (2020)
30. Momennejad, I.: Learning structures: predictive representations, replay, and generalization. Curr. Opin. Behav. Sci. **32**, 155–166 (2020)
31. Momennejad, I., Russek, E.M., Cheong, J.H., Botvinick, M.M., Daw, N.D., Gershman, S.J.: The successor representation in human reinforcement learning. Nat. Hum. Behav. **1**(9), 680–692 (2017)
32. Parr, T., Friston, K.J.: Generalised free energy and active inference. Biol. Cybern. **113**(5–6), 495–513 (2019)
33. Parr, T., Markovic, D., Kiebel, S.J., Friston, K.J.: Neuronal message passing using mean-field, bethe, and marginal approximations. Sci. Rep. **9**(1), 1–18 (2019)
34. Piray, P., Daw, N.D.: Linear reinforcement learning in planning, grid fields, and cognitive control. Nat. Commun. **12**(1), 1–20 (2021)
35. Rawlik, K.C.: On probabilistic inference approaches to stochastic optimal control (2013)

36. Smith, R., Friston, K.J., Whyte, C.J.: A step-by-step tutorial on active inference and its application to empirical data. J. Math. Psychol. **107**, 102632 (2022)
37. Stachenfeld, K.L., Botvinick, M.M., Gershman, S.J.: The hippocampus as a predictive map. Nat. Neurosci. **20**(11), 1643–1653 (2017)
38. Todorov, E.: General duality between optimal control and estimation. In: 2008 47th IEEE Conference on Decision and Control, pp. 4286–4292. IEEE (2008)
39. Todorov, E.: Efficient computation of optimal actions. Proc. Natl. Acad. Sci. **106**(28), 11478–11483 (2009)
40. Toussaint, M.: Probabilistic inference as a model of planned behavior. KI **23**(3), 23–29 (2009)
41. Tschantz, A., Millidge, B., Seth, A.K., Buckley, C.L.: Control as hybrid inference. arXiv preprint arXiv:2007.05838 (2020)
42. Tschantz, A., Millidge, B., Seth, A.K., Buckley, C.L.: Reinforcement learning through active inference. arXiv preprint arXiv:2002.12636 (2020)
43. Whittington, J.C., McCaffary, D., Bakermans, J.J., Behrens, T.E.: How to build a cognitive map: insights from models of the hippocampal formation. arXiv preprint arXiv:2202.01682 (2022)
44. Whittington, J.C., et al.: The tolman-eichenbaum machine: unifying space and relational memory through generalization in the hippocampal formation. Cell **183**(5), 1249–1263 (2020)

Learning Policies for Continuous Control via Transition Models

Justus Huebotter$^{(\boxtimes)}$ (ID), Serge Thill (ID), Marcel van Gerven (ID),
and Pablo Lanillos (ID)

Donders Institute, Radboud University, Nijmegen, The Netherlands
justus.huebotter@donders.ru.nl

Abstract. It is doubtful that animals have perfect inverse models of their limbs (e.g., what muscle contraction must be applied to every joint to reach a particular location in space). However, in robot control, moving an arm's end-effector to a target position or along a target trajectory requires accurate forward and inverse models. Here we show that by learning the transition (forward) model from interaction, we can use it to drive the learning of an amortized policy. Hence, we revisit policy optimization in relation to the deep active inference framework and describe a modular neural network architecture that simultaneously learns the system dynamics from prediction errors and the stochastic policy that generates suitable continuous control commands to reach a desired reference position. We evaluated the model by comparing it against the baseline of a linear quadratic regulator, and conclude with additional steps to take toward human-like motor control.

Keywords: Continuous neural control · Policy optimization · Active inference

1 Introduction

Using models for adaptive motor control in artificial agents inspired by neuroscience is a promising road to develop robots that might match human capabilities and flexibility and provides a way to explicitly implement and test these models and its underlying assumptions.

The use of prediction models in motor planning and control in biological agents has been extensively studied [12,15]. Active Inference (AIF) is a mathematical framework that provides a specific explanation to the nature of these predictive models and is getting increased attention from both the neuroscience and machine learning research community, specifically in the domain of embodied artificial intelligence [5,13]. At the core of AIF lies the presence of a powerful generative model that drives perception, control, learning, and planning all based on the same principle of free energy minimization [7]. However, learning these generative models remains challenging. Recent computational implementations harness the power of neural networks (deep active inference) to solve a variety of tasks based on these principles [13].

© The Author(s), under exclusive license to Springer Nature Switzerland AG 2023
C. L. Buckley et al. (Eds.): IWAI 2022, CCIS 1721, pp. 162–178, 2023.
https://doi.org/10.1007/978-3-031-28719-0_12

While the majority of the state of the art in deep AIF (dAIF) is focused on abstract decision making with discrete actions, in the context of robot control continuous action and state representations are essential, at least at the lowest level of a movement generating hierarchy. Continuous control implementations of AIF, based on the original work from Friston [7], is very well suited for adaptation to external perturbations [21] but it computes suboptimal trajectories and enforces the state estimation to be biased to the preference/target state [13]. New planning algorithms based on optimizing the expected free energy [18] finally uncouple the action plan from the estimation but they suffer from complications to learn the generative model and the preferences, specially for generating the actions.

In this paper, we revisit policy optimization using neural networks from the perspective of predictive control to learn a low-level controller for a reaching task. We show that by learning the transition (forward) model, during interaction, we can use it to drive the learning of an amortized policy. The proposed methods are not entirely novel, but instead combine aspects of various previous methods for low-level continuous control, active inference, and (deep) reinforcement learning. This is an early state proof-of-concept study aimed at understanding how prediction networks can lead to successful action policies, specifically for motor control and robotic tasks.

First, we summarize important related research and then go on to describe a modular neural network architecture that simultaneously learns the system dynamics from prediction errors and the stochastic policy that generates suitable continuous control commands to reach a desired reference position. Finally, we evaluated the model by comparing it against the baseline of a linear quadratic regulator (LQR) in a reaching task, and conclude with additional steps to take towards human-like motor control.

2 Related Work

This work revisits continuous control and motor learning in combination with system identification, an active direction of research with many theoretical influences. As the body of literature covering this domain is extensive, a complete list of theoretical implications and implementation attempts goes beyond the scope of this paper. Instead, we want to highlight selected examples that either represent a branch of research well or have particularly relevant ideas.

Motor learning and adaptation has been studied extensively in humans (for recent reviews please see [12,15]). Humans show highly adaptive behavior to perturbations in simple reaching tasks and we aim to reproduce these capabilities in artificial agents. While simple motor control can be implemented via optimal control when the task dynamics are known [12], systems that both have learning from experience and adaptation to changes have had little attention [2,6]. However, the assumption that the full forward and inverse model are given is not often met in practice and hence these have to be learned from experience [26].

Initial experiments in reaching tasks for online learning of robot arm dynamics in spiking neural networks inspired by optimal control theory have shown promising results [10].

Recently, the most dominant method for control of unspecified systems in machine learning is likely that of deep reinforcement learning (dRL) where control is learned as amortized inference in neural networks which seek to maximize cumulative reward. The model of the agent and task dynamics is learned either implicitly (model-free) [14] or explicitly (model-based) [8,27] from experience. The advantage of an explicit generative world model is that it can be used for planning [22], related to model predictive control, or generating training data via imagining [8,27]. Learning and updating such world models, however, can be comparatively expensive and slow. Recently, there has been a development towards hybrid methods that combine the asymptotic performance of model-free with the planning capabilities of model-based approaches [23]. Finally, model-free online learning for fast motor adaptation when an internal model is inaccurate or unavailable [2] shows promising results that are in line with behavioral findings in human experiments and can account for previously inexplicable key phenomena.

The idea of utilizing a generative model of the world is a core component of AIF, a framework unifying perception, planning, and action by jointly minimizing the expected free energy (EFE) of the agent [1,7,13]. In fact, here this generative model entirely replaces the need for an inverse model (or policy model in RL terms), as the forward model within the hierarchical generative model can be inverted directly by the means of predictive coding. This understands action as a process of iterative, not amortized, inference and is hence a strong contrast to optimal control theory, which requires both forward and inverse models [11]. Additionally, the notion of exploration across unseen states and actions is included naturally as the free energy notation includes surprise (entropy) minimization, a notion which is artificially added to many modern RL implementations [8,14,27]. Also, AIF includes the notion of a global prior over preferred states which is arguably more flexible than the reward seeking of RL agents, as it can be obtained via rewards as well as other methods such as expert imitation. Recently, the idea of unidirectional flow of top-down predictions and bottom-up prediction errors has been challenged by new hybrid predictive coding, which extends these ideas by further adding bottom-up (amortized) inference to the mix [24], postulating a potential paradigm shift towards learned habitual inverse models of action.

Recent proof-of-concept AIF implementations have shown that this framework is capable of adaptive control, e.g. in robotic arms [19] via predictive processing. In practice, most implementations of AIF by the machine learning community use neural networks to learn approximations of the probabilistic quantities relevant in the minimization of the EFE, named deep active inference. Using gradient decent based learning, these forward models can be used to directly propagate the gradients of desired states with respect to the control signals (or policy) [3,4,8,9,17,27]. Input to such policies is commonly given as either fully observable internal variables (related to proprioception) [3,4,25], visual observations directly [14] or a learned latent representation of single [8,9,27] or mixed sensory input [16,21]. This, however, makes use of amortized inference with bottom-up perception and top-down control [3,4,9,17,25] and is hence in some contrast to the predictive nature of the original AIF theory and more closely related to deep RL.

In summary, AIF postulates a promising approach to biologically plausible motor control [1,7], specifically for robotic applications [5]. The minimization of an agent's free energy is closely related to other neuroscientific theories such as the Bayesian brain hypothesis and predictive coding. Adaptive models can be readily implemented when system dynamics are known [6,20]. Unknown generative models of (forward and, if needed, inverse) dynamics may be learned from various perceptive stimuli through experience in neural networks via back propagation or error [3,4,8,9,17,23,27] or alternative learning methods [10,24,25]. This can be extended to also learn priors about preferred states and actions [3,4,8,9,14,23,27]. Generative models (and their priors) can then be utilized for perception, action, planning [9,22], and the generation of imagined training data [8,27].

In this work, we draw inspiration from these recent works. We are learning a generative model for a low-level controller with unknown dynamics from fully observable states through interaction. One component learns the state transitions, which in turn, similar to [8,27], is used to generate imagined training data for an amortized policy network. The prior about preferred states is assumed to be given to this low-level model and hence no reward based learning is applied.

3   Model

We consider a fully observable but noisy system with unknown dynamics. We formalize this system as an Markov Decision Process (MDP) in discrete time $t \in \mathbb{Z}$. The state of the system as an n-dimensional vector of continuous variables $\boldsymbol{x}_t \in \mathbb{R}^n$. Likewise, we can exert m-dimensional control on the system via continuous actions $\boldsymbol{u}_t \in \mathbb{R}^m$. We aim to learn a policy that can bring the system to a desired goal state $\tilde{\boldsymbol{x}} \in \mathbb{R}^n$, which is assumed to be provided by an external source. If the system dynamics were known, we could apply optimal control theory to find $\boldsymbol{u}_t^*$ for each point in time $t \in [0, \infty)$. However, the system dynamics are unknown and have to be learned (system identification). The dynamics of the system are learned via interaction and from prediction errors by a transition model v. This transition model is used to train in parallel a policy model π to generate the control actions. Both models are schematically summarized in Fig. 1.

3.1   Transition Model

The dynamics of the system are described by

$$\boldsymbol{x}_{t+1} = \boldsymbol{x}_t + f(\boldsymbol{x}_t,\ \boldsymbol{u}_t,\ \boldsymbol{\zeta}_t), \tag{1}$$

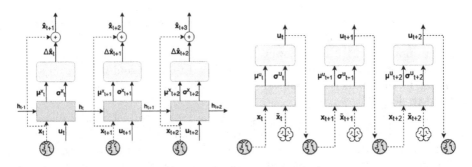

Fig. 1. Transition model (left) and policy model (right) workflow over three time steps. The policy network (orange) takes a state x and target $\tilde{x}$ as input from external sources to generate a control action u. The recurrent transition network (green) predicts the change to the next state Δx based on state x and control u. The gray box is a Gaussian sampling process (Color figure online).

where ζ is some unknown process noise. Further, any observation y cannot be assumed to be noiseless and thus

$$y_t = x_t + \xi_t, \tag{2}$$

where ξ is some unknown observation noise. As f is unknown, we want to learn a function g that can approximate it as

$$g(y_t,\ u_t,\ \phi) \approx f(x_t,\ u_t,\ \zeta_t), \tag{3}$$

by optimizing the function parameters ϕ. We hence define a state estimate $\hat{x}$ as

$$\hat{x}_t \sim \mathcal{N}(\hat{\mu}_t^x,\ \hat{\sigma}_t^x), \tag{4}$$

where the superscript x indicates not an exponent but association to the state estimate and

$$\hat{\mu}_t^x = y_{t-1} + \hat{\mu}_t^{\Delta x}. \tag{5}$$

In turn, both $\hat{\mu}_t^{\Delta x}$ and $\hat{\sigma}_t^x = \hat{\sigma}_t^{\Delta x}$ are outputs of a learned recurrent neural network (transition network) with parameters ϕ as

$$\hat{\mu}_t^{\Delta x},\ \hat{\sigma}_t^{\Delta x} = g(y_{t-1},\ u_{t-1},\ \phi). \tag{6}$$

To maintain differentiability to the state estimate we apply the reparametrization trick in Eq. (4). Further, we summarize the steps from Eq. (4)–6 (the transition model v, see Fig. 1 left) as

$$\hat{x}_t = v(y_{t-1},\ u_{t-1},\ \phi). \tag{7}$$

The optimal transition function parameters ϕ^* are given by minimizing the Gaussian negative log-likelihood loss

$$\mathcal{L}_v = \frac{1}{2T} \sum_{t=1}^{T} \left(\log\left(\max\left(\hat{\sigma}_t^x,\ \epsilon \right) \right) + \frac{(\hat{\mu}_t^x - y_t)^2}{\max\left(\hat{\sigma}_t^x,\ \epsilon \right)} \right), \tag{8}$$

and

$$\phi^* = \operatorname*{argmin}_{\phi} \mathcal{L}_v, \tag{9}$$

where ϵ is a small constant to avoid division by zero and the added constant has been omitted.

3.2 Policy Model

The actor is given by the policy π_θ that gives a control action $\boldsymbol{u}$ for a given current state $\boldsymbol{x}$ and target or preferred state $\tilde{\boldsymbol{x}}$ as

$$\pi(\boldsymbol{u}_t \mid \boldsymbol{x}_t, \ \tilde{\boldsymbol{x}}_t, \ \boldsymbol{\theta}), \tag{10}$$

where $\boldsymbol{x}_t$ can be either an observation from the environment $\boldsymbol{y}_t$ or an estimate from the transition network $\hat{\boldsymbol{x}}_t$ and

$$\boldsymbol{u}_t \sim \mathcal{N}(\boldsymbol{\mu}_t^u, \ \boldsymbol{\sigma}_t^u). \tag{11}$$

Here, $\boldsymbol{\mu}^u$ and $\boldsymbol{\sigma}^u$ are given by a function approximator that is a neural network with parameters θ (see Fig. 1 right). We aim to find the optimal policy π^* so that

$$\pi^* = \operatorname*{argmin}_{u} \sum_{t=1}^{T} (\boldsymbol{x}_t - \tilde{\boldsymbol{x}}_t)^2 . \tag{12}$$

However, as $\boldsymbol{x}_t$ is non-differentiable with respect to the action, we instead use the transition model estimate $\hat{\boldsymbol{x}}_t$. This also allows to find the gradient of the above loss with respect to the action u by using backpropagation through the transition network and the reparametrization trick. Policy and transition network are optimized by two separate optimizers as to avoid that the policy loss pushes the transition network to predict states that are the target state, which would yield wrong results.

While the above formulation in principle should find a system that is able to minimize the distance between the current state estimate $\hat{\boldsymbol{x}}$ and the target $\tilde{\boldsymbol{x}}$, in practice there are some additional steps to be taken into account to learn a suitable policy. As the state contains information about position and velocity, so does the target state. If the target state is a fixed position, the target velocity is given as zero. However, optimizing the system in a matter where the loss increases as the system starts moving, there is a strong gradient towards performing no action at all, even if this means that the position error will remain large throughout the temporal trajectory. To overcome this issue, we introduce a target gain vector $\tilde{\boldsymbol{x}}_g$, which weighs the relevance of each preference state variable. For instance, when the velocity of the system is non-important we set to 1 where x is a representing a position encoding and 0 for every velocity. The weighted policy loss becomes:

$$\mathcal{L}_\pi = \frac{1}{T} \sum_{t=1}^{T} \tilde{\boldsymbol{x}}_g \, (\hat{\boldsymbol{x}}_t - \tilde{\boldsymbol{x}})^2 . \tag{13}$$

The offline training procedure for both transition and policy networks is summarized in Algorithm 1 below, as well as Algorithm 2 & B in the Appendix B and C.

Algorithm 1. Offline training of transition and policy networks

---
1: Input: a differentiable transition parametrization $v(\hat{x}'|y, u, \phi)$,
2: a differentiable policy parametrization $\pi(u|x, \tilde{x}, \theta)$,
3: a task environment providing $(y', \tilde{x}'|u)$
4: Initialize transition parameters $\phi \in \mathbb{R}^d$ and policy parameters $\theta \in \mathbb{R}^{d'}$
5: Initialize a memory buffer of capacity M
6: **loop** for I iterations:
7:    Play out E episodes of length T by applying $u \sim \pi(y, \tilde{x}, \theta)$ at each step and save to memory
8:    Update transition network parameters for n_v batches of size N_v sampled from memory
9:    Update policy network parameters for n_π batches of size N_π sampled from memory

---

4   Results

Here we summarize the key results of this research. For a more detailed description of the task please refer to appendix Appendix A. To evaluate the performance of the trained models in comparison to an LQR baseline we have established a reaching task inspired by experiments conducted in humans and robots in previous research [6,12]. The agent is presented eight equidistant targets in sequence for $T = 200$ steps, while starting at the center position $x_{t0} = [0, 0, 0, 0]$. Initially, each target

Fig. 2. Eight equidistant targets (blue) are presented to the agent in sequence, starting from the center position (red) each time. (Color figure online)

is 0.7 units of distance removed from the center with offsets of 45° (Fig. 2). In one case these targets are stationary, or alternatively rotate in a clockwise motion with an initial velocity of 0.5 perpendicular to the center-pointing vector. To test agent performance under changed task dynamics, we offset the rotation angle γ during some evaluations, which influences the direction of the acceleration as given by control u (see Eq. (24)). To quantify the performance of target reaching, we measure the Euclidean distance between the current position $[x_1, x_2]$ and the target position $[\tilde{x}_1, \tilde{x}_2]$ at each step t, so that performance is defined as

$$J = \Delta t \sum_{t=1}^{T} \sqrt{(x_{1,t} - \tilde{x}_{1,t})^2 + (x_{2,t} - \tilde{x}_{2,t})^2}. \tag{14}$$

Results in Fig. 3 show that both the transition model and policy model are able to quickly learn from the environment interactions. The task of reaching

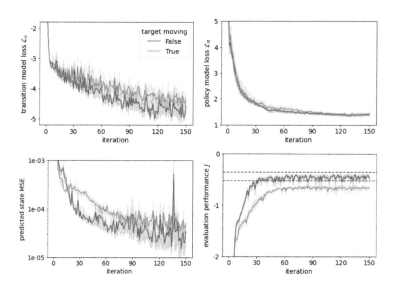

Fig. 3. Model performance improves during learning. The transition model shows better predictions when the target is stationary. The policy closely approaches but never reaches the LQR baseline scores for both stationary (red dotted line) and moving targets (green dotted line). (Color figure online)

stationary targets only is easier to conduct with predicted state mean squared error lower and a higher evaluation task performance. For both tasks, the model performance approached but never fully reached the optimal control baseline of LQR – for implementation details of the baseline please refer to appendix Appendix D).

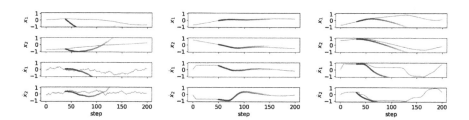

Fig. 4. Auto-regressive transition model predictions (blue to yellow) for 100 time steps over the true state development (green) are poor at the beginning of training (left), but can closely follow the true state development at the end of the training (center). Perturbing the action with a rotation angle $\gamma = 60°$ induces a mismatch between state prediction and true trajectory (right) (Color figure online).

Figure 4 shows auto-regressive predictions of the transition model when provided with some initial states and the future action trajectory. The model initially failed to make sensible predictions, but the final trained model closely predicts the true state development. When applying a rotational perturbation to the input control of $\gamma = 60°$ (Fig. 4(right)) these predictions start to diverge from the true state, as the model has capabilities for online adaptation.

The policy model is initially unable to complete the reaching task, but has a strong directional bias of movement (data not shown). After just 20 iterations (200 played episodes and 600 policy weight updates) we observe that the policy model can partially solve target reaching for both stationary and moving targets (Fig. 5 A & E respectively). At the end of training the model generated trajectories (B & F) closely match those of the LQR baseline (C & G). Applying perturbations results in non-optimal trajectories to the target (D & H). Once these perturbations become too large at around $\gamma = \pm 90°$, neither LQR nor the learned models can solve the tasks. However, the learned models closely track the performance of the LQR. This failure is a result of both policy and transition model being learned entirely offline and the inference being completely amortized. We believe that a more predictive coding based implementation of AIF as suggested by [1,7] and demonstrated by [20] would allow the system to recover from such perturbations. In future iterations of this research, we aim to extend both the transition and policy models by an adaptive component that can learn online from prediction errors to recover performance similar to [6,10] and match adaptation similar to that described in humans [12].

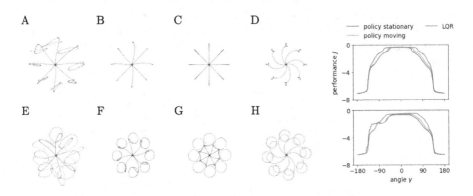

Fig. 5. Example trajectory plots from the evaluation task for stationary targets (top row) and moving targets (bottom row) show the improvement during learning from iteration 20 (A & E) to iteration 150 (B & F). The LQR baseline performs the target reaching optimally (C & D), but struggles when the control input u is rotated by $\gamma = 60°$ (D & H). The graph on the right shows that both models learned on stationary as well as moving targets perform close to the LQR under different perturbation conditions, but no model can reach the targets when the rotation becomes larger than $\gamma = 90°$.

5   Conclusion

Here, we show that a low-level motor controller and its state dynamics can be learned directly from prediction error via offline learning. Furthermore, it has similar capabilities to LQR to absorb rototranslation perturbations. However, as neither model has any means of online adaptation, they fail to show the behavioral changes described in humans [12] or control approaches [6]. In future research, hope to take steps towards human-like online motor adaptation as described in [12, 15]. AIF proposes a specific implementation of prediction error-driven motor action generation [1, 7, 20], but computational implementations in dRL and dAIF based on offline learning in neural networks often lack these online adaptation capabilities. In future iterations of this research, we aim to address this gap . Specifically, we propose to combine the offline learning of our model with model-free adaptation, such as e.g. presented in [2, 6].

Our implementation is based on some underlying assumptions. There are two kinds of input to the system that come from other components of a cognitive agent which we do not explicitly model. First, the position of the agent effector in relation to some reference frame (e.g. its base joint) is provided to the low-level controller in Cartesian coordinates. This information would have to be obtained through an integration of visual, proprioceptive, and touch information. Second, the target position of this effector is provided in the same coordinate system. This information would likely be generated by a motor planning area where abstract, discrete action priors (e.g. grasp an object) are broken down into a temporal sequence of target positions. Integrating our method with models of these particular systems is not part of this work but should be addressed in future research.

Acknowledgements. This research was partially supported by the Spikeference project, Human Brain Project Specific Grant Agreement 3 (ID: 945539).

Appendix

A   Task Description

The state of the 2d plane environment is given as

$$x = [x_1, x_2, \dot{x}_1, \dot{x}_2]. \tag{15}$$

Further, the desired target state is given as

$$\tilde{x} = [\tilde{x}_1, \tilde{x}_2, \dot{\tilde{x}}_1, \dot{\tilde{x}}_2]. \tag{16}$$

When we only care about the final position in the state, then the target gain is

$$\tilde{x}_g = [\tilde{x}_{g1}, \tilde{x}_{g2}, \dot{\tilde{x}}_{g1}, \dot{\tilde{x}}_{g2}] = [1, 1, 0, 0]. \tag{17}$$

The desired target state as well as it's target gain are currently provided by the task itself, but later should be provided by some higher level cognitive mechanism.

Further, the action influences the state by

$$\boldsymbol{u} = [u_1, \, u_2] \propto [\ddot{x}_1, \ddot{x}_2], \tag{18}$$

where $u_i \in [-u_{max}, u_{max}]$.

Following the forward Euler for discrete time steps with step size Δt we also get the environment dynamics as

$$\hat{x}_{i,t+1} \sim \mathcal{N}(x_{i,t} + \Delta t \dot{x}_{i,t}, \, \zeta_x), \tag{19}$$

and then clip the computed value based on the constrains

$$x_{i,t+1} = \begin{cases} x_{max} & \text{if } \hat{x}_{i,t+1} > x_{max} \\ \hat{x}_{i,t+1} & \text{if } x_{max} > \hat{x}_{i,t+1} > x_{min} \\ x_{min} & \text{if } \hat{x}_{i,t+1} < x_{min} \end{cases} \tag{20}$$

Doing the same for velocity and acceleration we get

$$\hat{\dot{x}}_{i,t+1} \sim \mathcal{N}(\dot{x}_{i,t} + \Delta t \ddot{x}_{i,t}, \, \zeta_{\dot{x}}), \tag{21}$$

and

$$\dot{x}_{i,t+1} = \begin{cases} \dot{x}_{max} & \text{if } \hat{\dot{x}}_{i,t+1} > \dot{x}_{max} \\ \hat{\dot{x}}_{i,t+1} & \text{if } \dot{x}_{max} > \hat{\dot{x}}_{i,t+1} > \dot{x}_{min} \\ \dot{x}_{min} & \text{if } \hat{\dot{x}}_{i,t+1} < \dot{x}_{min} \end{cases} \tag{22}$$

as well as

$$\hat{\ddot{x}}_{i,t+1} \sim \mathcal{N}(\kappa u'_{i,t}, \, \zeta_{\ddot{x}}), \tag{23}$$

where κ is some real valued action gain and u' may be subject to a rotation by the angle γ as

$$\boldsymbol{u}' = \boldsymbol{u} * \begin{bmatrix} \cos\gamma, \, -\sin\gamma \\ \sin\gamma, \, \cos\gamma \end{bmatrix}. \tag{24}$$

Finally,

$$\ddot{x}_{i,t+1} = \begin{cases} \ddot{x}_{max} & \text{if } \hat{\ddot{x}}_{i,t+1} > \ddot{x}_{max} \\ \hat{\ddot{x}}_{i,t+1} & \text{if } \ddot{x}_{max} > \hat{\ddot{x}}_{i,t+1} > \ddot{x}_{min} \\ \ddot{x}_{min} & \text{if } \hat{\ddot{x}}_{i,t+1} < \ddot{x}_{min} \end{cases} \tag{25}$$

where $\zeta = [\zeta_x, \zeta_{\dot{x}}, \zeta_{\ddot{x}}]$ is some Gaussian process noise parameter and the maximum and minimum values are the boundaries of space, velocity, and acceleration respectively. In the normal case $\zeta = [0, 0, 0]$, so that there is no process noise unless explicitly mentioned otherwise. Here, we can see that updating the state $\boldsymbol{x}$ by following Eq. (19) to Eq. (23) in this order, it takes three steps for any control signal to have an effect on the position of the agent itself. This is why it is necessary to use a RNN as the transition model to grasp the full relationship between control input and state dynamics.

Finally, the environment adds some observation noise $\boldsymbol{\xi} = [\xi_x, \xi_{\dot{x}}]$ to the state before providing it back to the controller, as mentioned in Eq. (2), so that

$$\boldsymbol{y} = [y_1, y_2, \dot{y}_1, \dot{y}_2], \tag{26}$$

with

$$y_{i,t} \sim \mathcal{N}(x_{i,t}, \xi_x), \tag{27}$$

$$\dot{y}_{i,t} \sim \mathcal{N}(\dot{x}_{i,t}, \xi_{\dot{x}}). \tag{28}$$

B Training Algorithms

The following two algorithms describe in more detail the offline learning of the transition network (Algorithm 2) and policy network (Algorithm 3) that correspond to lines 8 and 9 of Algorithm 1 respectively. For a summary please refer to Fig. 6).

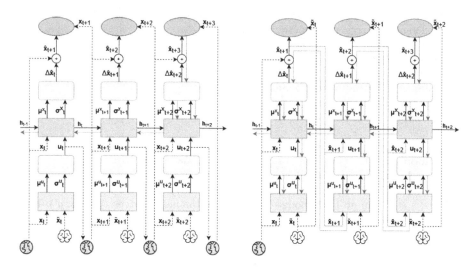

Fig. 6. Transition model learning (left) and policy model learning (right) use different algorithms. The transition model directly tries to predict the change in state and the gradients (red arrows) can directly flow from the loss computation (red) through the sampling step (gray) and to the recurrent model parameters (green). In case of the policy model update, the procedure is more involved. In order to obtain gradients with respect to the action, the models jointly roll out an imagined state and action sequence in an auto-regressive manner. The gradients have to flow from its own loss function (purple) through the transition model to reach the policy parameters (orange). This assumes that the transition model is sufficiently good at approximating the system dynamics. (Color figure online)

Algorithm 2. Updating of transition network parameters

1: Input: a differentiable transition parametrization $v(\hat{x}'|y, u, \phi)$,
2: a memory buffer object containing episodes,
3: a loss function $\mathcal{L}_v$,
4: a learning rate α_v
5: **loop** for n_v batches:
6:     Sample N_v episodes of length T from memory
7:     $L \leftarrow 0$
8:     **loop** for every episode e in sample (this is done in parallel):
9:         **loop** for every step (y, u, y') in e:
10:             Predict next state $\hat{x}' = v(y, u, \phi)$
11:             Evaluate prediction and update loss $L \leftarrow L + \mathcal{L}_v(\hat{x}', y')$
12:     $\phi \leftarrow \phi + \alpha_v \nabla_\phi \frac{L}{TN}$ (using Adam optimizer)
13: Return: ϕ

Algorithm 3. Updating of policy network parameters

1: Input: a differentiable transition parametrization $v(\hat{x}'|y, u, \phi)$,
2: a differentiable policy parametrization $\pi(u|x, \tilde{x}, \theta)$,
3: a memory buffer object containing episodes,
4: a loss function $\mathcal{L}_\pi$,
5: a learning rate α_π,
6: a number of warm-up steps w and unroll step r
7: **loop** for n_π batches:
8:     Sample N_π episodes of length T from memory
9:     $L \leftarrow 0$
10:     $n_{rollouts} \leftarrow \lfloor \frac{T}{w} \rfloor$
11:     **loop** for every episode e in sample (this is done in parallel):
12:         **loop** for every rollout in $n_{rollouts}$:
13:             Reset hidden state of transition and policy networks
14:             Warm up both models by providing the next w steps $(y, \tilde{x}, u, y')$ from e
15:             Predict next state $\hat{x}' = v(y, u, \phi)$
16:             **loop** for r steps:
17:                 Predict next hypothetical action $\hat{u} = \pi(\hat{x}', \tilde{x}, \theta)$
18:                 Predict next hypothetical state $\hat{x}' = v(y, \hat{u}, \phi)$
19:                 Evaluate hypothetical trajectory and update loss $L \leftarrow L + \mathcal{L}_\pi(\hat{x}', \tilde{x})$
20:     $\theta \leftarrow \theta + \alpha_\pi \nabla_\theta \frac{L}{Nrn_{rollouts}}$ (using Adam optimizer)
21: Return: θ

C   Training Parameters

The parameters to reproduce the experiments are summarized in Table 1. Training was conducted continuously over 1,500 episodes of 4 s each, making the total exposure to the dynamics to be learned 300,000 steps or 100 min. During this process, both models were updated a total of 4,500 times.

Table 1. Hyperparamters used to obtain data shown in results section.

Parameter	Value
Task	
Episode steps T	200
Episodes per iteration E	10
Iterations I	150
Time step [s] Δt	0.02
Memory size M	1500
Rotation angle [deg] γ	0.0
Acceleration constant κ	5.0
Process noise std. ζ	0.001
Observation noise std. ξ	0.001
Position range x_{max}	1.0
Velocity range $\dot{x}_{max}$	1.0
Control range u_{max}	1.0
Transition model	
Hidden layer size (MLP)	256
Learning rate α_υ	0.0005
Batches per iteration n_υ	30
Batch size N_υ	1024
Policy model	
Hidden layer size (GRU)	256
Learning rate α_π	0.0005
Batches per iteration n_π	30
Batch size N_π	1024
Warmup steps w	30
Unroll steps r	20

D   LQR Baseline

To compare the learned model with an optimal control theory-based approach, we implemented and hand-tuned a linear quadratic regulator (LQR) [11]. We used the Python 3 control library for the implementation. The input matrices describe system dynamics A, control influence B, as well as state cost Q and control cost R and were specified as follows:

$$A = \begin{bmatrix} 0 & 0 & 1 & 0 \\ 0 & 0 & 0 & 1 \\ 0 & 0 & 0 & 0 \\ 0 & 0 & 0 & 0 \end{bmatrix}, \qquad B = \begin{bmatrix} 0 & 0 \\ 0 & 0 \\ \kappa & 0 \\ 0 & \kappa \end{bmatrix},$$

$$Q = \begin{bmatrix} 1 & 0 & 0 & 0 \\ 0 & 1 & 0 & 0 \\ 0 & 0 & 0.1 & 0 \\ 0 & 0 & 0 & 0.1 \end{bmatrix}, \qquad R = \begin{bmatrix} 0.1 & 0 \\ 0 & 0.1 \end{bmatrix}. \tag{29}$$

This results in the control gain matrix K as

$$K = \begin{bmatrix} 3.16227766 & 0. & 1.50496215 & 0. \\ 0. & 3.16227766 & 0. & 1.50496215 \end{bmatrix}. \tag{30}$$

Controlling the task described in Appendix A to go from the initial state $x = [-0.5, 0.5, 0, 0]$ to the target state $\tilde{x} = [0.5, -0.5, 0, 0]$ results in the state evolution as shown in Fig. 7.

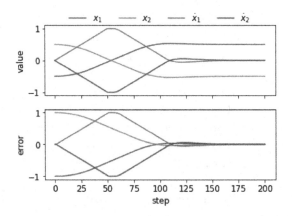

Fig. 7. State dynamics under LQR control show that initially, velocity is increased towards the target at the maximum rate, before it plateaus and declines at the same maximum rate. The tuned controller only has minimal overshoot at the target position.

References

1. Adams, R.A., Shipp, S., Friston, K.J.: Predictions not commands: active inference in the motor system. Brain Struct. Funct. **218**, 611–643 (2013). https://doi.org/10.1007/s00429-012-0475-5, http://link.springer.com/10.1007/s00429-012-0475-5
2. Bian, T., Wolpert, D.M., Jiang, Z.P.: Model-free robust optimal feedback mechanisms of biological motor control. Neural Comput. **32**, 562–595 (2020). https://doi.org/10.1162/neco_a_01260, http://www.ncbi.nlm.nih.gov/pubmed/31951794
3. Catal, O., Nauta, J., Verbelen, T., Simoens, P., Dhoedt, B.: Bayesian policy selection using active inference (2019). http://arxiv.org/abs/1904.08149

4. Catal, O., Verbelen, T., Nauta, J., Boom, C.D., Dhoedt, B.: Learning perception and planning with deep active inference. pp. 3952–3956. IEEE (2020). https://doi.org/10.1109/ICASSP40776.2020.9054364, https://ieeexplore.ieee.org/document/9054364/

5. Costa, L.D., Lanillos, P., Sajid, N., Friston, K., Khan, S.: How active inference could help revolutionise robotics. Entropy, **24**, 361 (2022). https://doi.org/10.3390/e24030361, https://www.mdpi.com/1099-4300/24/3/361

6. DeWolf, T., Stewart, T.C., Slotine, J.J., Eliasmith, C.: A spiking neural model of adaptive arm control. Proc. Roy. Soc. B Biol. Sci. **283**, 20162134 (2016). https://doi.org/10.1098/rspb.2016.2134, https://royalsocietypublishing.org/doi/10.1098/rspb.2016.2134

7. Friston, K.: What is optimal about motor control? Neuron, **72**, 488–498 (2011). https://doi.org/10.1016/j.neuron.2011.10.018, https://linkinghub.elsevier.com/retrieve/pii/S0896627311009305

8. Hafner, D., Lillicrap, T., Ba, J., Norouzi, M.: Dream to Control: Learning Behaviors by Latent Imagination. In:ICLR 2020 Conference, vol. 1, pp. 1–10. (2019). http://arxiv.org/abs/1912.01603

9. van der Himst, O., Lanillos, P.: Deep Active Inference for Partially Observable MDPs. In: Verbelen, T., Lanillos, P., Buckley, C.L., De Boom, C. (eds) Active Inference. IWAI 2020. Communications in Computer and Information Science, vol 1326, pp. 61–71 Springer, Cham (2020). https://doi.org/10.1007/978-3-030-64919-7_8, https://link.springer.com/10.1007/978-3-030-64919-7_8

10. Iacob, S., Kwisthout, J., Thill, S.: From models of cognition to robot control and back using spiking neural networks. In: Vouloutsi, V., Mura, A., Tauber, F., Speck, T., Prescott, T.J., Verschure, P.F.M.J. (eds.) Biomimetic and Biohybrid Systems. Living Machines 2020. Lecture Notes in Computer Science, vol 12413, 176–191, Springer, Cham (2020). https://doi.org/10.1007/978-3-030-64313-3_18, https://link.springer.com/10.1007/978-3-030-64313-3_18

11. Kalman, R.E.: Contributions to the theory of optimal control. Bol. Soc. Mat. Mexicana **5**(2), 102–119 (1960)

12. Krakauer, J.W., Hadjiosif, A.M., Xu, J., Wong, A.L., Haith, A.M.: Motor learning (2019). https://doi.org/10.1002/cphy.c170043, https://onlinelibrary.wiley.com/doi/10.1002/cphy.c170043

13. Lanillos, P., et al.: Active inference in robotics and artificial agents: survey and challenges, pp. 1–20 (2021). http://arxiv.org/abs/2112.01871

14. Lee, A.X., Nagabandi, A., Abbeel, P., Levine, S.: Stochastic latent actor-critic: deep reinforcement learning with a latent variable model. In: Advances in Neural Information Processing Systems, vol. 33, pp. 741–752. Curran Associates, Inc. (2020). https://proceedings.neurips.cc/paper/2020/file/08058bf500242562c0d031ff830ad094-Paper.pdf

15. McNamee, D., Wolpert, D.M.: Internal models in biological control. Annu. Rev. Control, Robot. Auton. Syst. **2**, 339–364 (2019). https://doi.org/10.1146/annurev-control-060117-105206, https://www.annualreviews.org/doi/10.1146/annurev-control-060117-105206

16. Meo, C., Franzese, G., Pezzato, C., Spahn, M., Lanillos, P.: Adaptation through prediction: multisensory active inference torque control (2021). http://arxiv.org/abs/2112.06752

17. Millidge, B.: Deep active inference as variational policy gradients. J. Math. Psychol. **96**, 102348 (2020). https://doi.org/10.1016/j.jmp.2020.102348, https://linkinghub.elsevier.com/retrieve/pii/S0022249620300298

18. Millidge, B., Tschantz, A., Buckley, C.L.: Whence the expected free energy? Neural Comput. **33**, 447–482 (2021). https://doi.org/10.1162/neco_a_01354, https://direct.mit.edu/neco/article/33/2/447/95645/Whence-the-Expected-Free-Energy
19. Oliver, G., Lanillos, P., Cheng, G.: An empirical study of active inference on a humanoid robot. IEEE Trans. Cogn. Dev. Syst. **14**, 462–471 (2022). https://doi.org/10.1109/TCDS.2021.3049907, https://ieeexplore.ieee.org/document/9316712/
20. Pio-Lopez, L., Nizard, A., Friston, K., Pezzulo, G.: Active inference and robot control: a case study. J. Roy. Soc. Interface, **13**, 20160616 (2016). https://doi.org/10.1098/rsif.2016.0616, https://royalsocietypublishing.org/doi/10.1098/rsif.2016.0616
21. Sancaktar, C., van Gerven, M.A.J., Lanillos, P.: End-to-end pixel-based deep active inference for body perception and action, pp. 1–8. IEEE (2020). https://doi.org/10.1109/ICDL-EpiRob48136.2020.9278105, https://ieeexplore.ieee.org/document/9278105/
22. Traub, M., Butz, M.V., Legenstein, R., Otte, S.: Dynamic action inference with recurrent spiking neural networks (2021). https://doi.org/10.1007/978-3-030-86383-8_19, https://link.springer.com/10.1007/978-3-030-86383-8_19
23. Tschantz, A., Millidge, B., Seth, A.K., Buckley, C.L.: Control as hybrid inference (2020). http://arxiv.org/abs/2007.05838
24. Tschantz, A., Millidge, B., Seth, A.K., Buckley, C.L.: Hybrid predictive coding: inferring, fast and slow (2022). http://arxiv.org/abs/2204.02169
25. Ueltzhöffer, K.: Deep active inference. Biol. Cybern. **112**, 547–573 (2018). https://doi.org/10.1007/s00422-018-0785-7, http://link.springer.com/10.1007/s00422-018-0785-7
26. Wolpert, D., Kawato, M.: Multiple paired forward and inverse models for motor control. Neural Netw.s **11**(7–8), 1317–1329 (1998). https://doi.org/10.1016/S0893-6080(98)00066-5, https://linkinghub.elsevier.com/retrieve/pii/S0893608098000665
27. Wu, P., Escontrela, A., Hafner, D., Goldberg, K., Abbeel, P.: DayDreamer: world models for physical robot learning (c), 1–15 (2022). http://arxiv.org/abs/2206.14176

Attachment Theory in an Active Inference Framework: How Does Our Inner Model Take Shape?

Erica Santaguida[(✉)] and Massimo Bergamasco[(✉)]

Institute of Mechanical Intelligence, Scuola Superiore Sant'Anna, Pisa, Italy
{erica.santaguida,massimo.bergamasco}@santannapisa.it

Abstract. Starting from the Attachment theory as proposed by John Bowlby in 1969, and from the scientific literature about developmental processes that take place early in life of human beings, the present work aims at exploring a possible approach to attachment, exploiting some aspects deriving from the Active Inference Theory. We describe how, from the prenatal stage until around the second year of life, the sensory, relational, affective and emotional dynamics could interplay in the formation of priors that get rigidly integrated into the internal generative model and that possibly affect the entire life of the individual. It is concluded that the presented qualitative approach could be of interest for experimental studies aiming at giving evidence on how Active Inference could sustain attachment.

Keywords: Attachment theory · Active inference · Internal working models · Mind · Perception

1 Introduction

Attachment behaviors have been observed in many animal species. Bowlby in his pioneering work directly observed such relational dynamics in human beings (Bowlby 1969). According to Bowlby, the attachment bond between the infant and the primary caregiver is a particular relationship that is established, during the first years of life, through dyadic interactive exchanges (see Fig. 1).

The attachment bond probably originates in the prenatal period, from the representational sphere of the parents (Cranley 1981). In fact, before baby's birth, parents' fantasies and representations regarding the baby and the new paternal/maternal role influence the attachment bond that will come to be defined within the second year of life. Attachment bonding is naturally guided by instinctive infant's behaviors aiming at maintaining proximity with the caregiver but also by adult's instinctive caregiving behaviors. Such a bond has the primary evolutionary function of inducing protection of the infant both from internal and external disturbances. Attachment behaviors result in a structured *attachment system* that is activated in the case of perceived external or internal threats, in order to regulate homeostatic proximity to the attachment figure (Simonelli and Calvo 2016). The attachment system is one of the core systems driving motivation,

© The Author(s), under exclusive license to Springer Nature Switzerland AG 2023
C. L. Buckley et al. (Eds.): IWAI 2022, CCIS 1721, pp. 179–191, 2023.
https://doi.org/10.1007/978-3-031-28719-0_13

other being the exploratory, affiliative, and fear-attention systems. The flow of activation of the attachment system is not continuous but reaches highest levels when stressing conditions are present. Every human being develops during her life an attachment bond, that, depending on the quality of reiterative interactions, results in a specific form of attachment.

Fig. 1. Visual, tactile, auditory interaction between mother and infant.

An attachment style is defined by the end of the second year, and it refers to the ways in which the subject finds and seeks for a sense of security in her caregiver and how she uses the security base (the caregiver) to freely explore the world (Bowlby 1969). Mary Ainsworth (Ainsworth et al. 1978) identified three attachment styles: secure attachment, insecure avoidant attachment, and insecure anxious-resistant attachment. A secure attachment style is established when the children's attempt to reach for external regulation are generally met. Otherwise, when the caregiver is perceived as systematically unavailable or unresponsive to the children's requests, then the attachment style is geared towards an avoidant organization, which is characterized by avoidance of proximity seeking behavior and oriented to exploration. The attachment style is said to be anxious-resistant when the attachment behaviors have unpredictable effects. In this case the caregiver is incoherent and inconsistent in her responses to the children's help requests and this makes the balance between attachment and exploration behaviors lean toward attachment, as the caregiver does not provide enough security.

The attachment system is a control system that in order to be efficient needs to be well informed of environmental characteristics. A further analysis carried out by Bowlby (Bowlby 1988) considers how early interactive behaviors and their consequences on the relational environment are processed as mental representations. Such representations, named by Bowlby as Internal Working Models (IWM), are given by repeated patterns of attachment behaviors and caregiver's response, and allow the infant to organize internal models about the relationship with the attachment figure (e.g. as safe or not safe), about the Self (e.g., worthy or not worthy of caring attentions), and about the characteristics of the significant other (e.g., available or not available). For instance, repeated experiences of inconsistent parenting, can induce in the child an ambivalent representation of the other and the relationship, since the caregiver is seen as unpredictable and often incoherent in regulating the child's states, then it can foster a negative self-representation character-ized by scarce confidence on personal emotions regulation skills. These representations usually induce an excess in attachment behaviors at the expenses of exploring behaviors, being that the child is overly occupied in finding regulation from her caregiver. IWM contain information on the characteristics of the components of the dyadic relationship, their relative position in space, their capacity to act, and the most likely modes of action depending on the context. They guide perception and interpretation of events and have the fundamental function of allowing the system to predict future events and prevent inconvenient situations (Bowlby 1980). The more accurate the prediction, the better suited the model to drive behavioral response given certain environmental conditions. In the IWM theorization, Bowlby starts from Craik's formulation (1943) of internal models as dynamical representations of world's phenomena, that can change over time and that are not exact copies of external events but representations concerning structural characteristics useful to analyze convenience and affordability of alternative environ-mental conditions. The ultimate objective of internal models is preventing undesirable conditions and being able to choose optimal behavioral strategies. According to Bowlby (Bowlby 1980), IWM are developed by following Piaget's concepts of assimilation and accommodation, previously used to understand sensorimotor development (Piaget 1936). In short, IWM can schematically assimilate new representations dependent on novel repeated experiences, furthermore IWM can be adapted to new environmental or internal contingencies such as those related to cognitive development. Thus, IWM can be defined as non-rigid cognitive structures, however they progressively activate in a more and more automatic manner in order to optimize resources and make them available for dealing with new situations. For example, IWM resulting from a neglect-characterized attachment will be strengthened through repeated interactions with the primary care-giver (e.g. the mother), and without the subject being aware of it, she will interpret, in the context of meaningful future relationships, social cues of abandonment even where these are not present. Even if these structures are dynamic, they tend to be preserved as they exert unconscious and automatic influence on thoughts, behaviors, and feelings.

Transactions between the mother and the infant directly influence the neural wiring of the brain, with particular effect on the imprinting of circuits that activate when dyadic synchrony is experienced (e.g. in the frontolimbic area) (Schore 2021; Choi et al 2018; Perlini et al. 2019; Ran and Zhang 2018; Petrowski et al. 2019). Salient elements of the interaction that determine its affective value, are constituted by tactile (e.g. affective touch), olfactory and taste (e.g. maternal milk and body smell), visual (particularly inputs coming from face-to face interactions), and interoceptive stimuli (Schore 2021; Fotopoulou and Tsakiris 2017). Positive experiences of dyadic interactions let the child find a secure base in her caregiver (i.e. the caregiver is perceived as comforting and responsive) and constitute a fundamental element for the development of skills such as: affective regulation capacity, motivational control of goal directed behavior (Tremblay and Schultz 1999), regulation of autonomic reactivity to social stimuli, and homeostatic and allostatic regulation (Fotopoulou and Tsakiris 2017). Homeostasis concerns the need of restoring internal balance following a dysregulation (Modell et al. 2015). At a more sophisticated temporal depth, another process named allostasis predicts and anticipates needs, allowing certain parameters to change with the objective of preventing dysregulation before they occur (Sterling 2014). We sustain the idea that these functions are preserved and refined later in life and depend on early experience of attachment. Homeostatic and allostatic regulation work in parallel with attachment behaviors, and this is more evident when we observe that when a toddler feels uncomfortable, for instance while experiencing a sense of pain, her first reaction will be most probably to cry and seek for maternal proximity, thus external regulation (e.g. her mother's reaction). Attachment is a vital function (Bowlby 1969), since it is a system that is activated to regulate caregiver proximity, and which has a role not only for the affective-relational sphere but also for the sustenance of basic biological functions. The caregiver, in fact, acts as an external regulator of these functions and the social signals used by the child (such as crying) very often have the fundamental role of favoring the regulation of body parameters that need to remain within a specific narrow range.

2　Active Inference and Attachment Development

Active Inference is a theoretical framework that aims at understanding how self-determined biological systems can move, perceive and behave in the environment (Friston 2010). Mathematical foundations of Active Inference lay on the principle of minimization of free energy (Friston 2006), which allows the organism to adaptively contain the dispersion of free energy that is given by the interaction with the environment (Friston 2009). The Bayesian model provides a statistical interpretation of the free energy minimization principle. In this sense, the minimization of free energy is explained as the reduction of the discrepancy between predicted (priors or beliefs) and observed states. According to Active Inference, the prediction error is solved by uploading predictions (i.e. perceptual inference) or by selectively sampling the predicted outcomes through action (peripheral and autonomic reflexes and instrumental responses) (i.e. active inference) (Friston 2010; Pezzulo et al. 2015). The brain learns a generative model of the world (Pezzulo et al. 2015; Friston 2010; Friston et al. 2009; Helmholtz 1866), which is said to be generative as it can be updated, through action-perception loops, as the environmental statistics change. The generative model is described as "a construct used to draw inferences about the causes of the data (i.e. use observations to derive inferred states)" (Parr et al. 2022), and is endowed with hierarchical and temporal depth (Parr et al. 2022). In the action-perception loop modeling, the Active Inference theory, besides the generative model, contextually introduces the concept of generative process. The generative process represents the modality through which sensory data can be generated in the brain starting from their true causal structure. Variational free energy must be kept in balance during fluctuations of the model's priorities of acquiring epistemic value (through prediction error) and being preserved (then maintain and induce preferred states) (Pagnoni, Guareschi 2021).

In what follows, it is described how interoceptive and exteroceptive inputs and reflexes or voluntary actions that take place in an interpersonal environment, collaborate in the shaping of a generative model of attachment. Hierarchical and temporal depth of generative models imply that different cerebral processes related to internal regulation and control of the external environment act in an integrated and coordinated manner. Furthermore, increasingly complex processes are plausibly thought to develop from a simple, ancestral mechanism of minimization of the prediction error by the means of the action-perception loop (Pezzulo et al. 2022).

Interoceptive, proprioceptive, and exteroceptive predictions are processed at different hierarchy and temporal depth and they are integrated at the highest hierarchical level, which is referable to the activity of the prefrontal cortex (Pezzulo et al. 2015). Higher levels of the hierarchy contextualize the functioning of the lower levels, allowing the integration of multimodal inputs and predictions that work at different time scales. Homeostatic needs are thought to be the most ancestral and essential drives of behavior and perception (Pezzulo et al. 2015).

From the scientific literature, it is demonstrated that the Active Inference theory could be exploited for studying affective aspects of child development. According to (Fotopoulou and Tsakiris 2017), early interaction gives rise to the formation of the "minimal affective selfhood", which is described as "the feeling qualities associated with being an embodied subject" and that "are fundamentally shaped by embodied interactions with other people in early infancy and beyond". The minimal affective selfhood comes to be constituted as a result of embodied mentalization processes, which depend on the mind's ability to organize sensorimotor and multisensory signals derived from internal and external inputs, to infer their causes and predict them on the basis of experience and interaction with each other. The other, therefore, acts as an external homeostatic regulator and the interaction with her allows the achievement of an embodied mentalization not only of one's own body but also of other bodies (such as that of the CG) (co-embodied mentalization. See Fotopoulou, Tsakiris 2017). This insight is of interest for the discussion of attachment underpinnings, as it allows to interpret the minimal part of the selfhood as an embodied construct that thrives only in condition of interaction with significant ones (in perinatal stage, mostly the mother). Thus, we assume that the shaping of a generative model originates from bodily cues that start to occur during the prenatal period. Furthermore, interpersonal reiterate dyadic exchanges which are consolidated in a style of attachment over the first years of life, are considered as essential elements that allow and characterize the internal (embodied) model, which will influence the subject for the rest of her life.

A recent work (Ciaunica et al. 2021), according to the principles of Active inference, argues that the homeostatic functions of the individual develop in an environment of co-embodiment between fetus and mother (intended as a unique condition of shared external and bodily environment). This condition allows a homeostatic co-regulation, tuned by feedback systems that serve when certain parameters go out of the range of acceptability, falling into what is defined as discrepancy. At this point, endocrine, immunological, and autonomic processes, and motor reflexes work in order to re-establish internal equilibrium. Authors refer to co-regulation as the bidirectional influence on mother's and fetus' systems. The fetus is dependent on the homeostatic capacities of the maternal system and co-regulation constitutes her first prior (Ciaunica et al. 2021). According to Barrett, Quigley, Barrett et al. (2016) allostasis is a predictive process that allows the correct energetic balance, and enteroception is centrally involved in the process as it signals the consequences of allostasis through inputs encoded as valence of circumstances.

In the present work the Active Inference framework is used to describe in general the ontogenetic evolution of internal models in a context of interaction with the primary attachment figure. The infant's system is the frame of reference for the subsequent discussion (see Fig. 2).

Fig. 2. The child's model is taken into account as frame of reference. It performs an adaptive control loop on the environment and, in particular, on the mother. The discrepancy given by the difference between prediction and observation is reduced and solved by the model updating (e.g. by updating beliefs about the mother, the body and the world) or through action on the environment (e.g. crying and other proximity reaching behaviors). Adapted from Parr, Pezzulo, Friston, 2022.

The generative model of attachment is believed to begin its development before birth, during intrauterine life. As above mentioned, the organization of the attachment system is supported by the experience-dependent development and consequent wiring of areas of the central nervous system, with particular reference to the fronto-limbic areas (Schore 2021). It is of interest that in (Barrett et al. 2016) the limbic system and the orbitofrontal cortex have been understood as areas centrally involved in fundamental functions for the persistence of the organism, as they are linked to the minimization of free energy in an Active Inference framework. By considering the temporal development of the generative model of attachment it is possible to analyze Active Inference theory aspects in different phases of the child development. The fetus comes into contact with the world and, in particular, with the mother, starting from the period the fetus spends in the intrauterine environment. We assume that humans' representations of their primary caregiver (i.e., biological mother) arise before birth over homeostatically (Ciaunica et al. 2021), genetically and multimodal experience-driven intrauterine inputs. The in-utero environment has peculiar physical and chemical characteristics that influence the transmission of sensorial information to the fetus. Exteroceptive inputs coming from the intrauterine (e.g., temperature and thermal variations, pressure, mother's voice) and extrauterine (e.g., external voices, mother's food ingestion, lights modifications, mother's touching of the womb) environment are transmitted to the fetus' system passing through biological barriers by which the fetus is contained (such as amniotic liquid and placenta) and through immature and rapidly evolving sensory organs (see Fig. 3).

Maternal evaluation of the contingencies influences fetus' interoceptive information through (Gilles et al. 2018) hormones secretion. In other words, when the mother experiences conditions that she perceives as stressing, the hormone secretion is transmitted to the fetus through the blood. During the in-utero phase of life, the fetus already has

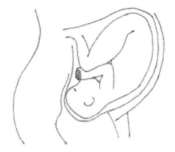

Fig. 3. Fetus contained in a biological membrane (utero) and (amniotic) liquid.

a Bayesian brain which can enact action-perception loops for the adaptive control of the environment. In fact, the fetus adaptively elaborates its own organization by tuning homeostatic values according to information received from the shared environment (e.g. through maternal hormone secretion) and according to genetically inherited information. The increasingly complex organism of the fetus begins to exert control over the internal (her own organism) and external (shared) environment through motor and autonomic reflexes for the minimization of free energy and to obtain an epistemic value of the milieu (e.g., by self-touching). These actions will act as a prompt for the elaboration of articulated behavioral strategies, which will respond to the same essential mechanism. Behavioral strategies are managed by centers of high complexity (e.g. the preforontal cortex) (Gu et al. 2013) which receive top-down and bottom-up multimodal signals and that aim to reduce prediction errors inherent to different hierarchical layers.

In addition to genetically and homeostatically driven priors, in-utero multisensory experiences inform the fetus' system about the probability that specific events appear in the environment. Embodied beliefs about such probabilities will be maintained also in the extrauterine environment and will influence the homeostatic strategies and the attachment bond (Arguz Cildir et al. 2020). An example of this dynamics concerns the longitudinal study of Dutch children born during the Dutch famine (Vaiserman and Lush chak 2021). The children of mothers who severely suffered from hunger in the gestational period had different health consequences depending on the extra-uterine environment they found at birth. Newborns experiencing hunger conditions, consistent with their mothers' condition during gestation period, demonstrated fewer adaptive problems (e.g., physical and mental illnesses) with respect to newborns appropriately fed after birth. We speculate that such a condition is indicative of a learning process about environmental parameters so that priors that are generated during the gestational period are used for adaptive control to environmental circumstances. We assume that a high discrepancy between homeostatic beliefs generated during gestation and conditions observed after birth, if persistent, is capable of provoking disturbances during the ontogenetic evolution.

Birth induces strong discrepancies in the newborn's system on multiple domains and induces a process of rapid upload of beliefs concerning observations, which is expressed through neuronal plasticity (Parr et al. 2022). We hypothesize that the interaction between fetus and mother gives rise to priors that will be used for the adaptive formation of attachment bond. The newborn faces several new conditions at the same time entering the extra-utero environment: air, that requires breathing, instead of liquid new exteroceptive stimuli such as lower temperature, no continuous tactile contact with the mother, brighter lights, louder sounds, etc. Therefore, birth constitutes a strong push for the system to adapt to new existential challenges, such as breathing and seeking nourishment. A slower process of learning is still active over time and is particularly oriented towards the acknowledgment and control of the social environment, primarily in order to survive. The act of crying is an example of control action of the social environment that the newborn carries out. This action follows a high dysregulation given by the changing environment and through this active behavior the newborn seeks, and generally finds, the caregiver proximity for external regulation. With the birth event the baby needs to put in place (new) behavioral patterns of proximity seeking in order to obtain the caregiver regulation action. The mother's voice perceived by the fetus during gestation will be used as prior of a preferred state once the baby is born and the sound of her voice will be associated with pleasant interoceptive sensations. It was experimentally (Kisilevsky et al. 2003) demonstrated that while fetus can recognize their mother's voice, newborns prefer it over other females' voices. Thus, mother's voice perception can be assumed as one of the Bayes-optimal conditions where the utility function is maximized.

Once the baby is born, she can act upon the environment through behaviors such as the implementation of the reflex of sucking the mother's breast, or the fixation of the face of the mother (Bushnell 2001). The last condition is given by the innate instinct to create a social connection with the "significant other" by directing the attention towards an indicator of human emotions (i.e. the face and its expressions and micro expressions), that is the foundation of social interaction and bonding. It is possible that the expectations on the caregiver diverge from the newborn's actual observation. Caregiving attitudes are generally activated in the neo-caregiver. The interaction, therefore, is induced by instinctive priors that are already included in the control system of the members of the mother-child dyad. It is possible that the pleasant interoceptive sensations the newborn experiences while hearing her mother's voice, smelling her odor, tasting her milk, are linked to the reduction of discrepancy from the internal priors. The absence of the natural mother at the time of birth could negatively affect the attachment bond and the general development. During the first months of life the generative model acquires new dynamics based on experiences and these acquisitions go hand in hand with physical, cognitive and sensory systems development.

Starting from the second month of life, as described by previous observational and neuroscientific works (Trevarten and Aitken 2001; Schore 2021), the infant is able to put in act face-to-face proto-interactive exchanges with the mother. Dyadic transactions are characterized by multimodal elements: visual, given by the observation of maternal expressions and movements, tactile, given by the exchange of warmth and pressure, auditory, such as the mother's voice and prosody, olfactory, such as the mother's skin odor, and interoceptive, such as hunger (Fig. 4).

Embodied models of the interaction are informed by exteroceptive and interoceptive inputs coming from dyadic interactions. Exchanges that follow a synchronized dynamic and harmonious course between the various sensory modalities result in the massive hiring of the orbitofrontal cortex and limbic system areas. Positive exchanges activate

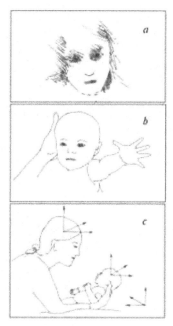

Fig. 4. a) 3PP of the synchronized dyadic interaction seen from an external perspective; b) 1PP of the newborn observing the mother. The sensory organs are not yet developed; however, there is a tendency to stare at the face; c) 1PP of the mother, who uses the face connotations, sounds and gestures of the child in the multisensory exchange.

pleasant interoceptive sensations and keep the free energy in balance. The synchronized and harmonious trend between the various sensory modalities during transactions characterizes the quality of the interaction and depends on one hand on the caregiver's readiness and sensitivity, and on the other hand on the child's predisposition and adaptivity. The repetition of certain patterns allows the infant's system to predict the expected availability and efficacy of the caregiver. During the first year of life embodied IWM arise in the individual. IWM integrate multimodal information concerning representations of the causes of "relational" events and concerning behavioral strategies learned in order to maintain the proximity of the caregiver within an acceptable range. Behavioral attachment strategies impact on underlying homeostatic parameters. The IWM are therefore complex generative models, which integrate interoceptive and exteroceptive elements, learned during repeated interactions with the caregiver from which representations are extracted, regarding salient characteristics of the relationship, such as self- efficacy and responsiveness of the caregiver (generalized to other agents of the environment) to their emotional and physical needs. IWM are associated with cognitive schemas and influence emotional regulation both voluntarily (intentional behaviors) and involuntary (reflexes, habitual behaviors and conditions) (Long et al. 2020).

(Cittern et al. 2018) have formulated the attachment theory in an Active Inference Framework, providing a computational demonstration of how different attachment styles are established following exposure to certain patterns of interaction with the caregiver.

The minimization of free energy is then considered as the basic principle that guides the organization of the organism, starting from homeostatic tuning to mental representations concerning the mother, the self, the relationship and the world.

In continuity with (Ciaunica et al. 2021) we assume that homeostasis is the first prior of the generative model and the values to be kept stable are largely learned during the co-embodiment period. Just like concentric rings of a tree, more complex models are learned on the basis of simpler ones during development in a context of interaction with the primary caregiver. A model with higher complexity and temporal depth is the allostatic one, which allows the fetus and then the child to predict physiological needs and act in advance on the environment (Fig. 5).

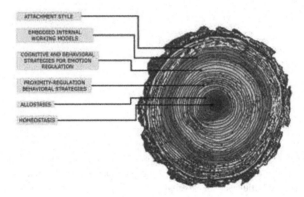

Fig. 5. Simpler models are nuclear and they are necessary for the formation of more peripheric and complex models. Externals layers are endowed with more hierarchical and temporal depth and they contain internal layers.

The attachment style of an individual constitutes an articulated generative model that will be used in the contextualization (Pezzulo et al. 2015) and interpretation of social cues for the entire life span. The predictions related to the generative model of attachment are capable of flexibilization over the course of life (e.g. Lange et al. 2021) however, they tend to be resistant to change. Resistance to change is mainly due to the fact that these representations and the associated behavior strategies are put in place in an automated and unconscious manner, as well explained by the attachment priming effect (Norman et al. 2015). In this sense, neuroscientific literature interestingly highlights that the areas involved in the recurrence of IWM overlap those involved in the default mode network (Long et al. 2020).

In summary, the generative model of attachment takes shape since the gestational period, during which homeostatic priors are established. Homeostasis could be, in fact, a first step for affective development, and it is strictly influenced by the biological mother and her reactivity to the environment. Action-perception loops are already employed in the intrauterine space, and they are used to keep the organism in a state of free energy minimization. Homeostatic values and autonomic and sensorimotor loops are the core feature from which attachment behavioral strategies and embodied representations, steam. The generative model of attachment is devoted to the optimal organization of an organism in its social environment which primarily serves to guarantee survival. Attachment style is the result following repeated experiences in the social context and "is held to characterize human beings from the cradle to the grave." (Bowlby 1980, p.154).

3   Conclusions

In the present work, development of the attachment bond over time, in an Active Inference framework, have been analyzed. The Active Inference framework provides a model which could be of interest for the study of functions that underlie bodily and representational functions, such attachment bonding starting from the prenatal stage. Following this hypothesis, experience and genetically-driven priors shape the generative model and influence the individual's functioning over life. The generative model of the fetus is compared with the statistics of the extrauterine environment, and the degree of discrepancy influences its adaptive capacity to the environment. In other words, contingencies experienced during pregnancy that induced changes in the sensorium, thus probably in the generative model, give rise to discrepancy when extra-uterine conditions are characterized by different observations. The discrepancy leads to adaptive gaps that need to be overcome. The characteristics of the social environment (i.e. of the primary caregiver) shape, through repeated interactions, the internal model of the child. The internal homeostatic organization and then representations of early multisensory relational events are firmly established at the base of the generative model and serve as a basic structure for subsequent acquisitions. Many questions remain open and require further analytical and experimental work. How is it possible to observe the assumptions that the child has before birth? At what moment does the search for proximity really emerge in the child? Considering it true that the baby, in the first months of life, does not distinguish herself from the mother, what are the boundaries of her Markov blanket? The present work lacks computational demonstrations of the presented concepts; however, it is in the authors' interest to deepen the analytic work in terms of formal modeling in the next future.

References

Ainsworth, M.S., Blehar, M.C., Waters, E., Wall, S.: Patterns of Attachment: A Psychological Study of the Strange Situation. Lawrence Erlbaum.s, Oxford, England (1978)

Arguz Cildir, D., Ozbek, A., Topuzoglu, A., Orcin, E., Janbakhishov, C.E.: Association of prenatal attachment and early childhood emotional, behavioral, and developmental characteristics: a longitudinal study. Infant Ment. Health J. 41(4), 517–529 (2020)

Barrett, L.F., Quigley, K.S., Hamilton, P.: An Active Inference theory of allostasis and enteroception in depression. Philos. Trans. R. Soc. B 371(1708), 2016001 (2016)

Bowlby, J.: Attachment and Loss: Attachment, vol. 1. Basic Books, New York (1969)

Bowlby, J.: Attachment and Loss: Loss, Sadness and Depression, vol. 3. Basic books, New York (1980)

Bowlby, J.: Open Communication and Internal Working Models: Their Role in the Development of Attachment Relationship. University of Nebraska Press, Lincoln (1988)

Bushnell, I.: Mother's face recognition in newborn infants: learning and memory. Infant Child Dev. (2001)

Choi, E.J., Taylor, M.J., Hong, S.B., Kim, C., Yi, S.H.: The neural correlates of attachment security in typically developing children. Brain Cogn. **124**, 47–56 (2018)

Ciaunica, A., Constant, A., Preissl, H., Fotopoulou, K.: The first prior: from co-embodiment to co-homeostasis in early life. Conscious. Cogn. **91**, 103117 (2021)

Cittern, D., Nolte, T., Friston, K., Edalat, A.: Intrinsic and extrinsic motivators of attachment under active inference. PLoS One **13**(4), e0193955 (2018)

Craik, K.: The Nature of Explanation. Cambridge University Press, Cambridge (1943)

Cranley, M.S.: Development of a tool for the measurement of maternal attachment during pregnancy. Nurs. Res. **5**, 281–284 (1981)

Fotopoulou, A., Tsakiris, M.: Mentalizing homeostasis: the social origins of interoceptive inference. Neuropsychoanalysis **19**(1), 3–28 (2017)

Friston, K., Kilner, J., Harrison, L.: A free energy principle for the brain. J. Physiol. Paris **100**, 1–3 (2006)

Friston, K., Daunizeau, J., Kiebel, S.J.: Reinforcement learning or active inference? PLoS ONE **4**, e6421 (2009)

Friston, K.: The free-energy principle: a unified brain theory? Nat. Rev. Neurosci. **11**(2), 127–138 (2010)

Gilles, M., et al.: Maternal hypothalamuspituitary-adrenal (HPA) system activity and stress during pregnancy: effects on gestational age and infant's anthropometric measures at birth. Psychoneuroendocrinology **94**, 152–161 (2018)

Gu, X., Hof, P.R., Friston, K.J., Fan, J.: Anterior insular cortex and emotional awareness. J. Comp. Neurol. **521**, 3371–3388 (2013)

Helmholtz, H.V.: Concerning the perceptions in general. In: Southall, J.P.C. (Ed.), New York (1866)

Kisilevsky, B.S., et al.: Effects of experience on fetal voice recognition. Psychol. Sci. J. **14**(3), 220–224 (2003)

Lange, J., Goerigk, S., Nowak, K., Rosner, R., Erhardt, A.: Attachment style change and working alliance in panic disorder patients treated with cognitive behavioral therapy. Psychotherapy **58**(2), 206 (2021)

Long, M., Verbeke, W., Ein-Dor, T., Vrtička, P.: A functional neuro-anatomical model of human attachment (NAMA): insights from first-and second-person social neuroscience. Cortex **126**, 281–321 (2020)

Modell, H., Cliff, W., Michael, J., McFarland, J., Wenderoth, M.P., Wright, A.: A physiologist's view of homeostasis. Adv. Physiol. Educ. **39**, 259–266 (2015)

Norman, L., Lawrence, N., Iles, A., Benattayallah, A., Karl, A.: Attachment-security priming attenuates amygdala activation to social and linguistic threat. Soc. Cogn. Affect. Neurosci. **10**(6), 832–839 (2015)

Pagnoni, G., Guareschi, F.T.: Meditative in-action: an endogenous epistemic venture. Preprint (2021)

Parr, T., Pezzulo, G., Friston, K.: Active Inference. The Free Energy Principle in Mind, Brain, and Behavior. The MIT Press, Cambridge, Massachusetts (2022)

Perlini, C., et al.: Disentangle the neural correlates of attachment style in healthy individuals. Epidemiol. Psychiatr. Sci. **28**(4), 371–375 (2019)

Petrowski, K., Wintermann, G.B., Hübner, T., Smolka, M.N., Donix, M.: Neural responses to faces of attachment figures and unfamiliar faces: associations with organized and disorganized attachment representations. J. Nerv. Ment. Dis. **207**(2), 112–120 (2019)

Pezzulo, G., Parr, T., Friston, K.: The evolution of brain architectures for predictive coding and active inference. Philos. Trans. R. Soc. B **377**(1844), 20200531 (2022)

Pezzulo, G., Rigoli, F., Friston, K.: Active inference, homeostatic regulation and adaptive behavioural control. Prog. Neurobiol. **134**, 17–35 (2015)

Piaget, J.: La naissance de l'intelligence chez l'enfant, Delachaux et Niestlé, Neuchatel-Paris (1936)

Ran, G., Zhang, Q.: The neural correlates of attachment style during emotional processing: an activation likelihood estimation meta-analysis. Attach. Hum. Dev. **20**(6), 626–633 (2018)

Schore, A.N.: The interpersonal neurobiology of intersubjectivity. Front. Psychol. **12**, 648616 (2021)

Simonelli, A., Calvo, V.: L'Attaccamento: Teoria e Metodi di Valutazione, Carocci Editore, Roma (2016)

Sterling, P.: Homeostasis vs allostasis: implications for brain function and mental disorders. JAMA Psychiat. **71**(10), 1192–1193 (2014)

Tremblay, L., Schultz, W.: Relative reward preference in primate orbitofrontal cortex. Nature **398**, 704–708 (1999)

Trevarthen, C., Aitken, K.J.: Infant intersubjectivity: research, theory, and practice. J. Child Psychol. Psychiatry **42**, 3–48 (2001)

Vaiserman, A., Lushchak, O.: DNA methylation changes induced by prenatal toxicmetal exposure: an overview of epidemiological evidence. Environ. Epigenetics **7**(1), dvab007 (2021)

Capsule Networks as Generative Models

Alex B. Kiefer[1,2](✉), Beren Millidge[1,3], Alexander Tschantz[1],
and Christopher L. Buckley[1,4]

[1] VERSES Research Lab, Culver City, USA
akiefer@gmail.com
[2] Monash University, Melbourne, Australia
[3] MRC Brain Network Dynamics Unit, University of Oxford, Oxford, England
[4] Sussex AI Group, Department of Informatics, University of Sussex,
Brighton, England

Abstract. Capsule networks are a neural network architecture specialized for visual scene recognition. Features and pose information are
extracted from a scene and then dynamically routed through a hierarchy
of vector-valued nodes called 'capsules' to create an implicit scene graph,
with the ultimate aim of learning vision directly as inverse graphics.
Despite these intuitions, however, capsule networks are not formulated
as explicit probabilistic generative models; moreover, the routing algorithms typically used are ad-hoc and primarily motivated by algorithmic
intuition. In this paper, we derive an alternative capsule routing algorithm utilizing iterative inference under sparsity constraints. We then
introduce an explicit probabilistic generative model for capsule networks
based on the self-attention operation in transformer networks and show
how it is related to a variant of predictive coding networks using Von-
Mises-Fisher (VMF) circular Gaussian distributions.

1 Introduction

Capsule networks are a neural network architecture designed to accurately
capture and represent part-whole hierarchies, particularly in natural images
[17,18,39], and have been shown to outperform comparable CNNs at visual
object classification, adversarial robustness, and ability to segment highly over-
lapping patterns [18,39]. A capsule network comprises layers of 'capsules' where
each capsule represents both the identity and existence of a visual feature as well
as its current 'pose' (position, orientation, etc.) relative to a canonical baseline.

This approach is heavily inspired by the concept of a scene graph in computer
graphics, which represents the objects in a scene in precisely such a hierarchical
tree structure where lower-level objects are related to the higher-level nodes by
their pose. The capsule network aims to flexibly parameterize such a scene graph
as its generative model and then perform visual object recognition by inverting
this generative model [16,17] to infer 3D scene structure from 2D appearances.

It is argued that the factoring of scene representations into transformation-
equivariant capsule activity vectors (i.e. vectors that change linearly with

C. L. Buckley et al. (Eds.): IWAI 2022, CCIS 1721, pp. 192–209, 2023.
https://doi.org/10.1007/978-3-031-28719-0_14

translation, rotation, etc.) and invariant pose transformation matrices is more flexible and efficient than the representation used in convolutional neural networks, where activities in higher layers are merely invariant to changes in viewpoint. In addition to arguably providing a better scene representation, capsule networks can use agreement between higher-level poses and their predictions based on lower-level poses to solve the binding problem of matching both the 'what' and the 'where' of an object or feature together.

Capsule networks are in part motivated by the idea that 'parse-trees' of the object hierarchy of a scene must be constructed at run-time, since they can be different for different images. Crucially, it is assumed that this dynamically constructed parse-tree must be sparse and almost singly connected - each low-level capsule or feature can be matched to only one high-level parent. This is because in natural scenes it is sensible to assume that each feature only belongs to one object at a time – for instance, it is unlikely that one eye will belong to two faces simultaneously. In [39], it is proposed to dynamically construct these parse-trees by an algorithm called 'routing by agreement' whereby low-level capsules are assigned to the high-level capsule whose pose matrix most closely matches their pose matrix under certain transformations.

While capsule networks appear to be a highly efficient architecture, invented using deep insights into the nature of visual scenes, there are, nevertheless, many elements of the construction that appear relatively ad-hoc. There is no construction of an explicit probabilistic generative model of the network. Moreover, it is unclear why the routing algorithm works and how it is related to other frameworks in machine learning. Indeed, some research [31,36] suggests that typical routing algorithms do not perform well which suggests that the goals of routing are better attained in some other way.

In this paper we propose a probabilistic interpretation of capsules networks in terms of Gaussian mixture models and VMF (circular Gaussian) distributions, which applies the self-attention mechanism used in modern transformer networks [15,47]. We argue that fundamentally, the purpose of the original routing-by-agreement algorithm of [39] is to approximate posterior inference under a generative model with the particular sparsity structure discussed above. We first demonstrate in experiments that we can achieve routing-like behaviour using sparse iterative inference, and show in addition that even in the original implementation of dynamic routing in capsules [39], sparsity of the top-level capsules is enforced via the margin loss function alone when iterative routing is turned off. This loss function can be interpreted as implementing a low-entropy prior on digit classes. We then write down a principled top-down generative model for capsules networks that provides a plausible description of the model that routing attempts to approximately invert. Overall, our results aim to provide a clear and principled route toward understanding capsule networks, and interpreting the idea of routing as fundamentally performing sparse iterative inference to construct sparse hierarchical program trees at runtime – a method that can be implemented in many distinct ways.

2   Capsule Networks

A capsule network comprises a hierarchical set of layers each of which consists of a large number of parallel capsules. In practice, several non-capsule layers such as convolutional layers are often used to provide input data preprocessing. We do not consider non-capsule layers in this analysis.

Each capsule j in a layer receives an input vector s_j consisting of a weighted combination of the outputs of the capsules i in the layer below, multiplied by their respective affine transformation matrices $\mathbf{T}_{i,j}$, which define the invariant relationships between the poses represented by i and j. The input from capsule i is denoted $\hat{u}_{j|i} = \mathbf{T}_{i,j}v_i$, where v_i is the output activity of capsule i after its input has been passed through a 'squash' nonlinearity defined as $f(\mathbf{x}) = \frac{||\mathbf{x}||^2}{1+||\mathbf{x}||^2} \cdot \frac{\mathbf{x}}{||\mathbf{x}||}$. The higher-level capsule then weights the contributions from its low-level input capsules by weighting coefficients $c_{i,j}$ which are determined by iterative routing. To obtain the output of the capsule, all its inputs are weighted and summed and then the output is fed through the nonlinear activation function f. The forward pass of a capsule layer can thus be written as,

$$v_{(l)_j} = f(\sum_i c_{i,j} \mathbf{T}_{i,j} v_{(l-1)_i}) \tag{1}$$

The core algorithm in the capsule network is the routing-by-agreement algorithm which iteratively sets the agreement coefficients $c_{i,j}$:

$$b_{i,j}^k = b_{i,j}^{k-1} + (\mathbf{T}_{i,j}\mathbf{v}_{(l-1)_i})^T \mathbf{v}_{(l)_j}^{k-1}$$
$$\mathbf{c}_i^k = \sigma(\mathbf{b}_i^{k-1}) \tag{2}$$

where k is the iteration index of the routing algorithm, $\sigma(\mathbf{x})$ is the softmax function such that $\sigma(\mathbf{x})_i = \frac{\exp(x_i)}{\sum_j \exp(x_j)}$, and $\mathbf{b}_i^k$ are the logit inputs to the softmax at iteration k, which act as log priors on the relevance of lower-level capsule i's output to all the higher-level capsules. These are initialized to 0 so all capsules are initially weighted equally.

The routing algorithm weights the lower-level capsule's contribution to determining the activities at the next layer by the dot-product similarity between the input from the low-level capsule and the higher-level capsule's output at the previous iteration. Intuitively, this procedure will match the pose of the higher-level capsule and the 'projected pose' $\hat{u}_{j|i} = \mathbf{T}_{i,j}\mathbf{v}_{(l-1)_i}$ from the lower-level capsule, so that each lower-level capsule will predominantly send its activity to the higher-level capsule whose pose best fits its prediction. In addition to matching parts to wholes, this procedure should also ensure that only higher-level capsules that receive sufficiently accurate pose-congruent 'votes' from the capsules below are activated, leading to the desired sparsity structure in the inferred scene representation.

3   Sparse Capsule PCN

Intuitively, the goal of routing is to match the poses of higher- and lower-level capsules and thus to construct a potential parse tree for a scene in terms of relations between higher-level 'objects' and lower-level 'features'. Crucially, this parse tree must be highly sparse such that, ideally, each lower-level feature is bound to only a single high-level object. To represent uncertainty, some assignment of probability to other high-level capsules may be allowed, but only to a few alternatives.

We argue that all of this is naturally accommodated if we interpret routing as implementing *sparse iterative inference*, where the sparsity constraints derive from an implicit underlying generative model. This is because the fundamental goal of routing is to obtain a 'posterior' over the capsule activations throughout the network given the input as 'data'. Unlike standard neural networks, this posterior is not only over the classification label at the output but over the 'parse tree' comprising activations at the intermediate layers. Taking inspiration from Predictive Coding Networks (PCNs) [3,6,12,27], we can imagine the parse tree posterior as being inferred in a principled way through iterative variational inference [2,48] applied to the activities at each layer during a single stimulus presentation.

The idea of using variational inference to perform capsule routing is also explored and shown to be very effective in [38]. Most closely related to our aims here, [42] propose a full generative model and variational inference procedure for capsules networks, focusing instead on the E-M routing version of capsules [18] in which existence is explicitly represented using a distinct random variable. They show that performing iterative inference to further optimize solutions at test time leads to improved digit reconstructions for rotated MNIST digits. [28] also proposes a generative model that aims to capture the intuitions behind capsule networks, and likewise derives a variational inference scheme for inverting this model.

There are various ways to achieve sparsity. [36] investigated unsupervised versions of capsules networks, and found that while routing in the CapsNet architecture did not produce the intended effects (i.e. sparse activations at each capsule layer and feature equivariance) without supervision at the output layer, these properties could be restored by adding a sparsity constraint adapted from k-sparse autoencoders [24]. A 'lifetime sparsity constraint' that forces all capsules to be active a small fraction of the time was also found to be necessary to discourage solutions in which a small number of capsules are used to reconstruct the input and the rest are ignored (which interferes with the ability to learn the desired equivariances). We experiment with a simpler form of sparsity in combination with iterative PC inference, using an L1 penalty, which is known to encourage sparsity, as an additional regularizing term added to the free energy. In Appendix B, we demonstrate this effect on a toy two-layer network where sparse iterative inference routs all inputs through a specific intermediate layer, thus constructing a single preferred parse-tree.

3.1   Experiments

To test our interpretation of iterative routing-by-agreement as inference under sparsity constraints, we investigated the role of routing in the canonical 'CapsNet' capsules network proposed in [39]. This network, diagrammed in Fig. 1A, consists of a preliminary conventional convolutional layer, followed by a convolutional capsules layer, and a final layer whose ten capsules are meant to represent the presence of the digits 0–9 in input images. Three iterations of the routing-by-agreement algorithm are used between the two capsules layers.

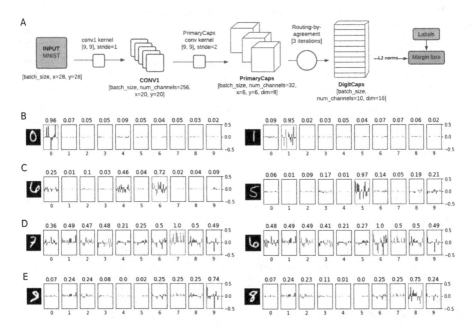

Fig. 1. A: CapsNet architecture. Input (left) is passed through Conv1, yielding 256 feature maps which provide the input to the first (convolutional) capsules layer, PrimaryCaps. This yields 1152 (32 dimensions × 6 * 6 output) 8-dimensional capsules, which are each connected to each of the 10 16-dimensional DigitCaps via their own transformation matrices. The L2 norms of the digit capsules are then compared with a one-hot encoding of the target. The auxiliary reconstruction network is not pictured. Rows B-E: Samples of DigitCaps activity vectors for test set examples under varying conditions. Red borders indicate correct digits and the number above each box is the corresponding vector norm. B: DigitCaps activities using a network trained with three iterations of dynamic routing. Sparsity is clearly enforced at the capsule level, though it is more extreme in some cases than others. C: Two random activity vectors from an otherwise identical network trained without routing. Note that the ambiguous image on the left predictably leads to less decisive capsule outputs (note also that this occurrence was not unique to the no-routing condition). D: Capsule network trained without routing, with 500 iterations of iterative inference performed in place of routing at inference time. E: Same as (D) but with an L1 regularization term on the capsule activities (i.e. $\Sigma_j \|\mathbf{v}_j\|$) added to the standard predictive coding (squared prediction error) loss function (Color figure online).

In the experiments reported in [39], the network is trained to classify MNIST digits via backpropagation, using a separate 'margin loss' function for each digit:

$$\mathcal{L}_k = \mathcal{T}_k \max\left(0, m^+ - \|\mathbf{v}_k\|\right)^2 + \lambda(1 - \mathcal{T}_k) \max\left(0, \|\mathbf{v}_k\| - m^-\right)^2 \qquad (3)$$

Here, $\mathcal{L}_k$ is the margin loss for digit k (0 through 9), $\mathcal{T}_k$ is a Boolean indicating the presence of that digit in the input image, $\|\mathbf{v}_k\|$ is the L2 norm of capsule output $\mathbf{v}_k$, and m^+ and m^- are thresholds used to encourage these vector norms to be close to 1 or 0, in the digit-present or digit-absent conditions, respectively. λ is an additional term used to down-weight the contribution of negative examples to early learning. In the full CapsNet architecture, this loss is combined with a downweighted reconstruction regularization term from an auxiliary reconstruction network used to encourage capsule activities to capture relevant features in the input.

In our experiments, we first trained a standard CapsNet on MNIST for 13 epochs, using the same hyperparameters and data preprocessing as in [39]. It is worth noting that, as a baseline for comparison, CapsNet does indeed produce sparse higher-level capsule activations (i.e. sparsity in the vector of L2 norms of the activity vectors of each of the 10 capsules in the DigitCaps layer). However, in Fig. 1C we show that training the same network architecture with 0 routing iterations (simply setting the higher-level capsule activities to the squashed sum of the unweighted predictions from lower layers) produces sparsity as well, suggesting that the transformation matrices learn to encode this sparsity structure based on the margin loss function alone. In addition to exhibiting similar higher-level capsule activities, the two networks also performed comparably (with $> 99\%$ accuracy on the test set) after training, though the no-routing network was initially slower to converge (see Appendix A).

To test whether the intuitions explored in the toy example in Appendix B would play out in a larger-scale architecture, we also tried using iterative inference in place of dynamic routing. In these experiments, we began with a forward pass through a trained CapsNet, clamped the target (label) nodes, and ran iterative inference for 500 iterations with a learning rate of 0.01, either with the standard squared-error predictive coding objective or with the standard PC loss plus L1 regularization applied to the final output vector of the CapsNet (see Appendix C). We found that, as in the toy experiment, standard iterative inference with an L1 penalty (in this case applied per-capsule) produced sparse outputs, while without the L1 penalty activity was more evenly distributed over the capsules, though the vector norms for the 'correct' capsules were still longest.

Overall, our findings on CapsNet are consistent with results reported in [36], which suggest that the sparsity seen at the output layer of CapsNet is attributable to its supervised learning objective alone and does not occur without this objective. Further confirming these results, we performed experiments on a modification of CapsNet with an intermediate capsule layer between PrimaryCaps and DigitCaps, and did not observe sparsity in the intermediate-layer activities despite comparable performance. Despite our largely negative findings, these experiments support our broader view that the main point of routing is to induce sparsity in the capsule outputs, and that this objective can be achieved by various means, including iterative inference in a predictive coding network. In the following section we propose an explicit generative model for capsules that produces the right kind of sparsity in a principled way.

4   A Generative Model for Capsules

To be fully consistent with the goal of learning vision as inverse computer graphics, a capsules network should be formulated as a top-down model of how 2D appearances are generated from a hierarchy of object and part representations, whose inversion recovers a sensible parse-tree. We now develop an explicit probabilistic generative model for the capsule network that achieves this which, interestingly, involves the self-attention mechanism used in transformer networks.

4.1   Attention and the Part-Whole Hierarchy

In recent years it has become increasingly clear that neural attention [15, 47] provides the basis for a more expressive class of artificial neural network that incorporates interactions between activity vectors on short timescales. As noted in [39], while conventional neural networks compute their feedforward pass by taking the dot products of weight vectors with activity vectors, neural attention relies on the the dot product between two activity vectors, thus producing representations that take short-term context into account. In particular, attention allows for the blending of vectors via a weighted sum, where the weights depend on dot-product similarities between input and output vectors. The core computation in neural attention can be written as,

$$\mathbf{Z} = \sigma(\mathbf{Q}\mathbf{K}^T)\mathbf{V} \tag{4}$$

with $\mathbf{Z}$ being the output of the attention block, $\mathbf{K}, \mathbf{Q}, \mathbf{V}$ being the 'Key', 'Query', and 'Value' matrices, and σ the softmax function, as above. Intuitively, the attention operation can be thought of as first computing 'similarity scores' between the query and key matrices and then normalizing them with the softmax. The similarity scores are then multiplied by the value matrix to get the output. In the transformer architecture [1, 47], these matrices are typically produced from a given input representation (e.g. a word embedding) via a learned linear transformation.

There is a tempting analogy between the capsule layer update and neural attention, since the output capsule activities are determined as a blend of functions of the inputs, using weights determined by applying the softmax function to dot-product similarity scores. A key difference, which seems to ruin the analogy, is that in routing-by-agreement, each of the weights that determine the output mixture comes from a distinct softmax, over the outputs of one lower-level capsule. Simply swapping out a neural attention module for the routing algorithm gives the wrong result however, since this enforces a 'single-child con-straint' where in the limit each higher-level object is connected to at most one lower-level part in the parse-tree.

It turns out however that the attention mechanism is precisely what is needed to naturally construct a *top-down* generative model of parse trees within a capsules architecture. Firstly, we note that we can aggregate the small trans-formation matrices $\mathbf{T}_{i,j}$ connecting input capsule i to output capsule j into a large tensor structure $\mathbf{W} = \begin{bmatrix} \mathbf{T}_{1,1} & \cdots & \mathbf{T}_{m,1} \\ \vdots & \ddots & \vdots \\ \mathbf{T}_{1,n} & \cdots & \mathbf{T}_{m,n} \end{bmatrix}$, for m lower-level and n higher-level capsules. Similarly, the N individual d-dimensional capsule vectors in a layer can be stacked to form an $N \times d$ matrix with vector-valued entries, $\mathbf{V}_{(l)} = [\mathbf{v}_{(l)_1}, \mathbf{v}_{(l)_2} \cdots \mathbf{v}_{(l)_N}]^T$, and the routing coefficients c_{ij} collected into a matrix $\mathbf{C}$ with the same shape as $\mathbf{W}$. We can then write the forward pass through a vector of capsules in a way that is analogous to a forward pass through a large ANN:

$$\mathbf{V}_{(l)} = f\Big[(\mathbf{C} \odot \mathbf{W})\mathbf{V}_{(l-1)}\Big] \tag{5}$$

Here, $\odot$ denotes element-wise multiplication and the nonlinearity f is also applied element-wise. The expression $\mathbf{W}\mathbf{V}_{(l-1)}$ should be read as a higher-level matrix-matrix multiplication in which matrix-vector multiplication is performed in place of scalar multiplication per element, i.e. $\mathbf{V}_{(l)_j} = \sum_{i=1}^{m} \mathbf{C}_{ji}\mathbf{W}_{ji}\mathbf{V}_{(l-1)_i}$. This term implements the sum of predictions $\sum_{i=1}^{m} \hat{\mathbf{u}}_{j|i}$ from lower-level capsules for the pose of each higher-level capsule j, where each transformation matrix is first scaled by the appropriate entry in the routing coefficient matrix $\mathbf{C}$.

We have argued that what the forward pass in Eq. 5 aims to implement is in effect posterior inference under a top-down generative model. To write down such a model, we first define $\tilde{\mathbf{W}}$ as the transpose of the original weight tensor $\mathbf{W}$, i.e. an $m \times n$ collection of transformations from n higher-level capsules to m lower-level capsules. Since each row of $\tilde{\mathbf{W}}$ collects the transformations from all higher-level capsules to the intrinsic coordinate frame of one lower-level capsule $\mathbf{v}_{(l-1)_i}$, we can then define a matrix $\hat{\mathbf{U}}_{(l-1)} = \begin{bmatrix} (\tilde{\mathbf{W}}_1 \odot \mathbf{V}_{(l)})^T \\ \vdots \\ (\tilde{\mathbf{W}}_m \odot \mathbf{V}_{(l)})^T \end{bmatrix}$ that contains all the predictions for the lower-level capsules, where matrix-vector multiplication is applied element-wise to the submatrices.

Setting $\mathbf{V} = \mathbf{K} = \hat{\mathbf{U}}_{(l-1)_i}$ and $\mathbf{Q} = \mathbf{V}_{(l-1)_i}$, we can then frame the top-down generation of one lower-level capsule vector within a capsules network as an instance of neural attention, where $\mathbf{V}_{(l)}^k$ is the matrix of capsule activities at layer l and iteration k, as above:

$$\mathbf{V}^k{}_{(l-1)_i} = \sigma(\mathbf{V}_{(l-1)_i}^{k-1} \hat{\mathbf{U}}_{(l-1)_i}^T)\hat{\mathbf{U}}_{(l-1)_i} \tag{6}$$

These updates are independent for each lower-level capsule but can clearly be vectorized by adding an extra leading dimension of size m to each matrix, so that an entire attention update is carried out per row.

The above can be viewed as an inverted version of routing-by-agreement in which the higher-level capsules cast 'votes' for the states of the lower-level capsules, weighted by the terms in the softmax which play a role similar to routing coefficients. There is also a relation to associative memory models [21, 26, 35] since we can think of the capsule network as associating the previous lower-level output (query) to 'memories' consisting of the predictions from the layer above, and using the resulting similarity scores to update the outputs as a blend of the predictions weighted by their accuracy.

Crucially, when used in this way for several recurrent iterations, neural attention encourages each row of the output matrix to be dominated by whichever input it is most similar to. Since the output in this case is the lower level in a part-whole hierarchy (where rows correspond to capsule vectors), this precisely enforces the single-parent constraint that routing-by-agreement aspires to.

In the routing-by-agreement algorithm, the routing logits b_{ij} are initially set to 0 (or to their empirical prior value, if trained along with the weights) and then accumulate the dot-product similarities during each iteration of routing. It is clear that the application of attention alone without the accumulation of log evidence over iterations should produce a routing-like effect, since there is a positive feedback loop between the similarities and the softmax weights. If one wanted to emulate routing-by-agreement more closely, the above could be supplemented with an additional recurrent state formed by the similarity scores of the previous iteration.

While the single-parent constraint alone is not sufficient to ensure that only lower-level capsules that are a good fit with some active higher-level capsule are activated, it is reasonable to expect that when no constant relationship between a higher- and lower-level entity exists in the data, training would encourage the weights of the corresponding transformation matrix to be close to 0, on pain of inappropriately activating a lower-level capsule, which would lead to an increase in the loss function (e.g. squared prediction error).

4.2   Probabilistic Generative Model

We now formulate the generative model sketched above explicitly in probabilistic terms, which therefore also doubles as a generative model of transformer attention (see Appendix D).

As remarked above, capsule networks can be seen as combining something like a standard MLP forward pass with additional vector-level operations. In particular, for a given lower-level capsule i, the attention mechanism can be interpreted as mixing the MLPs defined by each capsule pair (i, j) with the weights given by the attention softmax. If we interpret each MLP update in terms of a Gaussian distribution, as in PCNs (that is, where the uncertainty about the hidden state $\mathbf{V}_{(l-1)}$ is Gaussian around a mean given by the sum of weighted 'predictions'), we arrive at a mixture of Gaussians distribution over each lower-level capsule pose, whose mixing weights are determined by scaled dot-product similarity scores.

These similarity scores must be interpreted differently from the MLP-like part of the model, and we argue that these are correctly parametrized via a von Mises-Fisher (VMF) distribution with a mean of the pose input. The VMF implements the dot-product similarity score required by the routing coefficient in a probabilistic way, and can be thought of as parametrizing a distribution over vector angles between the output and input poses, which is appropriate since the purpose of the routing coefficients is to reinforce predictions that match a higher-level pose (where degree of match can be captured in terms of the angle between the pose and prediction vectors). Importantly, since it parametrizes an *angle* the distribution is circular since angles 'wrap-around'. The VMF distribution is a Gaussian defined on the unit hypersphere and so correctly represents this circular property.

Given the above, we can express the update in Eq. 6 above as a probabilistic generative model as follows:

$$p(\mathbf{V}_{(L)_i}|\hat{\mathbf{U}}_{(L)_i}) = \sum_j \left[\pi_{(i)_j} \mathcal{N}(\mathbf{V}_{(L)_i}; \hat{\mathbf{U}}_{(L)_{ij}}, \sigma_{ij}) \right]$$

$$\pi_{(i)} = Cat(n, \mathbf{p}_{(i)})$$

$$\mathbf{p}_{(i)} = \sigma\left[VMF(\mathbf{V}_{(L)_i}; \hat{\mathbf{U}}_{(L)_{i1}}, \kappa_{i1}) \ldots VMF(\mathbf{V}_{(L)_i}; \hat{\mathbf{U}}_{(L)_{in}}, \kappa_{in}) \right]$$

$$(7)$$

where $\pi_{(i)}$ are the mixing weights, σ_{ij} is the standard deviation of the Gaussian distribution over capsule i conditioned on higher-level capsule j, and κ_{ij} is the 'concentration parameter' of the corresponding VMF distribution, which determines how tightly probability mass is concentrated on the direction given by the mean $\hat{\mathbf{U}}_{(L)_{ij}}$.

The generative model then defines the conditional probability of an entire capsule layer given the predictions as the product of these per-capsule mixture distributions, mirroring the conditional independence of neurons within a layer in conventional MLPs:

$$p(\mathbf{V}_{(\mathbf{L})}|\hat{\mathbf{U}}_{(L)}) = \prod_i p(\mathbf{V}_{(L)_i}|\hat{\mathbf{U}}_{(L)_i}) \qquad (8)$$

It would be simpler to use the attention softmax itself directly to determine the mixing weights of each GMM, but using the probabilities returned by individual VMF distributions instead affords a fully probabilistic model of this part of the generative process, where not only the vector angle match but also the variance can be taken into account for each capsule pair i, j. It remains for future work to write down a process for inverting this generative model using variational inference.

5   Conclusion

In this paper, we have aimed to provide a principled mathematical interpretation of capsule networks and link many of its properties and algorithms to other better known fields. Specifically, we have provided a probabilistic generative model of the capsule network in terms of Gaussian and VMF distributions which provides a principled mathematical interpretation of its core computations. Secondly, we have shown how the ad-hoc routing-by-agreement algorithm described in [39] is related to self-attention. Moreover, we have demonstrated both in a toy illustrative example and through large-scale simulation of capsule networks how the desiderata of the routing algorithm can be achieved through a general process of sparse iterative inference.

Acknowledgements. Alex Kiefer and Alexander Tschantz are supported by VERSES Research. Beren Millidge is supported by the BBSRC grant BB/S006338/1 and by VERSES Research. CLB is supported by BBRSC grant number BB/P022197/1 and by Joint Research with the National Institutes of Natural Sciences (NINS), Japan, program No. 0111200.

Code Availability. Code for the capsules network is adapted from https://github.com/adambielski/CapsNet-pytorch and can be found at: https://github.com/exilefa ker/capsnet-experiments. Code reproducing the toy model experiments and figure in Appendix B can be found at: https://github.com/BerenMillidge/Sparse_Routing.

Appendix A: Convergence of CapsNet with and Without Routing

The following plots show the loss per epoch (plotted on a log scale for visibility) during training of a CapsNet architecture with 3 and 0 rounds of dynamic routing-by-agreement and without an auxiliary reconstruction net. The figure shows that after initially slower learning, CapsNet without routing converged to nearly the same test set loss as with routing (Fig. 2).

Interestingly, although classification performance was very similar across these networks, the test set accuracy for the four conditions (standard, no routing, routing without reconstruction loss, and neither routing nor reconstruction loss) were 99.23%, 99.34%, 99.32%, and 99.29% respectively. In this case at least, dynamic routing appears not to have led to improved accuracy, although it does lead to slightly lower values of the loss function both when using the full loss (margin + reconstruction) and when using margin loss alone.

This is consistent with the findings in [31] that iterative routing does not greatly improve the performance of capsules networks and can even lead to worse performance, though it is also consistent with Sabour et al. [39], who report a roughly 0.14% performance improvement using routing against a no-routing baseline.

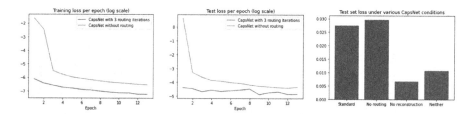

Fig. 2. Left: Training set loss during training of CapsNet under standard routing (3 iterations) and no-routing conditions. Middle: Test set loss. Right: Comparison of final losses across standard, no-routing, no-reconstruction, and no-routing no-reconstruction conditions. Note that the loss functions differ between the reconstruction and no-reconstruction conditions.

Appendix B: Toy Model of Routing as Sparse Iterative Inference

Capsule networks assume that a sparse parse tree representation of a stimulus is preferred and achieve this using the routing algorithm while an equivalent ANN would typically produce dense representations. To gain intuition for why sparse iterative inference may be able to achieve this result as well as capsule routing, we provide a simple illustrative example of how iterative inference with a sparsity penalty can result in routing-like behaviour. We consider a simple three-layer neural network with a single hidden layer and visible input and output layers. We fix both the top and bottom layers to an input or a target respectively. We can then infer the hidden layer activities which can both sufficiently 'explain' the output given the input. If we imagine the input layer of the network as representing features and the output as a classification label, then in the hidden layer we wish to uniquely assign the input features all to the best matching 'object'. We construct such a network with input size 3, hidden size 3, and output size 1, with input weights set to identity and the output weights set to a matrix of all 1 s. The network is linear although this is not necessary. Following the Gaussian generative model proposed for the capsule network, we implemented this network as a predictive coding network (PCN) and performed iterative inference by updating activities to minimize the variational free energy which can be expressed as a sum of squared prediction errors at each layer [6]. In additional to the standard iterative free energy, we also experimented with adding either a sparsity penalty (L1 regularisation) or L2 activity norm regularization to the network. We investigated the extent to which iterative inference with the sparsity penalty can reproduce the desired routing effect with only a single high-level feature being active, and found that it can (see Fig. 3).

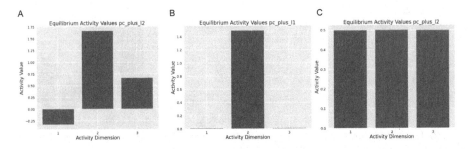

Fig. 3. A: Default behaviour of the outcome of iterative inference purely minimizing squared prediction errors. Probability mass is distributed between all three potential 'objects' in the hidden layer. B: Outcome of sparse iterative inference using an L1 penalty term in the objective function. All probability mass is concentrated on a single 'best' object so that a singly connected scene parse tree is constructed. C: Outcome of iterative inference with an L2 penalty term in the loss function which encourages probability mass to spread out. All objects have approximately equal probability of being selected.

Moreover, this sparsity penalty was necessary in this network in order for inference to exhibit routing-like behaviour. Without any regularization, iterative inference has a tendency to distribute probability mass between various high-level objects. This tendency is exacerbated with L2 regularisation which encourages the inference to spread probability mass as evenly as possible.

Interestingly, a similar intuition is applied in [18] where routing is explicitly derived as part of an EM algorithm with a clear probabilistic interpretation and where the MDL penalties derived for simply activating a capsule, which do not depend on its degree of activation, can perhaps also be thought of as effectively implementing a similar sparsity penalty which encourages the EM algorithm to assign capsule outputs to a single high-level capsule instead of spreading probability mass between them.

Appendix C: Iterative Inference Process for CapsNet

Our implementation of sparse iterative inference in place of capsule routing is the same in outline as that used for the toy model discussed in Appendix B, applied to the CapsNet architecture. That is, we minimize the sum of squared prediction errors per layer, which is equivalent to the variational free energy [27]. In this case the prediction error for the output layer is given by the difference between the prediction from the penultimate (PrimaryCaps) layer and the clamped target values. For the sparsity condition, we also add the capsule-level L1 sparsity penalty discussed in the caption of Fig. 1 at the output layer. Dynamic routing was turned off for this experiment, both at inference time and during training.

Appendix D: Relationship Between Capsule Routing and Attention

As noted above, our generative model of the capsule network can also describe the self-attention block in transformers, providing a fundamental building block towards building a full transformer generative model. Explicitly writing down such a generative model for the transformer architecture could enable a significantly greater understanding of the core mechanisms underlying the success of transformers at modelling large-scale sequence data as well as potentially suggest various improvements to current architectures.

This relationship to transformer attention is important because transformer attention is well-understood and found to be highly effective in natural language processing tasks [5,34] as well as recently in vision [11,32] and reinforcement learning [8,37,51]. Since capsule networks appear highly effective at processing natural scene statistics, this provides yet another example of the convergence of machine learning architectures towards a universal basis of attention mechanisms.

The basis of attention mechanisms can then be further understood in terms of associative memory architectures based on Modern Hopfield networks [21,35], as briefly discussed above. It has been found that sparsity of similarity scores is necessary for effective associative memory performance to prevent retrieved memories from interfering with each other [20,22,26]. The softmax operation in self-attention can be interpreted as a separation function with the goal of sparsifying the similarity scores by exponentially boosting the highest score above the others. Indeed, it is a general result that the capacity of associative memory models can be increased dramatically by using highly sparsifying separation functions such as high-order polynomials [10,22], softmaxes [35] and top-k activation functions [4].

An interesting aspect of our generative model is the use of VMF distributions to represent the dot-product similarity scores. Intuitively, this arises because the cosine similarity is 'circular' in that angles near $360°C$ are very similar to angles near 0. In most transformer and associative memory models, the update rules are derived from Gaussian assumptions which do not handle the wrap-around correctly and hence may be subtly incorrect for angles near the wrap-around point. By deriving update rules directly from our generative model, it is possible to obtain updates which handle this correctly and which may therefore perform better in practice. A second potential improvement relates to the VMF variance parameter κ. In transformer networks this is typically treated as a constant and set to $\frac{1}{\sqrt{d}}$. In essence, this bakes in the assumption that the variance of the distribution is inversely proportional to the data dimension. Future work could also investigate dynamically learning values of κ from data which could also improve performance.

One feature of routing-by-agreement not captured by iterative inference in standard PCNs is the positive feedback loop, in which low prediction error encourages even closer agreement between activities and predictions. This is similar to applying self-attention over time. A key distinction between attention as used in transformers and the routing mechanism in capsule networks is that the latter is iterative and can be applied sequentially for many iterations (although usually only 3–5), unlike in transformers where it is applied only once. Capsule networks therefore could provide ideas for improving transformer models by enabling them to work iteratively and adding the recurrent state that arises from the 'bias' term in the routing algorithm.

It has been proposed that highly deep networks with residual connections, a set of architectures that includes transformers, are implicitly approximating iterative inference using depth instead of time [14,19] which is a highly inefficient use of parameters. Instead, it is possible that similar performance may be obtained with substantially smaller models which can explicitly perform iterative inference similar to capsule networks. Some evidence for this conjecture comes from the fact that empirically it appears that large language models such as GPT2 [34] appear to perform most of their decisions as to their output tokens in their first few layers. These decisions are then simply refined over the remaining layers – a classic use-case for iterative inference.

The link between capsule routing and sparse iterative inference also has significant resonances in neuroscience. It is known that cortical connectivity and activations are both highly sparse (approximately only 1–5% neurons active simultaneously) [7,13,50] with even higher levels of sparsity existing in other brain regions such as the cerebellum [9,40,41]. Such a level of sparsity is highly energy efficient [43] and may provide an important inductive bias for the efficient parsing and representation of many input signals which are generated by highly sparse processes – i.e. dense pixel input is usually only generated by a relatively small set of discrete objects. Secondly, iterative inference is a natural fit for the ubiquitous recurrent projections that exist in cortex [23,25,44,46,49] and many properties of visual object recognition in the brain can be explained through a hybrid model of a rapid amortized feedforward sweep followed by recurrent iterative inference [45]. These considerations combine to provide a fair bit of evidence towards a routing-like sparse iterative inference algorithm being an integral part of cortical functioning. Moreover, it has been demonstrated many times in the sparse-coding literature that adding sparse regularisation on a variety of reconstruction and classification objectives can result in networks developing receptive fields and representations that resemble those found in the cortex [29,30,50].

Iterative inference is also important for enabling object discrimination and disambiguation in highly cluttered and occluded scenes because it can model the vital 'explaining away' [33] effect where inferences about one object can then inform parallel inferences about other objects. This is necessary in the case of occlusion since by identifying the occluder and implicitly subtracting out its visual features, it is often possible to make a much better inference about the occluded object [17]. It is therefore noteworthy, and suggestive of our hypothesis that routing can really be interpreted as iterative inference, that capsule networks perform much better at parsing such occluded scenes than purely feedforward models such as CNNs.

References

1. Bahdanau, D., Cho, K., Bengio, Y.: Neural machine translation by jointly learning to align and translate (2014). arXiv preprint arXiv:1409.0473
2. Beal, M.J.: Variational algorithms for approximate Bayesian inference. Technical report (2003)
3. Bogacz, R.: A tutorial on the free-energy framework for modelling perception and learning. J. Math. Psychol. **76**, 198–211 (2017)
4. Bricken, T., Pehlevan, C.: Attention approximates sparse distributed memory. arXiv preprint arXiv:2111.05498 (2021)
5. Brown, T.B., et al.: Language models are few-shot learners. arXiv preprint arXiv:2005.14165 (2020)
6. Buckley, C.L., Kim, C.S., McGregor, S., Seth, A.K.: The free energy principle for action and perception: a mathematical review. J. Math. Psychol. **81**, 55–79 (2017)
7. Buzsáki, G., Mizuseki, K.: The log-dynamic brain: how skewed distributions affect network operations. Nat. Rev. Neurosci. **15**(4), 264–278 (2014)
8. Chen, L., et al.: Decision transformer: reinforcement learning via sequence modeling. Adv. Neural. Inf. Process. Syst. **34**, 15084–15097 (2021)
9. De Zeeuw, C.I., Hoebeek, F.E., Bosman, L.W., Schonewille, M., Witter, L., Koekkoek, S.K.: Spatiotemporal firing patterns in the cerebellum. Nat. Rev. Neurosci. **12**(6), 327–344 (2011)
10. Demircigil, M., Heusel, J., Löwe, M., Upgang, S., Vermet, F.: On a model of associative memory with huge storage capacity. J. Stat. Phys. **168**(2), 288–299 (2017)
11. Dosovitskiy, A., et al.: An image is worth 16×16 words: transformers for image recognition at scale. arXiv preprint arXiv:2010.11929 (2020)
12. Friston, K.: A theory of cortical responses. Philos. Trans. Roy. Soc. B Biol. Sci. **360**(1456), 815–836 (2005)
13. Graham, D.J., Field, D.J.: Sparse coding in the neocortex. Evol. Nerv. Syst. **3**, 181–187 (2006)
14. Greff, K., Srivastava, R.K., Schmidhuber, J.: Highway and residual networks learn unrolled iterative estimation. arXiv preprint arXiv:1612.07771 (2016)
15. Gregor, K., Danihelka, I., Graves, A., Rezende, D., Wierstra, D.: Draw: a recurrent neural network for image generation. In: International Conference on Machine Learning, pp. 1462–1471. PMLR (2015)
16. Hinton, G.: How to represent part-whole hierarchies in a neural network. arXiv preprint arXiv:2102.12627 (2021)
17. Hinton, G.E., Krizhevsky, A., Wang, S.D.: Transforming auto-encoders. In: Honkela, T., Duch, W., Girolami, M., Kaski, S. (eds.) ICANN 2011. LNCS, vol. 6791, pp. 44–51. Springer, Heidelberg (2011). https://doi.org/10.1007/978-3-642-21735-7_6
18. Hinton, G.E., Sabour, S., Frosst, N.: Matrix capsules with EM routing. In: International Conference on Learning Representations (2018)
19. Jastrzbski, S., Arpit, D., Ballas, N., Verma, V., Che, T., Bengio, Y.: Residual connections encourage iterative inference. arXiv preprint arXiv:1710.04773 (2017)
20. Kanerva, P.: Sparse Distributed Memory. MIT Press, Cambridge (1988)
21. Krotov, D., Hopfield, J.: Large associative memory problem in neurobiology and machine learning. arXiv preprint arXiv:2008.06996 (2020)
22. Krotov, D., Hopfield, J.J.: Dense associative memory for pattern recognition. Advance in Neural Information Processing System, vol. 29, pp. 1172–1180 (2016)

23. Lamme, V.A., Roelfsema, P.R.: The distinct modes of vision offered by feedforward and recurrent processing. Trends Neurosci. **23**(11), 571–579 (2000)
24. Makhzani, A., Frey, B.J.: k-sparse autoencoders. CoRR abs/1312.5663 (2014)
25. Melloni, L., van Leeuwen, S., Alink, A., Müller, N.G.: Interaction between bottom-up saliency and top-down control: how saliency maps are created in the human brain. Cereb. Cortex **22**(12), 2943–2952 (2012)
26. Millidge, B., Salvatori, T., Song, Y., Lukasiewicz, T., Bogacz, R.: Universal hopfield networks: a general framework for single-shot associative memory models. arXiv preprint arXiv:2202.04557 (2022)
27. Millidge, B., Seth, A., Buckley, C.L.: Predictive coding: a theoretical and experimental review. arXiv preprint arXiv:2107.12979 (2021)
28. Nazábal, A., Williams, C.K.I.: Inference for generative capsule models. CoRR abs/2103.06676 (2021), https://arxiv.org/abs/2103.06676
29. Olshausen, B.A., Field, D.J.: Emergence of simple-cell receptive field properties by learning a sparse code for natural images. Nature **381**(6583), 607–609 (1996)
30. Olshausen, B.A., Field, D.J.: Sparse coding of sensory inputs. Curr. Opin. Neurobiol. **14**(4), 481–487 (2004)
31. Paik, I., Kwak, T., Kim, I.: Capsule networks need an improved routing algorithm. ArXiv abs/1907.13327 (2019)
32. Parmar, N., et al.: Image transformer. In: International Conference on Machine Learning, pp. 4055–4064. PMLR (2018)
33. Pearl, J.: Probabilistic Reasoning in Intelligent Systems: Networks of Plausible Inference. Morgan kaufmann, Burlington (1988)
34. Radford, A., Wu, J., Child, R., Luan, D., Amodei, D., Sutskever, I., et al.: Language models are unsupervised multitask learners. OpenAI Blog **1**(8), 9 (2019)
35. Ramsauer, H., et al.: Hopfield networks is all you need. arXiv preprint arXiv:2008.02217 (2020)
36. Rawlinson, D., Ahmed, A., Kowadlo, G.: Sparse unsupervised capsules generalize better. ArXiv abs/1804.06094 (2018)
37. Reed, S., Zolna, K., et al.: A generalist agent. arXiv preprint arXiv:2205.06175 (2022)
38. Ribeiro, F.D.S., Leontidis, G., Kollias, S.D.: Capsule routing via variational bayes. In: AAAI, pp. 3749–3756 (2020)
39. Sabour, S., Frosst, N., Hinton, G.E.: Dynamic routing between capsules. In: Advances in Neural Information Processing Systems, vol. 30 (2017)
40. Schweighofer, N., Doya, K., Lay, F.: Unsupervised learning of granule cell sparse codes enhances cerebellar adaptive control. Neuroscience **103**(1), 35–50 (2001)
41. Shepherd, G.M., Grillner, S.: Handbook of Brain Microcircuits. Oxford University Press, Oxford (2018)
42. Smith, L., Schut, L., Gal, Y., van der Wilk, M.: Capsule networks - a probabilistic perspective. CoRR abs/2004.03553 (2020). https://arxiv.org/abs/2004.03553
43. Sterling, P., Laughlin, S.: Principles of Neural Design. MIT Press, Cambridge (2015)
44. Theeuwes, J.: Top-down and bottom-up control of visual selection. Acta Physiol. (Oxf) **135**(2), 77–99 (2010)
45. Tschantz, A., Millidge, B., Seth, A.K., Buckley, C.L.: Hybrid predictive coding: Inferring, fast and slow. arXiv preprint arXiv:2204.02169 (2022)
46. VanRullen, R.: The power of the feed-forward sweep. Adv. Cogn. Psychol. **3**(1–2), 167 (2007)
47. Vaswani, A., Shazeer, N., et al.: Attention is all you need. In: Advances in Neural Information Processing Systems, pp. 5998–6008 (2017)

48. Wainwright, M.J., Jordan, M.I., et al.: Graphical models, exponential families, and variational inference. Found. Trends® Mach. Learn. **1**(1–2), 1–305 (2008)

49. Weidner, R., Krummenacher, J., Reimann, B., Müller, H.J., Fink, G.R.: Sources of top-down control in visual search. J. Cogn. Neurosci. **21**(11), 2100–2113 (2009)

50. Willmore, B.D., Mazer, J.A., Gallant, J.L.: Sparse coding in striate and extrastriate visual cortex. J. Neurophysiol. **105**(6), 2907–2919 (2011)

51. Zheng, Q., Zhang, A., Grover, A.: Online decision transformer. arXiv preprint arXiv:2202.05607 (2022)

Home Run: Finding Your Way Home by Imagining Trajectories

Daria de Tinguy$^{(\boxtimes)}$, Pietro Mazzaglia, Tim Verbelen, and Bart Dhoedt

IDLab, Department of Information Technology Ghent University - imec,
Technologiepark-Zwijnaarde 126, 9052 Ghent, Belgium
{daria.detinguy,pietro.mazzaglia,tim.verbelen,bart.dhoedt}@ugent.be

Abstract. When studying unconstrained behaviour and allowing mice
to leave their cage to navigate a complex labyrinth, the mice exhibit foraging behaviour in the labyrinth searching for rewards, returning to their
home cage now and then, e.g. to drink. Surprisingly, when executing such
a "home run", the mice do not follow the exact reverse path, in fact, the
entry path and home path have very little overlap. Recent work proposed
a hierarchical active inference model for navigation, where the low level
model makes inferences about hidden states and poses that explain sensory inputs, whereas the high level model makes inferences about moving
between locations, effectively building a map of the environment. However, using this "map" for planning, only allows the agent to find trajectories that it previously explored, far from the observed mice's behaviour.
In this paper, we explore ways of incorporating before-unvisited paths in
the planning algorithm, by using the low level generative model to imagine potential, yet undiscovered paths. We demonstrate a proof of concept
in a grid-world environment, showing how an agent can accurately predict a new, shorter path in the map leading to its starting point, using
a generative model learnt from pixel-based observations.

Keywords: Robot navigation · Active inference · Free energy
principle · Deep learning

1 Introduction

Humans rely on an internal representation of the environment to navigate, i.e.
they do not require precise geometric coordinates or complete mappings of the
environment; a few landmarks along the way and approximate directions are
enough to find our way back home [1]. This reflects the concept of a "cognitive
map" as introduced by Tolman [2], and matches the discovery of specific place
cells firing in the rodent hippocampus depending on the animal position [3] and
our representation of space [1].

Recently, Çatal et al. [4] showed how such mapping, localisation and path
integration can naturally emerge from a hierarchical active inference (AIF)
scheme and are also compatible with the functions of the hippocampus and

entorhinal cortex [5]. This was implemented on a real robot to effectively build a map of its environment, which could then be used to plan its way using previously visited locations [6].

However, while investigating the exploratory behaviour of mice in a maze, where mice were left free to leave their home to run and explore, a peculiar observation was made. When the mice decided to return to their home location, instead of re-tracing their way back, the mice were seen taking fully new, shorter, paths directly returning them home [7].

On the contrary, when given the objective to reach a home location, the hierarchical active inference model, as proposed by [4,6], can only navigate between known nodes of the map, unable to extrapolate possible new paths without first exploring the environment. To address this issue, we propose to expand the high level map representation using the expected free energy of previously unexplored transitions, by exploiting the learned low-level environment model. In other worlds, we enlarge the projection capabilities of architecture [6] to unexplored paths.

In the remainder of this paper we will first review the hierarchical AIF model [4], then explain how we address planning with previously unvisited paths by imagining novel trajectories within the model. As a proof of concept, we demonstrate the mechanism on a Minigrid environment with a four-rooms setup. We conclude by discussing our results, the current limitations and what is left to improve upon the current results.

2 Navigation as Hierarchical Active Inference

The active inference framework relies upon the notion that intelligent agents have an internal (generative) model optimising beliefs (i.e. probability distributions over states), explaining the causes of external observations. By minimising the surprise or prediction error, i.e, free energy (FE), agents can both update their model as well as infer actions that yield preferred outcomes [8,9].

In the context of navigation, Çatal et al. [4] introduced a hierarchical active inference model, where the agent reasons about the environment on two different levels. On the low level, the agent integrates perception and pose, whereas on the high level the agent builds a more coarse grained, topological map. This is depicted in Fig. 1.

The low level, depicted in blue, comprises a sequence of low-level action commands a_t and sensor observations o_t, which are generated by hidden state variables s_t and p_t. Here s_t encodes learnable features that give rise to sensory outcomes, whereas p_t encodes the agent's pose in terms of its position and orientation. The low level transition model $p(s_{t+1}|s_t, p_t, a_t)$ and likelihood model $p(o_t|s_t)$ are jointly learnt from data using deep neural networks [10], whereas the pose transition model $p(p_{t+1}|s_t, p_t, a_t)$ is instantiated using a continuous attractor network similar to [11].

At the high level, in red in the Figure, the agent reasons over more coarse grained sequences of locations l_τ, where it can execute a move m_τ that gives

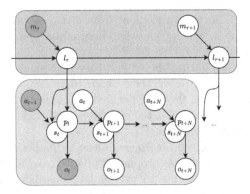

Fig. 1. Navigation as a hierarchical generative model for active inference [4]. At the lower level, highlighted in blue, the model entertains beliefs about hidden states s_t and p_t, representing hidden causes of the observation and the pose at the current timestep t respectively. The hidden states give rise to observations o_t, whereas actions a_t impact future states. At the higher level, highlighted in red, the agent reasons about locations l. The next location $l_{\tau+1}$ is determined by executing a move m_τ. Note that the higher level operates on a coarser timescale. Grey shaded nodes are considered observed.

rise to a novel location $l_{\tau+1}$. In practice, this boils down to representing the environment as a graph-based map, where locations l_τ are represented by nodes in the graph, whereas potential moves m_τ are links between those nodes. Note that a single time step at the higher level, i.e. going from τ to $\tau+1$, can comprise multiple time steps on the lower level. This enables the agent to first 'think' far ahead in the future on the higher level.

To generate motion, the agent minimizes expected free energy (EFE) under this hierarchical generative model. To reach a preferred outcome, the agent first plans a sequence of moves that are expected to bring the agent to a location rendering the preferred outcome highly plausible, after which it can infer the action sequence that brings the agent closer to the first location in that sequence. For a more elaborate description of the generative model, the (expected) free energy minimisation and implementation, we refer to [4].

3   Imagining Unseen Trajectories

As discussed in [4], minimising expected free energy under such a hierarchical model induces desired behaviour for navigation. In the absence of a preferred outcome, an epistemic term in the EFE will prevail, encouraging the agent to explore actions that yield information on novel (hidden) states, effectively expanding the map while doing so. In the presence of a preferred state, the agent will exploit the map representation to plan the shortest (known) route towards the objective. However, crucially, the planning is restricted to previously visited locations in the map. This is not consistent with the behaviour observed in mice [7], as these, apparently, can exploit new paths even when engaging in a goal-directed run towards their home.

In order to address this issue, we hypothesize that the agent not only considers previously visited links and locations in the map during planning, but also imagines potential novel links. A potential link from a start location l_A to a destination location l_B is hence scored by the minimum EFE over all plans π (i.e. a sequence of actions) generating such a trajectory under the (low level) generative model, i.e.:

$$G(l_A, l_B) = \min_{\pi} \underbrace{\sum_{k=1}^{H} D_{KL}\left[Q(s_{t+k}, p_{t+k}|\pi)Q(s_t|l_A) \| Q(s_{t+H}, p_{t+H}|l_B)\right]}_{\text{probability reaching} l_B \text{ from } l_A} \tag{1}$$
$$+ \underbrace{\mathbb{E}_{Q(s_{t+k})}\left[H(P(o_{t+k}|s_{t+k}))\right]}_{\text{observation ambiguity}}.$$

The first term is a KL divergence between the expected states to visit starting at location l_A and executing plan π, and the state distribution expected at location l_B. The second term penalizes paths that are expected to yield ambiguous observations.

We can now use $G(l_A, l_B)$ to weigh each move between two close locations (the number of path grows exponentially the further the objective is), even through ways not explored before, and plan for the optimal trajectory towards a goal destination. In the next section, we work out a practical example using a grid-world environment.

4   Experiments

4.1   MiniGrid Setup

The experiments were realised in a MiniGrid environment [12] of 2×2 up to 5×5 rooms, of sizes going from 4 to 7 tiles and having a random floor color chosen among 6 options : red, green, blue, purple, yellow and grey. Rooms are connected by a single open tile, randomly spawned in the wall. The agent has 3 possible actions at each time step: move one tile forward, turn 90° left or turn 90° right. It can't see through walls and can only venture into an open grid space. Note that the wall blocking vision is not really realistic and the agent can see the whole room if there is an open door in its field of view, thus even if part of the room should be masked by a wall (eg. Fig. 2C raw observation). It can see ahead and around in a window of 7 × 7 tiles, including its own occupied tile. The observation the agent receives is a pixel rendering in RGB of shape 3 × 56 × 56.

4.2   Model Training and Map Building

Our hierarchical generative model was set up in similar fashion as [4]. To train the lower level of the generative model, which consists of deep neural networks,

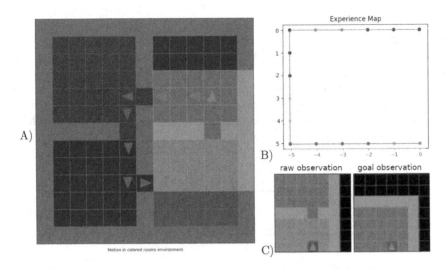

Fig. 2. MiniGrid test maze and associated figures, A) An example of the maze with a reachable goal (door open allowing shortcut) and the agent path toward a home-run's starting point, the transparent grey box correspond to the agent's field of view at the starting position. B) The topological map of the path executed in A as generated by the high level of our generative model, C) The currently observed RGB image as reconstructed by the agent's model at the end of path and the view at the desired goal position. (Color figure online)

we let an agent randomly forage the MiniGrid environments, and train those end to end by minimising the free energy on those sequences. Additional model details and training parameters can be found in Appendix A.

The high level map is also built using the same procedure as [4]. However, since we are dealing with a grid-world, distinct places in the grid typically yield distinct location nodes in the map, unless these are near and actually yield identical observations. Also, we found that predicting the effect of turning left or right was harder for neural networks to predict, yielding a higher surprise signal. However, despite these limitations, we can still demonstrate the main contribution of this paper.

4.3   Home Run

Inspired by the mice navigation in [7], we test the following setup in which the agent first explores a maze, and at some point is provided with a preference of returning to the start location. Figure 2 shows an example of a test environment and associated trajectories realised by the agent. At the final location, the agent is instructed to go back home, provided by the goal observation in Fig. 2C. Figure 2B illustrates the map generated by the hierarchical model.

First, we test whether the agent is able to infer whether it can reach the starting node in the experience map from the current location. We do so by

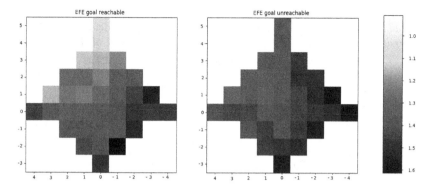

Fig. 3. Lowest expected free energy of each end position after 5 steps. The right figure shows the agent at position (0,0) facing the goal at position (0,5), as represented in Fig. 2A i). In the left figure, the door is open, therefore the goal is reachable, on the right figure the door is closed, the goal cannot be reached in 5 steps.

imagining all possible plans π, and evaluating the expected free energy of each plan over an average of $N = 3$ samples from the model. Figure 3 shows the EFE for all reachable locations in a 5 steps planning horizon. It is clear that in case the door is open, the agent expects the lowest free energy when moving forward through the door, expecting to reach the start node in the map. In case the path is obstructed (the door as in 2A, allowing a shortcut, is closed), it can still imagine going forward 5 steps, but this will result in the agent getting stuck against the wall, which it correctly imagines and reflects on the EFE.

However, the prior model learnt by the agent is far from perfect. When inspecting various imagined rollouts of the model, as shown in Fig. 4, we see that the model has trouble encoding and remembering the exact position of the door, i.e. predicting the agent getting stuck (top) or incorrect room colours and size (bottom). While not problematic in our limited proof of concept, also due to the fact that the EFE is averaged over multiple samples, this shows that the effectiveness of the agent will be largely dependent on the accuracy of the model.

To test the behaviour in a more general setting, we set multiple home-run scenarios, where the agent's end position is $d = 5, 6, 7, 9$ steps away from the start location. For each d, we sample at least 20 runs over 4 novel 2×2 rooms environment, with different room sizes and colours, similar to the train set, in which 10 have an open door between the start and goal, and 10 have not. We count the average number of steps required by the agent to get back home, and compare against two baseline approaches. First is the Greedy algorithm, inspired by [13], in which the agent greedily navigates in the direction of the goal location, and follows obstacles in the known path direction when bumping into one. Second is a TraceBack approach, which retraces all its steps back home, similar to Ariadne's thread. Our approach uses the EFE with a planning horizon of d to decide whether or not the home node is reachable based on a fixed threshold, and falls back to planning in the hierarchical model, which boils down to a TraceBack strategy (Table 1).

Table 1. Home run strategies and the resulting number of steps, for different distances d to home, and open versus closed scenarios. For small d our model correctly imagines the outcome. For $d = 9$ the agent infers an open door about 27% of the time.

	Open			Closed		
d	Greedy	TraceBack	Ours	Greedy	TraceBack	Ours
5	5	25	6.5	29.5	25	25
6	6	31	6	41	31	31
7	7	27	11.5	31.5	27	27
9	9	36	23.7	46	36	36

In case of small d (≤ 6), our approach successfully identifies whether the goal is reachable or not, even when the agent is not facing it, which results in a similar performance for a Greedy approach in the 'open' case, and a reverting to TraceBack in the 'closed' case. There is been only one exception in our test-bench at 5steps range issued by a reconstruction error on all samples (the occurrence probability is 0.04% as having a sample wrongly estimating the door position at 5steps is 33%). For $d = 7$ our model misses some of the shortcut opportunities, as the model's imagination becomes more prone to errors for longer planning horizons. For $d = 9$, the rooms are larger and the wall separating the two rooms is actually not visible to the agent. In this regime, we found the agent imagines about 27% of the time that it will be open, and takes the gamble to move towards the wall, immediately returning on its path if the wall is obstructed.

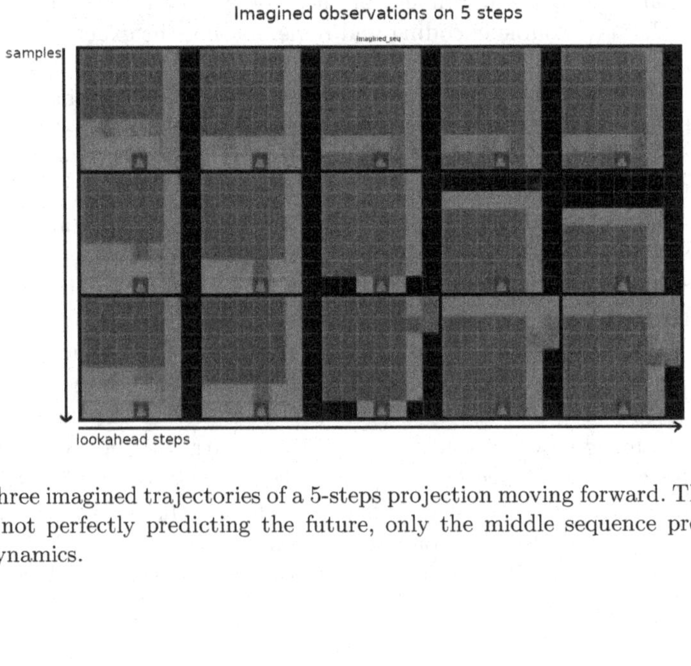

Fig. 4. Three imagined trajectories of a 5-steps projection moving forward. The trained model is not perfectly predicting the future, only the middle sequence predicts the correct dynamics.

5   Discussion

Our experiments show that using the EFE of imagined paths can yield more optimal, goal-directed behaviour. Indeed, our agent is able to imagine and exploit shortcuts when planning its way home. However, our current experimental setup is still preliminary and we plan to further expand upon this concept. For instance we currently arbitrarily set the point at which the agent decide to home-run. In a real experiment, the mice likely decide to go home due to some internal stimulus, e.g., when they get thirsty and head back home where water is available. We could further develop the experimental setup to incorporate such features and do a more extensive evaluation.

One challenge of using the Minigrid environment as an experimental setup [12] is the use of top view visual observations. Using a pixel-wise error for learning the low-level perception model can be problematic, as for example the pixel-wise error between a closed versus an open tile in the wall is small in absolute value, and hence it's difficult to learn for the model, as illustrated in Fig. 4. A potential approach to mitigate this is to use a contrastive objective instead, as proposed by [14].

Another important limitation of the current model is that it depends on the effective planning horizon of the lowest level model to imagine shortcuts. Especially in the Minigrid environment, imagining the next observation for a 90° turn is challenging, as it requires a form of memory of the room layout to correctly predict the novel observation. This severely limits the planning horizon of our current models. A potential direction of future work in this regard is to learn a better location, state and pose mapping. For instance, instead of simply associating locations with a certain state and pose, conditioning the transition model on a learnt location descriptor might allow the agent to learn and encode the shape of a complete room in a location node.

Other approaches have been proposed to address the navigation towards a goal by the shortest way possible in a biologically plausible way. For instance, Erdem et al. [15] reproduced the pose and place-cell principle of the rat's hippocampus with spiking neural networks and use a dense reward signal to drive goal-directed behaviour, with more reward given the closer the agent gets to the goal. Hence, the path with the highest reward is sought, and trajectories on which obstacles are detected are discarded. In Vegard et al. [13], the process is also bio-inspired, based on the combination of grid cell-based vector and topological navigation. The objective is now explicitly represented as a target position in space, which is reached by vector navigation mechanisms with local obstacle avoidance mediated by border cells and place cells. Both alternatives also adopt topological maps and path integration in order to reach their objective. However, both exhibit more greedy and reactive behaviour, whereas our model is able to exploit the lower level perception model to already predict potential obstacles upfront, before bumping into those.

6   Conclusion

In this paper we have proposed how a hierarchical active inference model can be used to improve planning by predicting novel, previously unvisited paths. We demonstrated a proof of concept using a generative model learnt from pixel based observations in a grid-world environment.

As future work we envision a more extensive evaluation, comparing shallow versus deep hierarchical generative models in navigation performance. Moreover, we aim to address several of the difficulties of our current perception model, i.e. the limitations of pixel-wise prediction errors, the limited planning horizon, and a more expressive representation for locations in the high level model. Ultimately, our goal is to deploy this on a real-world robot, autonomously exploring, planning and navigating in its environment.

Acknowledgment. This research received funding from the Flemish Government under the "Onder- zoeksprogramma Artificiële Intelligentie (AI) Vlaanderen" programme.

A   Model Details and Training

In this appendix, we provide some additional details on the training data, model parameters, training procedure and building the hierarchical map.

A.1   Training Data

To optimize the neural network models a dataset composed of sequences of action-observation pairs was collected by human demonstrations of interaction with the environment. The agent was made to move around from rooms to room, circle around and turn randomly. About 12000 steps were recorded in 39 randomly created environments having different room size, number of rooms, open door emplacements and floor colors, as well as the agent having a random starting pose and orientation. 2/3 of the data were used for training and 1/3 for validation. Then a fully novel environment was used for testing.

A.2   Model Parameters

The low level perception model is based on the architecture of [10], and is composed of 3 neural networks that we call: prior, posterior and likelihood.

The prior neural network consists in a LSTM layer followed with a variational layer giving out a distribution (i.e. mean and std).

The posterior model first consists of a convolutional network to compress sensor data. This data is then concatenated with the hot encoded action and the previous state, all of that is then processed by a fully connected neural network coupled with a variational layer to obtain a distribution.

The likelihood model performs the inverse of the convolutional part of the posterior, generating an image out of a given state sample.

The detailed parameters are listed in Table 2.

A.3   Training the Model

The low level perception pipeline was trained end to end on time sequences of 10 steps using stochastic gradient descent with the minimization of the free energy loss function [10]:

$$FE = \sum_t D_{KL}[Q(s_t|s_{t-1}, a_{t-1}, o_t)||P(s_t|s_{t-1}, a_{t-1})] - \mathbb{E}_{Q(s_t)}[\log P(o_t|s_t)]$$

The loss consists of a negative log likelihood part penalizing the error on reconstruction, and a KL-divergence between the posterior and the prior distributions on a training sequence. We trained the model for 300 epochs using the ADAM optimizer [16] with a learning rate of $1 \cdot 10-4$.

Table 2. Models parameters

	Layer	Neurons/Filters	Stride
Prior	Concatenation		
	LSTM	200	
	Linear	2*30	
Posterior	Convolutional	16	2
	Convolutional	32	2
	Convolutional	64	2
	Convolutional	128	2
	Convolutional	256	2
	Concatenation		
	Linear	200	
	Linear	2*30	
Likelihood	Linear	200	
	Linear	256*2*2	
	Upsample		
	Convolutional	128	1
	Upsample		
	Convolutional	64	1
	Upsample		
	Convolutional	32	1
	Upsample		
	Convolutional	16	1
	Upsample		
	Convolutional	3	1

A.4 Building the Map

The high level model is implemented as a topological graph representation, linking pose and hidden state representation to a location in the map. Here we reuse the LatentSLAM implementation [6] consisting of pose cells, local view cells and an experience map.

The pose cells are implemented as a Continuous Attractor Network (CAN), representing the local position x, y and heading θ of the agent. Pose cells represent a finite area, therefore the firing fields of a single grid cell correspond to several periodic spatial locations.

The local view cells are organised as a list of cell, each cell containing a hidden state representing an observation, the pose cell excited position, and the map's experience node linked to this view. After each motion, the encountered scene is compared to all previous cells observation by calculating the cosine distance between hidden state features. If the distance is smaller than a given threshold, then the cell corresponding to this view is activated, else a new cell is created.

The experience map contains the experience of the topological map. It gives an estimate of the agent global pose in the environment and link the pose cell position with the local view cell active at this moment. If those elements do not match with any existing node of the map, a new one is created and linked to the previous experience, else a close loop is operated and the existing experiences are linked together.

References

1. Peer, M., Brunec, I.K., Newcombe, N.S., Epstein, R.A.: Structuring knowledge with cognitive maps and cognitive graphs. Trends Cogn. Sci. **25**(1), 37–54 (2021). https://www.sciencedirect.com/science/article/pii/S1364661320302503
2. Tolman, E.C.: Cognitive maps in rats and men. Psychol. Rev. **55**(4), 189–208 (1948)
3. Milford, M.J., Wyeth, G.F., Prasser, D.: RatSLAM: a hippocampal model for simultaneous localization and mapping. In: IEEE International Conference on Robotics and Automation, 2004. Proceedings. ICRA 2004, vol. 1, 403–408 (2004)
4. Çatal, O., Verbelen, T., Van de Maele, T., Dhoedt, B., Safron, A.: Robot navigation as hierarchical active inference. Neural Netw. **142**, 192–204 (2021). https://www.sciencedirect.com/science/article/pii/S0893608021002021
5. Safron, A., Çatal, O., Verbelen, T.: Generalized simultaneous localization and mapping (g-SLAM) as unification framework for natural and artificial intelligences: towards reverse engineering the hippocampal/entorhinal system and principles of high-level cognition (2021). https://doi.org/10.31234/osf.io/tdw82
6. Çatal, O., Jansen, W., Verbelen, T., Dhoedt, B., Steckel, J.: LatentSLAM: unsupervised multi-sensor representation learning for localization and mapping. CoRR, vol. abs/2105.03265 (2021). https://arxiv.org/abs/2105.03265
7. Rosenberg, M., Zhang, T., Perona, P., Meister, M.: Mice in a labyrinth show rapid learning, sudden insight, and efficient exploration. ELife **10**, e66175 (2021). https://doi.org/10.7554/eLife.66175

8. Friston, K., FitzGerald, T., Rigoli, F., Schwartenbeck, P., Pezzulo, G.: Active inference and learning. Neurosci. Biobehav. Rev. **68**, 862–879 (2016). https://www.sciencedirect.com/science/article/pii/S0149763416301336

9. Kaplan, R., Friston, K.: Planning and navigation as active inference, vol. 12 (2017)

10. Çatal, O., Wauthier, S., De Boom, C., Verbelen, T., Dhoedt, B.: Learning generative state space models for active inference. Front. Comput. Neurosci. **14**, 574372 (2020). https://www.frontiersin.org/article/10.3389/fncom.2020.574372

11. Milford, M., Jacobson, A., Chen, Z., Wyeth, G.: RatSLAM: using models of rodent hippocampus for robot navigation and beyond. In: Inaba, M., Corke, P. (eds.) Robotics Research. STAR, vol. 114, pp. 467–485. Springer, Cham (2016). https://doi.org/10.1007/978-3-319-28872-7_27

12. Chevalier-Boisvert, M., Willems, L., Pal, S.: Minimalistic gridworld environment for openAI gym (2018). https://github.com/maximecb/gym-minigrid

13. Edvardsen, V., Bicanski, A., Burgess, N.: Navigating with grid and place cells in cluttered environments. Hippocampus **30**, 220–232 (2019)

14. Mazzaglia, P., Verbelen, T., Dhoedt, B.: Contrastive active inference. In: Advances in Neural Information Processing Systems, vol. 34 (2021). https://openreview.net/forum?id=5t5FPwzE6mq

15. Erdem, U.M., Hasselmo, M.: A goal-directed spatial navigation model using forward planning based on grid cells. Eur. J. Neurosci. **35**, 916–31 (2012)

16. Kingma, D.P., Ba, J.: Adam: A method for stochastic optimization. https://arxiv.org/abs/1412.6980

A Novel Model for Novelty: Modeling the Emergence of Innovation from Cumulative Culture

Natalie Kastel[1,2,3] and Guillaume Dumas[3,4](✉) (iD)

[1] Amsterdam Brain and Cognition Centre, University of Amsterdam, Science Park 904, 1098 XH Amsterdam, The Netherlands
Natalie.Kastel@PPSP.team
[2] Institute for Advanced Study, University of Amsterdam, Oude Turfmarkt 147, 1012 GC Amsterdam, The Netherlands
[3] Precision Psychiatry and Social Physiology Laboratory, CHU Sainte-Justine Research Center, Department of Psychiatry, Université de Montréal, Montréal, QC, Canada
Guillaume.Dumas@PPSP.team
[4] Mila - Quebec AI Institute, Université de Montréal, Montréal, QC, Canada

Abstract. While the underlying dynamics of active inference communication and cumulative culture have already been formalized, the emergence of novel cultural information from these dynamics has not yet been understood. In this paper, we apply an active Inference framework, informed by genetic speciation, to the emergence of innovation from a population of communicating agents in a cumulative culture. Our model is premised on the idea that innovation emerges from accumulated cultural information when a collective group of agents agree on the legitimacy of an alternative belief to the existing (or- status quo) belief.

Keywords: Active inference · Innovation · Communication · Cumulative culture · Cultural dynamics

1 Introduction

The dynamics underlying cultural evolution include the introduction of novel cultural information to a population (i.e., innovation), the transmission of established cultural information within a population (i.e., communication), and its change in prevalence (i.e., cumulative culture) (Kashima et al. 2019).

While there is a fast growing body of theoretical and empirical work on characterizing these dynamics (Aunger 2001; Buskell et al. 2019; Bettencourt et al. 2006; Creanza et al. 2017; Dawkins 1993; Dean et al. 2014; Dunstone and Caldwell 2018; Enquist et al. 2011; Gabora 1995; Heylighen and Chielens 2009; Kashima et al. 2019; Richerson et al. 2010; Stout and Hecht 2017; Weisbuch et al. 2009) mathematical models able to integrate this data into quantifiable models are scarce.

In 2015, Friston & Frith provide a quantitative model of joint communication and show that communication couples the internal states of active-inference agents and underwrites a minimal form of generalized synchrony between their internal states at a level of abstraction that allows us to characterize a statistical coupling even for agents operating with fundamentally different underlying neurobiological structures. Kastel and Hesp (2021) build on this and cast cultural transmission as a bi-directional process of communication. The idea is that when active inference agents communicate, they are able to understand each other by referring to their own generative model and inferring the internal state of the other from their behavior. This couples communicating agents in an action perception cycle of prediction and inference that induces a generalized synchrony between their internal states. Kastel & Hesp operationalise this generalized synchrony as a particular convergence between the internal states of interlocutors such that distinct belief states converge into one shared belief, and in that sense modify both of the original "parent" belief states.

The simulation of these local communication dynamics (and specifically the convergence and subsequent modification of the belief state of each communicating agent) also serves as the basis from which to build a full blown cumulative culture model (Kastel and Hesp 2021). Cumulative culture is an emerging and prominent theory of cultural evolution which describes cultural traits as being slightly modified with every transmission such that over time these modifications accumulate to bring about an adaptive culture. Though cumulative culture is a powerful theory in that it faithfully represents the complex nature of societal change, this complexity is exceptionally challenging to formalize in quantitative models. Kastel & Hesp provided an active-inference formalization of cumulative culture by casting it as the emergence of accumulated modifications to cultural beliefs from the local efforts of agents to converge on a shared narrative. As a proof of principle for this hypothesis, they simulate a population of agents that interchangeably engage in dialogue with each other over time. When a divergent belief state is introduced to a uniform population holding (variations of) a status quo belief, it spreads through it and brings about a cumulative collective behavior of separation and isolation between groups holding distinct beliefs.

While they provide a sufficient formulation of the way slight modifications to cultural information occur during communication (previously understood as transmission) and shown how the accumulation of these dynamics affect an entire population (i.e., cumulative culture), Kastel & Hesp did not provide an account of the way novel information (i.e., the hypothesized belief state) is introduced into a population to begin with.

Within a cumulative culture framework, innovation is interpreted as the emerging property of a complexity of exchanges between agents, as opposed to the result of the mental effort of an exceptionally skillful individual. Indeed, emerging theories have put forward the suggestion that inventors and entrepreneurs are not "the brains" behind a creative idea, but are the product of a collective cultural brain (Muthukrishna and Henrich 2016). Their ideas do not stand in competition or comparison with other agents in the population, but are better understood as a nexus for previously isolated ideas within it. This cumulative approach to cultural innovation is supported by empirical findings showing that innovation rates are higher in cultures with high sociality (i.e. large and highly interconnected populations that offer exposure to more ideas), transmission fidelity (i.e. better learning between agents) and transmission variance (i.e. a willingness to somewhat deviate from the accepted learned norms (Muthukrishna and Henrich 2016).

The theory of innovation as the emerging property of a complex cultural "brain" is compatible with the theory of cumulative culture and with empirical data, but it does not provide a specific account of the mechanism by which innovation may be achieved and novel cultural beliefs and practices introduced into a population.

This paper provides an active-inference based theoretical account of Innovation as the emergence of novel cultural information from a cumulative culture. This novel account of innovation derives inspiration from the way novelty emerges in biology, namely, through a process of genetic speciation.

2 The Emergence of Innovation

2.1 Speciation in Biology

We propose that it may prove constructive to draw on specific analogies between the way novel cultural beliefs and practices emerge within a culture and the emergence of a new species in the context of biological evolution, while remaining sensitive to points at which such analogies break down. In nature, speciation occurs when a group of organisms from a particular species are separated from their original population, thus encouraging the development of their own unique characteristics. These new characteristics increasingly differentiate the two population groups when their differences grow larger as the two groups reproduce separately due to their separate environments or characteristics. Across generations, genetic differences between the old and new group become so large that they are no longer able to create offspring (i.e. mechanisms of reproductive isolation), thus highlighting the status of the subgroup as an entirely new species in its own right (Rundle and Nosil 2005).

A classic example of speciation is that of the Galápagos finch. Different species of this bird inhabit separate environments, located on different islands of the Galápagos peninsula. Over time and numerous generations, separate populations of finches developed a variety of beak morphologies, each group's morphology appearing to be adapted specifically to the feeding opportunities available on their island. While one group had developed long and thin beaks, ideal for probing cactus flowers without getting injured by the cactus, other finches developed large and blunt beaks that were perfect for nut cracking. Due to the reproductive isolation of these birds (geographic based, in this case), they developed into separate species with their own unique features.

2.2 Innovation as Cultural Speciation

Before discussing possible similarities between biological speciation and cultural innovation, a crucial difference between them should be noted. Mechanisms of biological reproductive isolation prevent members of different species from producing offspring or, in edge cases, render such offspring sterile (Palumbi 1994). In contrast, cultural evolution frequently involves cross-talk between different branches of the cultural tree – as different cultures have tended to co-opt and refurbish each other's beliefs and practices. While horizontal gene transfer is exceedingly rare in biology, cultural evolution has experienced some of its greatest accelerations precisely due to transmission of beliefs and practices across diverse cultures.

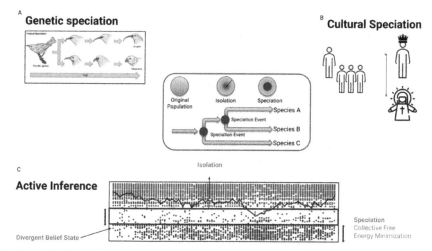

Fig. 1. A visual representation of speciation in genetics and its model in culture and active inference. Each model of speciation requires the existence of an original population that diverges through a process of group isolation such that each group develops its own unique characteristics and features. (A) genetic speciation in the galapagos finch. Geographic reproductive isolation broke up the original population of finches into those inhabiting separate islands with different selective pressures. Due to the differences in selective pressures, the separated populations developed a variety of beak morphologies that distinguished them from each other. (B) A model of speciation in religious practices. The divergence of early christians from the established jewish religion on the basis of differing interpretations of jewish scripture is modeled as a form of cultural isolation. While early Christians interpreted jewish eschatology as foretelling the arrival of a divine jewish seviour, conservative jews interpreted the same scripture as ascribing royalty to this envisioned liberator, but not divinity of any kind. As these separate streams of cultural beliefs and practices developed their own unique set of characteristics (i.e., traditions, beliefs and followers) they were no longer recognisable as part of the same religion at all and Christianity emerged as an established religion. (C) An active inference model of speciation in a cumulative culture (Kastel and Hesp 2021). When an intractably divergent belief state is introduced to a largely status quo population, locally parameterised efforts to minimize free energy bring about a self organized divergence in the population, which aligns with the process of reproductive isolation. Speciation is qualitatively observed in these simulations and is plausible under an active inference framework when representations within belief groups homogenize (i.e. shared expectations between agents emerge from a collective effort to minimize free energy). For detailed information on the methodology, and architecture of the generative model used to generate these simulations see Appendix A & B).

This crucial difference may be seen as a threat to a possible analogy between speciation and innovation because the former is made possible by virtue of a complete isolation between subgroups of a population, while such rigid isolation is not usually the case in culture. In theory this might lead to the logical conclusion that cultural speciation is simply not a possibility, because different branches of the cultural tree would not be able to maintain their characteristic integrity (as different species do) when external influences are so prevalent that they threaten any possibility for group level cultural stability.

Despite this undoubtedly logical concern, we have hard indisputable observational evidence of the existence of different nations, religions and cultural practices that have maintained their integrity for hundreds and even thousands of years in spite of cultural cross talk. This teaches us that novelty in culture is able to emerge despite perturbations from external forces to cultural 'bubbles' of communicative isolation.

For instance, Judaism is an example of a religion that has rather successfully maintained its cultural integrity despite being highly susceptible to external cultural and religious influences throughout history. That being said, practicing Judaism today is still unmeasurably different than it would have been 3000 years ago, a fact pointing to internal changes in culture while its characterizing features and socio-cultural boundaries remain at least partially intact. This can be well explained by an account of cumulative cultural dynamics (Kastel and Hesp 2021). Internal changes to the Jewish religion can be attributed to an accumulation of incremental modifications that occurred with every generation (and within generations) through the communication of religious practices within the community, with relatively minimal (though still unavoidable) blending and mixing with Roman or Christian religious traditions and beliefs.

According to our account so far, each transmitted belief is translated and fitted to a specific phenotype-congruent representation of that belief state on the receiving end of the exchange (Kashima et al. 2019). Individual "private" representations of cultural beliefs therefore fuse together to create new representations of old traditions, such that cultural reproduction consists in the facilitation of different subjective representations of the same belief state, where "sameness" is in turn derived completely from subjective interpretations inherent to each individual's communicative capacity (e.g., their language). Our account is therefore capable of capturing the way a culture evolves internally, without much need for (and even despite!) external influences.

Importantly, we suggest that it might also be possible to describe and even model the way innovation emerges from these dynamics, in a form of cultural speciation.

Our model leaves the notion of a cultural speciation purposefully abstract, Such that it may take on the form of any event, fashion, ideology, preference, language or behavior that, in time, separates between two identifiable streams of culture.

To give a concrete example of what might be meant by this, we return to the slightly thorny subject of religion, and specifically, the speciation (i.e., cultural differentiation) of Christianity in the context of Judaism. This discussion is intended as an illustration of divergence of belief and practices, without judging the value of their content per se, in light of a specific point of disagreement. It should not be taken as an exhaustive treatment of differences between Judaism and Christianity. Historically, early Christians diverged from the established Jewish religion, at least partly on the basis of differing interpretations of Jewish scripture (as referred to in panel (B) of Fig. 1). According to Jewish eschatology (i.e., Jewish scholars' interpretations of their scriptures with regard to the end of times) At the time, a Jewish king referred to as "messiah" (savior) and "son of David" (a descendent from the Davidic line) would rise to rule the Jewish people and bring them salvation from their hardships. Early Christians believed that Jesus of Nazareth fulfilled the criteria for being that promised savior (Lucass 2011, p. 4–6) A critical divergence between the interpretation of early Christians and conservative Jews, was that the characteristics of the "messiah" – according to the interpretation of conservative Jews – did not (and for religious reasons, could not) include divinity of any kind but merely plenty of charisma and leadership that would lead to royalty. The emergence of these distinct interpretations, or incompatible representations for the belief in the prophecy of a messiah could be seen as the speciation event that accelerated the differentiation of Christianity from Judaism (Fig. 1B). As one example of direct behavioral incompatibility stemming from these divergent interpretations, early Christians believed that Jesus had "fulfilled" the religious law brought by Moses and quickly discarded adherence to, e.g., Jewish dietary laws, circumcision, and sacrificial practices (see, e.g., the Council of Jerusalem described in the Book of Acts, chapter 15; estimated to have taken place around 50 AD). Cultural reproductive isolation followed as these separate streams of cultural beliefs and practices developed their own unique set of characteristics, traditions, beliefs and followers – until they were no longer recognisable as part of the same religion at all and Christianity emerged as an established religion (Boyarin 1999, p. 1–22). An interesting recent example of further "speciation event" in this regard is the recent (20th century) emergence of "Messianic Judaism", a syncretic Christian movement that mixes adherence to Judaic laws with acceptance of Jesus Christ as the messiah (Melton 2005).

This is only one example of the type of speciation that might take place within a cultural arena, and it represents a direct analogy to one of four types of biological speciation (Rundle and Nosil 2005). Our particular example corresponds to sympatric speciation, which occurs when genetic polymorphism causes two groups from the same species to evolve differently until they can no longer interbreed and are considered separate species. Our Galápagos finch example, on the other hand, was an example of allopatric biological speciation, in which a particular geographic barrier prevents groups of the same species from interbreeding until they undergo genotypic divergence to the point of reproductive isolation, where they are also considered different species. Our simulations focus on the sympatric form of cultural speciation, although we assume that parallels can be made for all four types of biological speciation.

2.3 Innovation in Active Inference

In active inference, the attunement of interlocutor's generative models on the microscale translates over time and with multiple encounters into collective free energy minimisation on the macroscale. Kastel and Hesp (2021) (Fig. 1C) show that simulations of cumulative culture are aligned with this premise when locally parameterised efforts to minimize free energy by individual agents bring about a self organized separation in the population when an intractably status-quo-divergent belief is introduced. In other words, it would appear that simulations cumulative cultural dynamics imitate reproductive isolation when separate belief groups (i.e. blue and red in these simulations) diverge and communicate in observable isolation from one another. Cultural speciation, while not specifically observed in these simulations, is plausible under an active inference framework when representations within belief groups homogenize (i.e. shared expectations between agents emerge).

While this paper is limited to theorizing about the emergence of innovation from a cumulative culture diffusion, there is a great deal of potential for future modeling work in this field. Such work should include at least two added components to our theoretical account in order to provide a consistent and complete model of the emergence of innovation from the cumulative culture dynamics provided by Kastel and Hesp (2021).

First, both reproductive (i.e., communicative) isolation and speciation need to be simulated in a formalized model of active inference. Communicative isolation between groups holding divergent beliefs should be quantifiable and should emerge naturally from local (agent based) free energy minimization. Similarly, cultural speciation should be operationalised and simulated as collective free energy minimization that brings about the emergence of shared representations within groups.

Secondly, and most importantly, A novel belief state should naturally emerge from these dynamics as opposed to being synthetically introduced into the population, as was done in our simulations. This should involve the design of a new paradigm for simulating the natural emergence of a sufficiently divergent belief state from the dynamics of cumulative culture, namely, from the accumulation of modifications to cultural information. What this means is that contrary to being mechnichally placed in the population in a manner that allows belief states to remain abstract, a belief state that emerges from several modifications on it will need to refer back to the complex content (namely, the manifold of alterations) that brought it about.

For the emerging belief state to be "sufficiently divergent" from a status quo population, it needs to be dissimilar to the status quo belief to a degree that is large enough that it creates a desired level of isolation between belief-groups (i.e. each group maintains its integrity), but not so large that communication with the first agent holding this belief becomes completely impossible. Note that the latter condition is simply the assumption in active inference for the possibility of communication. Namely, that for agents to arrive at a hermeneutic resolution and be able to understand each other, they must employ sufficiently similar generative models (Friston and Frith 2015). When this is not the case, and agents employ intractably dissimilar cultural beliefs, they will not be able to refer to their own generative model to infer the internal state of another from their behavior.

We arrive at an interesting, Goldilocks precondition for the emergence of a novel belief from cumulative cultural dynamics. A novel belief that emerges from the mixing and merging of beliefs over time should be neither too divergent, nor too similar to the status quo belief. When the former is the case, the agent that suddenly emerges with a potentially novel belief, has in his mind an idea so unique and exclusive that it is incomprehensible and unrelatable to other members of the population. On the other hand, if the latter is the case, and an agent emerges with a belief that is too similar to the status quo, his belief does not differ enough from the status quo to be isolated from it as a separate stream.

In conclusion, the theory of innovation we have discussed defines exactly the difference between a belief state that is only slightly modified during communication, and one that is considered novel. For cultural beliefs and practices to be slightly modified such that they continue to evolve, they need only comply with the hermeneutic condition and allow for communication between agents carrying this information to exchange ideas. This conclusion is derived from the theories and formulations of communication and cumulative culture brought forward in Kastel and Hesp (2021). Innovation, however, seems to have harsher requirements. It needs to comply with both the hermeneutic condition (i.e., needs to be sufficiently similar to the status quo) as well as an "isolation condition" (i.e., needs to be sufficiently divergent from the status quo). This conclusion is derived from theories brought forward in this paper, which provides a solid foundation on which to build a complete account of cultural dynamics.

A formalized account of innovation as an emergent property from the cumulative dynamics presented in this proposal would bring the circular dynamics of a complex culture to a satisfying close. Under such an account, not only would cumulative culture naturally emerge from a complex communication network of agents (as shown in Kastel and Hesp (2021)), but innovation would emerge from cumulative culture and underlie communication in a repeating, recursive loop that is the hallmark of complex dynamical systems.

3   Conclusions

We discuss a possible theory of innovation as the emergent property of cumulative cultural dynamics. We suggest that innovation emerges when gradual modifications to cultural information spontaneously produce a "sufficiently divergent" belief state that meets a goldilocks condition of being neither too similar, nor too conflicting with the status quo in the population. If the former is not met, communication with the agent holding this belief will not result in coordination with members of the existing population, and the alternative belief will not propagate. If the latter is not met, its propagation will not create the level of isolation necessary between belief groups for each group to maintain its integrity to be considered novel at all.

Appendix A - Methodology for Simulating the Dynamics of Cumulative Culture

A.1 Simulating the Local Dynamics of Communication

In our model, cultural transmission is cast as the mutual attunement of actively inferring agents to each other's internal belief states. This builds on a recent formalization of communication as active inference (Friston and Frith 2015) which resolves the problem of hermeneutics, (i.e., provides a model for the way in which people are able to understand each other rather precisely despite lacking direct access to each other's internal representations of meaning) by appealing to the notion of generalized synchrony as signaling the emergence of a shared narrative to which both interlocutors refer to. In active inference, this shared narrative is attained through the minimisation of uncertainty, or (variational) free energy when both communicating parties employ sufficiently similar generative models. We build on this to suggest that having sufficiently similar generative models allows communicating agents to recombine distinct representations of a belief (expressed as generative models) into one synchronized, shared model of the world (Fig. 2). When we simulate the belief-updating dynamics between interacting agents, the cultural reproduction of a particular idea takes the form of a specific convergence between their respective generative models.

Under this theory, the elementary unit of heritable information takes the form of an internal belief state, held by an agent with a certain probability. When we simulate the belief-updating dynamics between interacting agents, a reproduced cultural belief is carried by the minds (or generative models) of both interlocutors as a site of cultural selection, where it may be further reproduced through the same process. Our simulations of communication involve two active inference agents with distinct generative models and belief claims that engage in communication over a hundred time steps.

A.2 Simulating the Global Dynamics of Cumulative Culture

Cultural beliefs and practices spread within a society through communication, a process which we have referred to as the local dynamics of cumulative culture. This description is appropriate because the accumulated outcomes of each (local) dyadic interaction collectively determine the degree to which an idea is prevalent in a culture. Moving from local communication dynamics to a degree to which an idea is prevalent in a cumulative culture is what we refer to as the global dynamics of cumulative culture.

In our simulations of a cumulative culture, 50 active inference agents simultaneously engage in local dyadic communication as shown in our first simulation, such that 25 couples are engaged in conversation at every given time step. At the first time step, all agents have relatively similar belief states- referred to as the status quo. When we introduce an agent holding a divergent belief state to that of the status quo in the population, it propagates through it via pseudo-random engagements of agents in dialogue. In a simulated world of actively inferring agents, their individual mental (generative) models are slightly modified with every interlocutor they encounter, as their distinct representations converge to a shared narrative (Constant et al. 2019). The attunement of interlocutor's to each other's generative models on the microscale thus translates over time and with multiple encounters into collective free energy minimisation on the macroscale.

Appendix B - Generative Model Architecture, Factors and Parameters

In our simulations, agents attempt to convince each other of a cultural belief by utilizing generative models that operate with local information only. For the establishment of such generative models, we will formulate a partially observed Markov decision process (MDP), where beliefs take the form of discrete probability distributions (for more details on the technical basis for MDP'S under an active inference framework, see Hesp 2019).

Under the formalism of a partially observed Markov decision process, active inference entails a particular structure. Typically, variables such as agent's hidden states (x, s), observable outcomes (o) and action policies (u) are defined, alongside parameters (representing matrices of categorical probability distributions).

B.1 Perceptual Inference

The first level of this generative model aims to capture how agents process belief claims they are introduced to through conversation with other agents. The perception of others' beliefs (regarded in active inference as evidence) requires prior beliefs(represented as likelihood mapping A1 about how hidden states (s1) generate sensory outcomes (o).

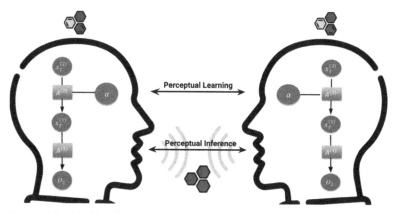

Fig. 2. Communication Coupling Parameters. Our model defines two groups of parameters that couple the internal states of agents: Learning and inference. Perceptual learning (A2) is the learning of associations between emotional valence and belief states that guide the long term actions of our agents who hold and express beliefs. This learning happens at slow time scales, accumulating across multiple interactions and used to modify models over extended periods of exchange. Perceptual Inference (A1) – namely, sensitivity to model evidence – operates on fast time scales and is direct and explicit to agents during dialogue. Importantly, we hypothesized that without precise evidence accumulation, agents would be insensitive to evidence regarding the belief state of the other, and their internal states would not converge.

Specifically, our agents predict the likelihood of perceiving evidence toward a particular expressed belief, given that this belief is "the actual state of the world". Parameterizing an agent's perception of an interlocutor's expression of belief in terms of precision values can be simply understood as variability in agents' general sensitivity to model evidence. High precisions here correspond to high responsiveness to evidence for a hidden state and low precisions to low responsiveness to evidence. Precisions for each agent were generated from a continuous gamma distribution which is skewed in favor of high sensitivity to evidence on a population level (See Fig. 2 & Fig. 3: Perception).

B.2 Anticipation

At this level, our generative model specifies agents' beliefs about how hidden states (detailed in Appendix A2) evolve over time. State transition probabilities [B1] define a particular value for the volatility of an agent's meeting selection (s2) and belief expression (s1) [B1]. For each agent, this precision parameter is sampled from a gamma distribution, determining the a priori probability of changing state, relative to maintaining a current state. Note that belief states themselves are defined on the continuous range <0, 1> (i.e., as a probability distribution on a binary state), such that multiplication tends to result in a continuous decay of confidence over time in the absence of new evidence (where the rate of decay is inversely proportional to the precision on B) (See Fig. 3: Anticipation).

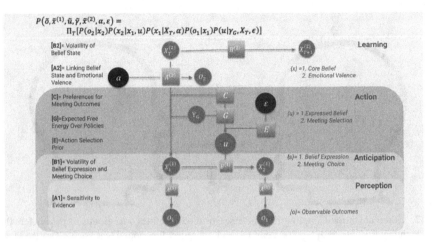

Fig. 3. A generative model of communication. Variables are visualized as circles, parameters as squares and concentration parameters as dark blue circles. Visualized on a horizontal line from left to right-states evolve in time. Visualized on a vertical line from bottom to top- parameters build to a hierarchical structure that is in alignment with cognitive functions. Parameters are described to the left of the generative model and variables are described on the right.

B.3 Action

After perceiving and anticipating hidden belief states in the world, our agents carry out deliberate actions biased towards the minimum of the expected free energy given each action (a lower level generative model for action is detailed in Appendix A4 and A5). At each time point, a policy (U) is chosen out of a set of possible sequences for action. In our simulations, two types of actions are allowed: selecting an agent to meet at each given time point (u2) and selecting a specific belief to express in conversation (u1). The first allowable action holds 50 possible outcomes (one for each agent in the simulation) while the second is expressed on the range <0, 1>, where the extremes correspond to complete confidence in denying or supporting the belief claim, respectively. Each policy under the G matrix specifies a particular combination of action outcomes weighted by its expected negative free energy value and a free energy minimizing policy is chosen (See Fig. 3: Action).

Voluntary Meeting Selection. While the choice of interlocutor is predetermined in a dyad, our multi-agent simulations required some sophistication in formulating the underlying process behind agents' selection for a conversational partner (s3) at each of the hundred time points. Building on previous work on active inference navigation and planning (Kaplan and Friston 2018), agents' meeting selection in our model is represented as a preferred location on a grid, where each cell on the grid represents a possible agent to meet.

We demonstrate the feasibility of incorporating empirical cultural data within an active inference model by incorporating (1) confirmation bias through state-dependent preferences [C], biasing meeting selection through the risk component of expected free energy (G) and (2) novelty seeking through the ambiguity component of expected free energy. The first form of bias reflects the widely observed phenomenon in psychology research that people's choices tend to be biased towards confirming their current beliefs (Nickerson 1998). The second form of bias reflects the extent to which agents are driven by the minimisation of ambiguity about the beliefs of other agents, driving them towards seeking out agents they have not met yet. For a detailed account on the process of meeting selection in these simulations, the reader is referred to Kastel and Hesp 2021.

B.4 Perceptual Learning

On this level agents anticipate how core belief states (specified in Appendix A1) might change over time [B2] (Fig. 2.3). This is the highest level of cognitive control, where agents experience learning as a high cognitive function (higher level generative model is detailed in Appendix A3). By talking with other simulated agents and observing their emotional and belief states, our agents learn associations between EV and beliefs via a high level likelihood mapping [A2], (updated via concentration parameter α). The Updating of core belief, based on beliefs expressed by other agents, is detailed in Appendix A7. This learning is important because it provides our agents with certainty regarding the emotional value they can expect from holding the alternative belief to the status quo, which has low precision at the beginning of the simulation (before the population is introduced to an agent proclaiming this belief). The prior P(A) for this likelihood mapping is specified in terms of a Dirichlet distribution.

References

Aunger, R.: Darwinizing culture: the status of memetics as a science (2001)

Buskell, A., Enquist, M., Jansson, F.: A systems approach to cultural evolution. Palgrave Commun. **5**(1), 1–15 (2019)

Bettencourt, L.M., Cintron-Arias, A., Kaiser, D.I., Castillo- Chávez, C. The power of a good idea: quantitative modeling of the spread of ideas from epidemiological models. Phys. A: Stat. Mech. Appl. **364**, 513–536 (2006)

Boyarin, D.: Dying for God: Martyrdom and the Making of Christianity and Judaism. Stanford University Press (1999)

Creanza, N., Kolodny, O., Feldman, M.W.: Cultural evolutionary theory: how culture evolves and why it 562 matters. Proc. Natl. Acad. Sci. **114**(30), 7782–7789 (2017)

Dawkins, R.: Viruses of the mind. Dennett Crit. Demystifying Mind **13**, e27 (1993)

Dunstone, J., Caldwell, C.A.: Cumulative culture and explicit metacognition: a review of theories, evidence and key predictions. Palgrave Commun. **4**(1), 1–11 (2018)

Enquist, M., Ghirlanda, S., Eriksson, K.: Modeling the evolution and diversity of cumulative culture. Philos. Trans. Roy. Soc. B: Biol. Sci. **366**(1563), 412–423 (2011)

Friston, K., Frith, C.: A duet for one. Conscious. Cogn. **36**, 390–405 (2015)

Gabora, L.: Meme and variations: A computational model of cultural evolution. In: 1993 Lectures in Complex Systems, pp. 471–485. Addison Wesley (1995)

Gordon Melton, J.: Encyclopedia of Protestantism. Infobase Publishing (2005)

Heylighen, F., Chielens, K.: Evolution of culture, memetics. In: Meyers, R. (ed.) Encyclopedia of Complexity and Systems Science, pp. 3205–3220. Springer, New York (2009). https://doi.org/10.1007/978-0-387-30440-3_189

Kashima, Y., Bain, P.G., Perfors, A.: The psychology of cultural dynamics: What is it, what do we know, and what is yet to be known? Annu. Rev. Psychol. **70**, 499–529 (2019)

Kastel, N., Hesp, C.: Ideas worth spreading: a free energy proposal for cumulative cultural dynamics. In: Kamp, M., et al. (eds.) ECML PKDD 2021. Communications in Computer and Information Science, vol. 1524, pp. 784–798. Springer, Cham (2021). https://doi.org/10.1007/978-3-030-93736-2_55

Muthukrishna, M., Henrich, J.: Innovation in the collective brain. Philos. Trans. Roy. Soc. B: Biol. Sci. **371**(1690), 20150192 (2016)

Palumbi, S.R.: Genetic divergence, reproductive isolation, and marine speciation. Annu. Rev. Ecol. Syst. 547–572 (1994)

Richerson, P.J., Boyd, R., Henrich, J.: Gene-culture coevolution in the age of genomics. Proc. Natl. Acad. Sci. (Supplement 2), 8985–8992 (2010)

Rundle, H.D., Nosil, P.: Ecological speciation. Ecol. Lett. **8**(3), 336–352 (2005)

Stout, D., Hecht, E.E.: Evolutionary neuroscience of cumulative culture. Proc. Natl. Acad. Sci. **114**(30), 7861–7868 (2017)

Weisbuch, M., Pauker, K., Ambady, N.: The subtle transmission of race bias via televised nonverbal behavior. Science **326**(5960), 1711–1714 (2009)

Active Inference and Psychology of Expectations: A Study of Formalizing ViolEx

Dhanaraaj Raghuveer[(✉)] and Dominik Endres

Theoretical Cognitive Science, Department of Psychology,
Philipps Universität Marburg, Marburg, Germany
{dhanaraaj.raghuveer,dominik.endres}@uni-marburg.de

Abstract. Expectations play a critical role in human perception, cognition, and decision-making. There has been a recent surge in modelling such expectations and the resulting behaviour when they are violated. One recent psychological proposal is the ViolEx model. To move the model forward, we identified three areas of concern and addressed two in this study - Lack of formalization and implementation. Specifically, we provide the first implementation of ViolEx using the Active Inference formalism (ActInf) and successfully simulate all expectation violation coping strategies modelled in ViolEx. Furthermore, through this interdisciplinary exchange, we identify a novel connection between AIF and Piaget's psychology, engendering a convergence argument for improvement in the former's structure/schema learning. Thus, this is the first step in developing a formal research framework to study expectation violations and hopes to serve as a base for future ViolEx studies while yielding reciprocal insights into ActInf.

Keywords: ViolEx · Active inference · Prior expectations · Formalization · FEP · Accommodation-Assimilation

1 Introduction

Human lives are rife with expectations, predictions, and anticipations in a variety of domains, fields, and facets of life, ranging from perception & attention [20], cognition [9] to decision-making in economic and social spheres [12]. Expectations are also said to play a crucial role in many psychological sub-fields like Clinical Psychology (treatment expectations), Social Psychology (stereotypical expectations), Developmental Psychology (performance expectations), and Cognitive Psychology [21]. Given the importance of expectations in understanding human behaviour, cognition, and its interdisciplinary nature, several psychological models have been proposed to study it in depth [16,22].

Out of these, the Violated Expectations (ViolEx) model [21,25] has been composed to unify the cognitive processes and behavioural responses of several proposals in one comprehensive framework [21,22]. In its most recent formulation, the ViolEx 2.0 model postulates four coping approaches in the context of expectations and expectation violation - two cognitive responses for expectation violations (accommodation, immunization) and two anticipatory (re)actions (experimentation, assimilation).

Panitz et al. [21] provide a detailed description of these coping responses at three different resolution levels. Here, elaborating on the conceptual level will suffice. When a cognitive agent finds itself in a situation in the world, it derives a situation-specific expectation from some generalized expectations. The agent can take an anticipatory (re)action that leads to outcomes that confirm the agent's expectation and avoid expectation dis-confirmation (assimilation). Alternatively, the agent can also behave to obtain expectation-relevant information irrespective of whether the agent anticipates confirmation/dis-confirmation of expectation (experimentation). Once a situational outcome is observed, the agent compares it with its situation-specific expectation, and if there is a mismatch between the two, it can cognitively respond by either integrating the new information (accommodation) or not (immunization). Intuitively, one can think of this whole process as acting to gain enough information about the world (experimentation) to have precise expectations and then using the information to produce desired/expected outcomes (assimilation). At the cognitive level, after each action, the agent exercises belief updating to match the information (accommodation) or does not (immunization).

Despite the comprehensive aims of ViolEx 2.0, we have identified three interconnected areas of concern that need to be addressed to develop the model to its full potential. First, there is a lack of implementation of the ViolEx model. Implementing the model through computer simulations can help identify hypotheses that would not have been considered before, make the model's verbal assumptions mathematically precise, and show the theory's mechanistic claims [29]. This lack of implementation is because the current formulation of ViolEx does not have a formalization of its constructs, which is our second concern. Finally, it is unclear what the theoretical grounding of the ViolEx model is. Theoretical grounding means the underlying theory of cognition and the philosophical commitments on which the model bases its explanation, like Computational Theory of Mind [11,18], or Embodied Theory of mind [4,5]. Such grounding and commitment are crucial for experimental procedures [17], control groups setup [3] and drawing valid conclusions from scientific models.

This third concern is beyond the scope of this paper. Below, we attempt to address the first two concerns: formalize and implement ViolEx using Active Inference (ActInf), a corollary process theory of the Free Energy Principle (FEP). We first elucidate the reasons for using the ActInf formalism to implement ViolEx and observe that there are significant overlaps between the constructs of ViolEx and ActInf, even though they were developed from independent disciplines: ViolEx has a historical root in (Developmental) Psychology, while ActInf comes from Theoretical Neuroscience. We then situate the constructs within the formalism and provide two simple generative models to implement all four coping strategies of ViolEx [21,22].

Furthermore, we leverage the historical origins of ViolEx and its partial overlap with Piaget's psychology - Accommodation/Assimilation [16]. While the concept of assimilation is significantly different between the two, accommodation has considerable similarities, which we utilize to identify a gap in the ActInf formalism, thus highlighting the mutual benefit of this interperspectival study.

2   Rationale for Active Inference and Generative Models

The FEP [14] is a potentially unifying principle from theoretical neuroscience, which states that minimizing the information-theoretic free energy (upper bound on surprisal) w.r.t a generative model is all that an agent has to do to maintain its organization adaptively [8,15]. Based on the FEP, the process theory Active Inference was proposed to construct artificial agents that simulate neurally plausible cognitive processes and behavioural tasks [13]. Apart from the neuronal plausibility, ActInf offers two additional benefits for use in a ViolEx context - an action-perception loop [7,24], and explore-exploit trade-off [19,28]. Pinquart et al. [22] proposed that ViolEx accounted for action and exploratory behaviour better than other expectation-related models (which included Predictive Processing (PP)). However, unlike its predecessor PP, Bayesian models, and Reinforcement Learning, ActInf provides a unique solution to both the above arguments [28], making it a strong candidate to be used for formalizing ViolEx. In later sections, we will return to how these two advantages play out while elaborating on situating ViolEx in ActInf terms and highlighting the overlap now.

2.1   Situating ViolEx

An active inference agent is governed by two free-energy equations - one for perceptual inference and one for choosing actions, and a generative model w.r.t which the agent minimizes these FEs. The variational free energy (VFE) used for perceptual inference is given by $E_{q(s|o)}[\ln q(s|o) - \ln p(s,o)]$, decomposed as[1]

$$VFE = D_{KL}[q(s|o)||p(s)] - E_{q(s|o)}[\ln p(o|s)] \qquad (1)$$

where $q(s|o)$ is the approximate/variational posterior on the hidden states (s, percepts) and $p(s)$ is the prior belief on the hidden states derived from the generative model (also called the generative prior). $p(o|s)$ reflects the predictive accuracy of the agent's sensory states o, which is averaged over the updated model beliefs $q(s|o)$. $D_{KL} = E_{q(s|o)}\left[\ln \frac{q(s|o)}{p(s)}\right]$ is the Kullback-Leiber divergence between the variational posterior and the prior which is zero if and only if the two distributions are equal. With these definitions in mind, minimizing VFE is interpreted as updating my prior beliefs to an approximate posterior such that it maximizes my predictive accuracy for the observations while not being very divergent from my prior beliefs. One can observe the striking similarity between

---

[1] For a step-by-step guide on the decomposition of VFE and EFE, see Appendix.

this FE functional and the cognitive responses in ViolEx. Integrating new information and updating my belief to account for it (accommodation) corresponds to maximizing the accuracy term of VFE. At the same time, immunization refers to minimizing the divergence term D_{KL}, which constrains this update, leading to a discounting or even discarding sensory evidence.

The second free energy equation for inferring action is the Expected Free energy (EFE), which is a direct extension of the VFE into the future, calculated under the expectation of future observations and conditioned on a policy. It is given by $E_{q(o,s|\pi)}[\ln q(s|\pi) - \ln p(s,o)]$, decomposed as

$$EFE = -E_{q(o,s|\pi)}[\ln q(s|o) - \ln q(s|\pi)] - E_{q(o,s|\pi)}[\ln p(o)] \qquad (2)$$

$q(s|o)$ is the approximate posterior belief of hidden states under sensory inputs o, whereas $q(s|\pi)$ is the approximate prior belief of hidden states under a policy π. The first expectation term, called 'Epistemic value' in ActInf literature, forces the agent to choose actions that reduce the uncertainty in its state inference ($q(s|o)$) relative to its prior prediction ($q(s|\pi)$). Minimizing the first term thus implies that policies that give the agent maximal information about the current state of the world are preferred (i.e., actions that maximize the difference between prior and posterior beliefs about the states).

Actions that reduce state uncertainty are not the only option in Active Inference. When learning is activated, the agent should also choose actions that reduce model parameter uncertainty [10, 27]. This close coupling between learning and information-seeking/experimentation is why we try to control for learning when simulating accommodation-immunization, as elaborated in upcoming sections. For more on learning and parameter uncertainty, see Appendix.

Minimizing the second term leads to policies that maximize $p(o)$, which are the prior preferred observations of the agent. In this case, the meaning of 'prior' is interpreted as 'expected' and 'preferred' as desirable, thereby collapsing the distinction between expected and desired observations.

To further reinforce our point, experimentation in ViolEx is acting so as to "collect expectation-relevant information" ($E_{q(o,s|\pi)}[\ln q(s|o) - \ln q(s|\pi)]$) while Assimilation is acting to "bring about outcomes that conform with one's expectations" ($E_{q(o,s|\pi)}[\ln p(o)]$). Accommodation corresponds to updating my beliefs "to increase consistency with situational outcome" ($E_{q(s|o)}[\ln p(o|s)]$), while immunization is "minimizing the impact of evidence" ($D_{KL}[q(s|o)||p(s)]$) (quotes from [21]).

2.2   Generative Models

Lastly, these FE functionals are always minimized w.r.t a generative model. Specifying the suitable generative model is vital to simulating a particular real-world situation. To sum up ActInf agents succinctly, their objective is to infer policies and hidden states that lead to realizing their goals. The former is achieved by minimizing EFE, while the latter is by minimizing VFE. However, in order to calculate the EFE & VFE values (like $p(o|s), p(s), q(o|\pi), p(o)$), we need to

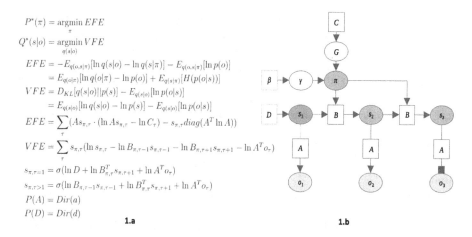

$$P^*(\pi) = \underset{\pi}{\arg\min} \, EFE$$

$$Q^*(s|o) = \underset{q(s|o)}{\arg\min} \, VFE$$

$$EFE = -E_{q(o,s|\pi)}[\ln q(s|o) - \ln q(s|\pi)] - E_{q(o,s|\pi)}[\ln p(o)]$$

$$= E_{q(o|\pi)}[\ln q(o|\pi) - \ln p(o)] + E_{q(s|\pi)}[H(p(o|s))]$$

$$VFE = D_{KL}[q(s|o)||p(s)] - E_{q(s|o)}[\ln p(o|s)]$$

$$= E_{q(s|o)}[\ln q(s|o) - \ln p(s)] - E_{q(s|o)}[\ln p(o|s)]$$

$$EFE = \sum_\tau (As_{\pi,\tau} \cdot (\ln As_{\pi,\tau} - \ln C_\tau) - s_{\pi,\tau} diag(A^T \ln A))$$

$$VFE = \sum_\tau s_{\pi,\tau}(\ln s_{\pi,\tau} - \ln B_{\pi,\tau-1}s_{\pi,\tau-1} - \ln B_{\pi,\tau+1}s_{\pi,\tau+1} - \ln A^T o_\tau)$$

$$s_{\pi,\tau=1} = \sigma(\ln D + \ln B^T_{\pi,\tau}s_{\pi,\tau+1} + \ln A^T o_\tau)$$

$$s_{\pi,\tau>1} = \sigma(\ln B_{\pi,\tau-1}s_{\pi,\tau-1} + \ln B^T_{\pi,\tau}s_{\pi,\tau+1} + \ln A^T o_\tau)$$

$$P(A) = Dir(a)$$

$$P(D) = Dir(d)$$

1.a

1.b

The agent must choose between two slots where only one yields a reward. In a stationary environment, the reward-yielding ('correct') slot remains constant, while in a non-stationary one, it switches midway (**encoded in the D matrix**). If the agent chooses the 'correct' slot, it wins a reward 90% of the time (**encoded in the A matrix**). Before choosing the slot, the agent can ask for a hint that says the 'correct' slot machine 90% of the time (**A matrix**). If the hint is availed, the reward is halved, else not (**encoded in C matrix**). This task is divided into 3 time steps; The first step is the 'start' stage, where the agent does not have to do anything. In the next step, the agent can choose whether to take a hint. In the final time step, the agent must choose one (either Left or Right) slot machine (**B matrix encodes all actions**).

1.c

Fig. 1. Equations of factorized generative model, Bayesian Network representation and description of the task used for experimentation-assimilation. Different model content (A, B, D) leads to different FE functionals, thus varying behaviour and cognitive response. The three subplots highlight the close link between a real-world task, Generative model, FE minimization and policy/state inference. Refer to the text for how accommodation-immunisation's model structure and content vary.

specify what the situation is formally, and this is what Generative models do. For example, in this study, we take the situation of a two-armed bandit task and specify it formally as a generative model. We then analyze the resulting behaviour patterns and cognitive responses of ActInf agents in those situations.

Owing to the prior work done in ViolEx, we were endowed with some initial predictors of ViolEx coping processes, which we implemented in the generative model. For example, uncertainty is the main predictor of anticipatory (re)actions according to [21]. While they do not specify whether uncertainty in the state, state-outcome mapping, or state-state transition, simulating all the above uncertainties is possible in ActInf [27]. To that end, we simulated the anticipatory (re)actions of ViolEx via the generative model in Fig. 1.

Here, s is the hidden state that causes the sensory observation o, under policy π. a and d are the Dirichlet concentration parameters acting as priors for Categorical A and D, respectively. A, D are parameters of the generative process (GP), but random variables from the agent's perspective [10], while a,d are parameters of the generative model (GM) that the agent uses. This distinction is important because learning (of ground-truths D, A, B) is possible only if d, a, and b are defined.[2] Every non-hierarchical ActInf agent will have the same generative model structure as in Fig. 1 with minor modifications [28]. However, the difference lies in the detail of what goes into each of the variables D (Initial prior), A (Likelihood), B (Actions), and C (Prior preference), which we call model content. Thus, by varying the model content, we can instantiate ViolEx predictors like uncertainty, information credibility, habits, disposition for risk, and many more. In the interest of space and relevance, we will not describe the different conditions and parameterizations used to simulate results for Experimentation-Assimilation[3] but constrain ourselves to Accommodation-Immunization.

Similar to anticipatory (re)actions, we gathered from ViolEx that the reliability of an information source is identified as one of the main predictors of an immunizing response. The hypothesis is that if the agent believes the information source's reliability is low, it will conveniently discard the evidence leading to possibly false inferences. Furthermore, an interesting parameter in ActInf literature implicitly present in the 'individual differences' part of ViolEx is habit learning [1]. Habit 'learning' is a peculiar kind of prior in that it is the only learning independent of the observed evidence but driven primarily by the frequency of the particular policy's selection. Given this dissociation from evidence, we hypothesized that entrenched habits could lead to immunization-like responses.

As mentioned earlier, most non-hierarchical ActInf agents have the same model structure as in Fig. 1 with minor changes, but the model content can vary significantly. The only change in the structure in our case is habit learning. This change yields two additional equations:

$$P(\pi) = \sigma(E - \gamma * EFE)$$
$$P(E) = Dir(e) \tag{3}$$

where E is the prior belief on policy, encoding habits, e, its Dirichlet concentration parameter that enables learning and σ, classical softmax function for normalization,. In model content, reward rate (X) was varied from 60% (low) to 90% (high) in GM and fixed at 100% in GP to simulate varying reliability of information source (varying precision of likelihood) [2] in agents' beliefs.

We precluded all learning except habits and hints made default but with perfectly random (50%) accuracy; thus, d, a, and b were either not defined or defined with high initial concentration, which prevents learning [28]. The above changes were made so that the agent's only source of information available for inference is the reward rate we control. Analyzing the effect of reliability in multiple information source setups would be interesting; however, we avoid it for simplification.

---

[2] For a detailed description of all the variables and belief-updating in the model, refer [28].

[3] See Appendix for different parameterizations, results and discussion of Experimentation-Assimilation.

Learning was prevented to study accommodation-immunization in isolation from experimentation-assimilation. We interpret accommodation as within-trial belief updating while experimentation as within & between-trial information seeking. This latter notion of between-trial information seeking is what learning is in ActInf agents. For a detailed analysis of how learning and experimentation are connected and thus our decision to prevent learning, we refer the reader to Appendix.[4]

3   Results and Discussion

3.1   Accommodation-Immunization

For accommodation-immunization, we simulated four different parameterizations of the generative model described in the previous section - stationary environment with high information reliability (reward-rate X = 90%) and no habits, non-stationary environment with low reliability (X = 60%) and no habits, non-stationary environment with low reliability (X = 60%) & habit learning and lastly, non-stationary environment with high reliability (X = 90%) and habit learning. We interpret accommodation as using the feedback received at the final time point of each trial (win or loss) to accurately infer the 'correct' slot and failure to do so as immunization. If this inference is false, it means the agent has failed to take the evidence into account, thus exhibiting immunization. We consider only the third time point because that is when the agent gets helpful information (see below) to accommodate, even though belief-updating happens at all three time steps.

Starting with our first simulation in Fig. 2.a, we see that the prior belief about the 'correct' slot machine is consistently 0.5 throughout the entire task as it should be. This belief is because the agent's only useful information source is the win/lose feedback (the hint is random, see Generative Models). Since we have prevented learning, as mentioned in the Generative Models section, it cannot carry information from previous trials either. Moreover, this helpful information comes after deciding on the 'correct' slot, so the prior should not be affected. This can also be observed in the action probabilities being equally likely for both Left and Right actions throughout the task, leading to chance-level performance. The Free Energy subplots reflect an agent's surprise if its action leads to a loss in that trial.

However, if proper accommodation of evidence happens in every trial, the posterior should point in the same direction as the actual 'correct' slot. We see this happening at near certainty (closer to 0 or 1, see 'Tracking beliefs' subplot) in the first simulation (Fig. 2.a) because the agent thinks the information source is highly reliable. The fact that the agent's beliefs are updated in the right direction by integrating the observed outcomes indicates ViolEx's accommodation.

---

[4] For a complete specification of our models and replication of plots, please visit https://github.com/danny-raghu/ViolEx_Simulation_ActInf. All results presented here were simulated using MATLAB R2021a and spm12 package.

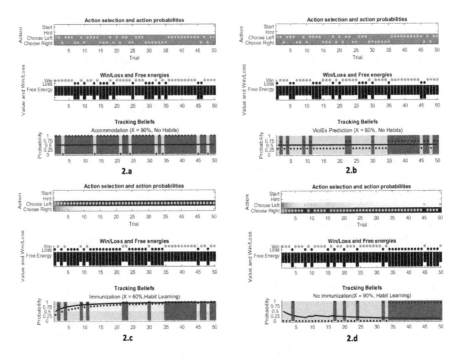

Fig. 2. The first and second subplot labels in each panel have the following interpretation: Cyan dots and shading in the 'Action' subplot correspond to the chosen actions and the probability of choosing the action, respectively; Darker the shade, higher the probability. Green and black dots are winning or losing in the trial, respectively. The third subplot tracks posterior (dots) and prior (line) beliefs on the 'correct' slot with subtitles depicting simulation parameterization (Refer to text). Dark Blue corresponds to the belief that the left slot is the 'correct' slot in the trial and Peach to the right. Light Blue bars represent false inference (>50% posterior prob for 'incorrect' slot) due to immunization in the trial. (Color figure online)

Moving on, we next tested the hypothesis derived from VioIEx about lower information credibility playing a role in immunization. Contrary to our expectations, we found that reducing the reliability of information (X = 60%) did not produce immunizing tendencies like false beliefs resistant to evidence. In our model, false beliefs have a greater than 50% posterior probability for a slot machine, while the evidence in that trial pointed to the other slot machine. This is depicted in Fig. 2.b. As we can see, while the posterior is uncertain (closer to ∼0.5) about the slot machine, it does not yet result in a belief that contradicts the observed evidence. Thus, one can see that the low reliability of information alone does not lead to immunization.

Third, we checked if keeping the reliability low as in the previous simulation but combining it with habit formation would have immunizing effects. When only habits are allowed to be learned, the policy sampling distribution changes as explained in Eq. 3, making the agent choose policies habitually (i.e., not as a consequence of sensory evidence o). This change in policy selection impacts its state inferences because the prior, $q(s) = \sum_\pi q(s, \pi) = \sum_\pi q(s|\pi) * P(\pi)$ (see below for details).

As Fig. 2.c illustrates, low reliability of the information, when combined with habits, leads to immunization. The mechanism behind this kind of cognitive response is intriguing. Firstly, habits are the only learning we have allowed in the agent and, therefore, are the sole info for policy selection. However, habits pay no regard to outcomes but are driven by the frequency of previous selections and can thus become unregulated. As habits strengthen, policy selection gets biased, which leads to a biased prior on hidden states ($q(s) \propto P(\pi)$). Once this prior becomes strong enough, if the evidence/likelihood is unreliable or of low precision [2], one can make robust false inferences, as shown in Fig. 2.c.

One crucial question could be why the agent here developed a habit of choosing left when right is the 'correct' slot. That is because of our information setup. The agent's choice in the first trial should have been left slot, which increased the likelihood of choosing left in the subsequent trial (since there is no other counteracting information source for policy). Thus, the habit of choosing left is purely random, reinforcing itself (incorrectly) because of a random choice in the first trial. We note, however, that habits may also accidentally lead to the correct behaviour, here: choosing the right slot. Whether or not this happens depends on the agent's initial (random) choice in our setup.

Another question is whether the agent would avoid immunization if it had better quality evidence, despite habits. We tested this in the final simulation (X = 90%), and as expected, there was no immunizing tendency (see Fig. 2.d). This result is also in line with the predictions from ViolEx, however, with a slight modification. Low reliability of information source in and of itself does not result in immunization, but in conjunction with developed habits, it can lead to such tendencies. Without habits, the priors for state inference do not get extremely precise/reliable compared to the precision/reliability of sensory evidence (likelihood) [2] leading to unbiased posterior inference or accommodation. More importantly, reliable information sources can avoid such false inferences even with developed habits.

3.2   Psychology of Expectations and AIF Formalism

As a psychological proposal for modelling Expectation and Expectation Violation, ViolEx has essential relevance in accounting for cognitive, social, clinical, and developmental issues like self-fulfilling prophecies, confirmation bias, cognitive dissonance, addiction, optimism bias and has strong empirical support for the model [6,16,21–23,25]. However, it lacks a formal framework to provide mechanistic explanations of how its constructs yield a particular cognitive or behavioural response, and Active Inference provides precisely this.

Reciprocally, the empirical evidence accrued by ViolEx and its partial overlap with Piaget's Psychology has some interesting contributions to the AIF formalism. Specifically, both in Piaget's model and ViolEx, there are two notions of accommodation. In ViolEx, these are Expectation (de)stabilization and Expectation change. The former refers to the belief updating of expectations already available to the organism (employed in this study). In contrast, the latter refers to changing the schema of expectations to introduce a novel and previously unavailable expectation into the organism's cognitive repertoire. This change in the structure of the existing schema to integrate a novel concept is accommodation proper in Piaget's model, sometimes called structure learning in Developmental Psychology [26].

This introduction of a novel category/concept was a puzzle for us when we tried to simulate all the properties of ViolEx, and realized that the fixed state-space of a dynamical POMDP (like the one used here) currently lacks the flexibility to model such emergent expectations [26]. Immunization in ViolEx also has a similar property, and we could only implement what is called Data-oriented immunization [21,22] but not Concept-oriented immunization (analogous breaking and conceptual restructuring of expectation schema).

4   Conclusion

We started our paper with three concerns about the current state of the ViolEx model and hoped to address two of those through the Active Inference framework. We showed the uncanny overlap between ViolEx and ActInf, even though they were developed in disparate fields, and argued for leveraging this overlap to formalize the constructs in ViolEx through ActInf. Through the formalization, we simulated all the core aspects of ViolEx, providing the first computational implementation of its core constructs and resulting behaviours.

We acknowledge that the generative models provided in this paper are idealizations and do not depict the complete picture of what could contribute to an assimilative behaviour or immunizing response. Also, the main goal of this study was not to test all the possible mediators of the constructs in ViolEx and provide a full-blown analysis of it. Instead, we aimed to suggest a suitable formalism that makes the constructs precise and simultaneously gives justice to the intricacies involved in ViolEx.

Furthermore, through this interdisciplinary exchange, we also identified room for extending the AIF formalism to flexibly model novel and emergent expectations, thus yielding mutual benefits to both ViolEx and Active Inference.

Acknowledgments. This work was supported by the DFG GRK-RTG 2271 'Breaking Expectations' project number 290878970.

A   Appendix

A.1   Detailed Decomposition of VFE and EFE

The first line of the equation below starts with the classical definition of VFE, as mentioned in the main text. The intermediate lines show how to derive the immunization-accommodation or complexity-accuracy decomposition step-by-step.

$$VFE = E_{q(s|o)}[\ln q(s|o) - \ln p(s,o)]$$
$$= E_{q(s|o)}[\ln q(s|o) - \ln (p(s) * p(o|s))]$$
$$= E_{q(s|o)}[\ln q(s|o) - \ln p(s) - \ln p(o|s)]$$
$$= E_{q(s|o)}[\ln q(s|o) - \ln p(s)] - E_{q(s|o)}[\ln p(o|s)] \qquad (4)$$
$$= E_{q(s|o)}\left[\ln \frac{q(s|o)}{p(s)}\right] - E_{q(s|o)}[\ln p(o|s)]$$
$$= D_{KL}[q(s|o)||p(s)] - E_{q(s|o)}[\ln p(o|s)]$$

Starting with the second line, we use the Product rule of probability to factorize the generative model into the generative prior on states and likelihood of observations, $p(s)$ and $p(o|s)$ respectively. Then we use the property of logarithms, $\ln (A * B) = \ln A + \ln B$ to split the factorised generative model in the third line. Finally, we gather up the variational posterior $(q(s|o))$ and generative prior $(p(s))$ together and once again use the property of logarithms, $\ln A - \ln B = \ln (A/B)$ to yield the D_{KL} term (see main text). Thus, the decomposition of immunization-accommodation, as pointed out in the main text, is achieved.

Like VFE, we start with the definition of EFE mentioned in the main text - a direct extension of the VFE into the future, calculated under the expectation of future observations and conditioned on a policy. We then derive the Experimentation-Assimilation or Epistemic-Pragmatic decomposition.

$$EFE = E_{q(o,s|\pi)}[\ln q(s|\pi) - \ln p(s,o)]$$
$$= E_{q(o,s|\pi)}[\ln q(s|\pi) - \ln (p(o) * p(s|o))]$$
$$= E_{q(o,s|\pi)}[\ln q(s|\pi) - \ln (p(o) * q(s|o))]$$
$$= E_{q(o,s|\pi)}[\ln q(s|\pi) - \ln p(o) - \ln q(s|o)] \qquad (5)$$
$$= E_{q(o,s|\pi)}[\ln q(s|\pi) - \ln q(s|o)] - E_{q(o,s|\pi)}[\ln p(o)]$$
$$= -E_{q(o,s|\pi)}[\ln q(s|o) - \ln q(s|\pi)] - E_{q(o,s|\pi)}[\ln p(o)]$$

In the second line, we factorize the generative model just like in VFE decomposition using the Product rule. However, unlike VFE, we factorize it into a generative prior on observations and posterior on states, $p(o)$ and $p(s|o)$, respectively. Next, we make an assumption that the variational posterior can approximate the generative posterior well enough, i.e.: $q(s|o) \approx p(s|o)$. These are the two key moves in EFE decomposition, and the remaining steps involve the property of logarithms and gatherings illustrated above to yield the experimentation-assimilation form.

To calculate the values of $q(o,s|\pi)$, $q(s|o)$ and $q(s|\pi)$, further assumptions and approximations are made. For example, $q(o|s)$ and $p(o|s)$ are assumed to be the same distribution encoded by the A matrix. This yields the divergence between variational posterior and variational prior,

$$\left[\ln\frac{q(s|o)}{q(s|\pi)}\right] = \left[\ln\frac{p(o|s)*q(s|\pi)}{q(o|\pi)*q(s|\pi)}\right]$$

$$= \left[\ln\frac{p(o|s)}{q(o|\pi)}\right]$$

$$q(o|\pi) = \sum_s q(o,s|\pi) \tag{6}$$

$$= \sum_s p(o|s)*q(s|\pi)$$

$$= A*s_\pi$$

These equations highlight the close relation between minimizing Free energy functionals, optimally inferring the policies/states and the generative model parameters like A, B, and D, as mentioned in the main text.

A.2   Learning and Parameter Uncertainty

As briefly mentioned in the main text, learning the parameters of the Generative Process, A (for example), is possible only if 'a' is defined in the Generative model. This is because a, b, and d are the Dirichlet concentration parameters that act as priors on the categorical distributions A, B, and D, respectively. Figure 1.a depicts this by equations, $P(A) = Dir(a), P(D) = Dir(d)$ and the choice of choosing Dirichlet distribution as the prior is because of its conjugacy with Categorical distributions [28]. By being a conjugate prior to the parameters of categorical distributions, updating the model parameters amounts to the simple addition of counts to the vector/matrix.

The learning equation used to update the model parameters is given by,

$$d_{trial+1} = \omega * d_{trial} + \eta * s_{\tau=1} \tag{7}$$

where, d is the initial prior on states given by, $d = p(s_{\tau=1}) = \begin{bmatrix} d_1 & d_2 \end{bmatrix}^T$, ω forgetting rate, and, η learning rate. Learning of model parameters mean just updating the concentration parameters of d, namely d_1 and d_2. An intuitive understanding of learning can be given using an example. Suppose that in the task described in the main text, the agent initially did not know which of the two slots was the 'correct' slot and so had an initial prior of $d_{trial=1} = \begin{bmatrix} 0.5 & 0.5 \end{bmatrix}^T$. However, at the end of the trial, the agent observes that left slot is the 'correct' slot with probability, $s_{\tau=1} = \begin{bmatrix} 1 & 0 \end{bmatrix}^T$. Assuming that both ω, and η is equal to 1, then the agent would have an updated initial prior at the next trial, $d_{trial=2} = \begin{bmatrix} 1.5 & 0.5 \end{bmatrix}^T$. If the agent observes left being the 'correct' slot for 8 consecutive trials, then $d_{trial=8} = \begin{bmatrix} 8.5 & 0.5 \end{bmatrix}^T$. This is what learning through adding counts means in Active Inference.

Note that with a higher initial concentration, the impact of additional count is way lower than when the concentration is lower. Upon normalising, even though $\begin{bmatrix} 5 & 5 \end{bmatrix}^T$, and $\begin{bmatrix} 50 & 50 \end{bmatrix}^T$ represent the same probability distribution, the addition of a count in the former changes the distribution more impactfully than in the latter. This is why initializing a model parameter with high concentration prevents learning [28]. This state of having a high initial concentration is called the 'saturated' state, where the agent has nothing to learn. Meanwhile, the state of low concentration is called the 'uncertain' state because these Dirichlet parameters are a kind of confidence estimate. The lower the Dirichlet counts, the less confident the agent is in its belief and vice versa. Thus, if the Dirichlet count for one state factor, say d_1 is low compared to the Dirichlet count for another, d_2, the agent actively seeks out observations that increase the count of d_1.

In order to actively seek out observations, this learning process has to be reflected in the EFE functional. This is precisely what happens when learning is activated in an agent,

$$
\begin{aligned}
EFE &= E_{q(o,s,\theta|\pi)}[\ln q(s,\theta|\pi) - \ln p(s,o,\theta)] \\
&= -E_{q(o,s,\theta|\pi)}[\ln q(s|o) - \ln q(s|\pi)] - E_{q(o,s,\theta|\pi)}[\ln p(o)] \\
&\quad - E_{q(o,s,\theta|\pi)}[\ln q(\theta|s,o,\pi) - \ln q(\theta)]
\end{aligned}
\tag{8}
$$

where θ could be either of a, d, b or all of them. Depending on the number of learning parameters, the number of terms in EFE increases. The connection between learning and EFE/experimentation is hinted at in the main text. Having elaborated on it, it becomes clear why learning and information-seeking behaviour are closely linked and why we chose to run our accommodation-immunization simulation without it.

A.3   Experimentation-Assimilation

The relevant question for experimentation-assimilation is which parts of the task mentioned in the main text parameterize uncertainty and risk-seeking. Any action that reduces uncertainty about the situation (by generating valid information) is regarded as experimentation in VioLEx [21]. There are several uncertainties for the agent here. 1) The uncertainty of finding the 'correct' slot (slot uncertainty) in the trial. 2) Uncertainty of not getting a reward even if one chooses the 'correct' slot (reward-rate = 90%). 3) Uncertain whether the hint-giver gives a correct hint (hint accuracy = 90%). The simulated agent does not know all these values (because a & d are defined in the GM, but A and D are random variables) but can learn by interacting with the environment and gathering information (experimentation). One can also simulate a lack of this experimentation by making the agent strongly prefer getting rewards w/o hints rather than with hints. To that extent, we construe taking a hint over directly choosing the slot machine as experimentation.

We used three different parameterizations of the above task in two conditions (Risk-seeking, Risk-averse). The agents in the first and second parameterization have only the slot uncertainty but operate in different (stationary/non-stationary) environments while having all other information. In the third one, the agents do not have information about hint accuracy but should experiment to learn about the uncertainty to perform well. This gradual increase of uncertainty should allow us to analyze the impact of uncertainty and disposition to risk in experimentation and assimilation.

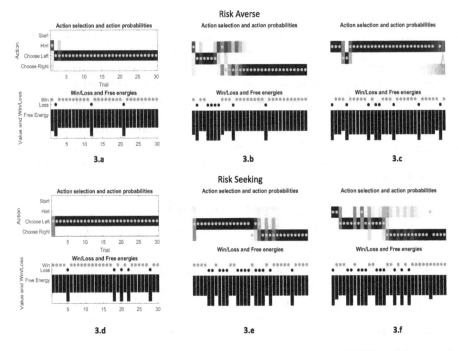

Fig. 3. The top and bottom rows correspond to Risk-averse and Risk-seeking conditions. From left to right, the uncertainties involved are 1) slot in a stationary environment, 2) slot in a non-stationary environment, and 3) slot and hint in a non-stationary environment. Labels of 'Action' and 'Free Energy' subplots in each panel have the same interpretation as in Fig. 2.

As mentioned above, we ran six simulations to explore how uncertainty and risk mediate experimentation-assimilation. Starting with a risk-averse agent with just the slot uncertainty (Fig. 3.a), we could see that the agent initially explores once and considers exploring briefly right after a loss. Since this is the least uncertain of the three environments, it quickly gains enough information about its environment to start assimilating.

However, when there is higher uncertainty due to a non-stationary setting, it has to be in experimentation mode for a more extended period before starting to assimilate. Experimentation mode here is hint action, or higher probability of taking hint action to collect information but not "random selection of behavioural alternatives" [21] as in RL. The qualitative difference in behaviour observed in the two cases shows that uncertainty, as hypothesized from ViolEx, does play a critical role in Experimentative behaviour. Finally, to drive home the point, we simulated hint uncertainty in the non-stationary environment, and the plot shows that experimentation increases even further (Fig. 3.c). Only in the last 3–4 trials does the agent have enough information to choose right (highlighted through the action probability).

To test our second hypothesis of risk-seeking as a potential mediator of assimilative behaviour, we simulated the same three conditions as above but with a higher difference between prior preference $(\ln p(o))$ for winning with and without hints.[5] This difference, in turn, affects the EFE value (see Eq. (2)), which changes the agent's policy π (Fig. 1 equations). We could garner from Fig. 3.a and 3.d that there is not a vast difference in observed behaviour between an RS and an RA agent when there is relatively small uncertainty. However, the difference in behaviour and performance becomes striking when uncertainty gradually increases, as depicted in Figs. 3.b, 3.c, 3.e, 3.f. Even when observing more losses than wins, the inherent risk-seeking tendency makes experimentation less likely than the risk-averse agent. These qualitative results show that while increased uncertainty leads to increased experimentation, strong individual traits like risk-seeking can counteract the effects of uncertainty to make assimilative strategies more or less likely.

References

1. Adams, R.A., Vincent, P., Benrimoh, D., Friston, K.J., Parr, T.: Everything is connected: inference and attractors in delusions. Schizophr. Res. (2021). https://doi.org/10.1016/j.schres.2021.07.032

2. Albarracin, M., Demekas, D., Ramstead, M.J., Heins, C.: Epistemic communities under active inference. Entropy **24**, 476 (2022). https://doi.org/10.3390/E24040476

3. Allen, J.W., Bickhard, M.H.: Stepping off the pendulum: why only an action-based approach can transcend the nativist-empiricist debate. Cogn. Dev. **28**, 96–133 (2013). https://doi.org/10.1016/j.cogdev.2013.01.002

4. Bickhard, M.: Troubles with computationalism. The Philosophy of Psychology, pp. 173–183 (2014). https://doi.org/10.4135/9781446279168.N13

5. Bickhard, M.H.: Interactivism: a manifesto. New Ideas Psychol. **27**, 85–95 (2009). https://doi.org/10.1016/J.NEWIDEAPSYCH.2008.05.001

6. Braun-Koch, K., Rief, W.: Maintenance vs. change of negative therapy expectation: an experimental investigation using video samples. Front. Psychiatry **13**, 474 (2022). https://doi.org/10.3389/FPSYT.2022.836227/BIBTEX

7. Bruineberg, J., Kiverstein, J., Rietveld, E.: The anticipating brain is not a scientist: the free-energy principle from an ecological-enactive perspective. Synthese **195**(6), 2417–2444 (2016). https://doi.org/10.1007/s11229-016-1239-1

8. Bruineberg, J., Rietveld, E.: Self-organization, free energy minimization, and optimal grip on a field of affordances. Front. Hum. Neurosci. **8**, 599 (2014). https://doi.org/10.3389/fnhum.2014.00599

9. Clark, A.: Whatever next? Predictive brains, situated agents, and the future of cognitive science. Behav. Brain Sci. **36**, 181–204 (2013). https://doi.org/10.1017/S0140525X12000477

10. Costa, L.D., Parr, T., Sajid, N., Veselic, S., Neacsu, V., Friston, K.: Active inference on discrete state-spaces: a synthesis. J. Math. Psychol. **99** (2020). https://doi.org/10.1016/j.jmp.2020.102447

11. Fodor, J.A.: The Language of Thought. Harvard University Press, Cambridge (1979)

---

[5] To be precise, the difference was increased from 2 in risk-averse agents to 2.5 in risk-seeking.

12. Friston, K.: The free-energy principle: a rough guide to the brain? Trends Cogn. Sci. **13**, 293–301 (2009). https://doi.org/10.1016/J.TICS.2009.04.005

13. Friston, K., FitzGerald, T., Rigoli, F., Schwartenbeck, P., Pezzulo, G.: Active inference: a process theory. Neural Comput. **29**, 1–49 (2017). https://doi.org/10.1162/NECO_a_00912

14. Friston, K., Stephan, K.E.: Free-energy and the brain. Synthese **159**, 417–458 (2007). https://doi.org/10.1007/s11229-007-9237-y

15. Friston, K., Thornton, C., Clark, A.: Free-energy minimization and the dark-room problem. Front. Psychol. **3**, 130 (2012). https://doi.org/10.3389/fpsyg.2012.00130

16. Gollwitzer, M., Thorwart, A., Meissner, K.: Editorial: psychological responses to violations of expectations. Front. Psychol. **8**, 2357 (2018). https://doi.org/10.3389/fpsyg.2017.02357

17. Hochstein, E.: How metaphysical commitments shape the study of psychological mechanisms. Theory Psychol. **29**, 579–600 (2019). https://doi.org/10.1177/0959354319860591

18. Hohwy, J.: The Predictive Mind. Oxford University Press, Oxford, January 2009. https://doi.org/10.1093/ACPROF:OSO/9780199682737.001.0001

19. Kiverstein, J., Miller, M., Rietveld, E.: The feeling of grip: novelty, error dynamics, and the predictive brain. Synthese **196**(7), 2847–2869 (2017). https://doi.org/10.1007/s11229-017-1583-9

20. Ouden, H.E.D., Kok, P., de Lange, F.P.: How prediction errors shape perception, attention, and motivation. Front. Psychol. **3** (2012). https://doi.org/10.3389/fpsyg.2012.00548

21. Panitz, C., et al.: A revised framework for the investigation of expectation update versus maintenance in the context of expectation violations: the ViolEx 2.0 model. Front. Psychol. **12** (2021). https://doi.org/10.3389/fpsyg.2021.726432

22. Pinquart, M., Endres, D., Teige-Mocigemba, S., Panitz, C., Schütz, A.C.: Why expectations do or do not change after expectation violation: a comparison of seven models. Conscious. Cogn. **89**, 103086 (2021). https://doi.org/10.1016/J.CONCOG.2021.103086

23. Pinquart, M., Koß, J.C., Block, H.: How do students react when their performance is worse or better than expected? **52**, 1–11 (2021). https://doi.org/10.1026/0049-8637/A000222, https://econtent.hogrefe.com/doi/10.1026/0049-8637/a000222

24. Ramstead, M.J., Kirchhoff, M.D., Friston, K.J.: A tale of two densities: active inference is enactive inference. Adapt. Behav. **28**, 225–239 (2020). https://doi.org/10.1177/1059712319862774

25. Rief, W., Glombiewski, J.A., Gollwitzer, M., Schubö, A., Schwarting, R., Thorwart, A.: Expectancies as core features of mental disorders. Curr. Opin. Psychiatry **28**, 378–385 (2015). https://doi.org/10.1097/YCO.0000000000000184

26. Rutar, D., de Wolff, E., van Rooij, I., Kwisthout, J.: Structure learning in predictive processing needs revision. Comput. Brain Behav. **5**, 234–243 (2022). https://doi.org/10.1007/S42113-022-00131-8/FIGURES/3, https://link.springer.com/article/10.1007/s42113-022-00131-8

27. Schwartenbeck, P., Passecker, J., Hauser, T.U., Fitzgerald, T.H., Kronbichler, M., Friston, K.J.: Computational mechanisms of curiosity and goal-directed exploration. eLife **8** (2019). https://doi.org/10.7554/ELIFE.41703

28. Smith, R., Friston, K.J., Whyte, C.J.: A step-by-step tutorial on active inference and its application to empirical data (2021). https://doi.org/10.31234/osf.io/b4jm6

29. Thagard, P.: Why cognitive science needs philosophy and vice versa. Top. Cogn. Sci. **1**, 237–254 (2009). https://doi.org/10.1111/j.1756-8765.2009.01016.x

AIXI, FEP-AI, and Integrated World Models: Towards a Unified Understanding of Intelligence and Consciousness

Adam Safron[1,2,3]([⊠])

[1] Center for Psychedelic & Consciousness Research, Department of Psychiatry & Behavioral Sciences, Johns Hopkins University School of Medicine, Baltimore, MD, USA
[2] Cognitive Science Program, Indiana University, Bloomington, IN, USA
[3] Institute for Advanced Consciousness Studies, Santa Monica, CA, USA
asafron@gmail.com

Abstract. Intelligence has been operationalized as both goal-pursuit capacity across a broad range of environments, and also as learning capacity above and beyond a foundational set of core priors. Within the normative framework of AIXI, intelligence may be understood as capacities for compressing (and thereby predicting) data and achieving goals via programs with minimal algorithmic complexity. Within the Free Energy Principle and Active Inference framework, intelligence may be understood as capacity for inference and learning of predictive models for goal-realization, with beliefs favored to the extent they fit novel data with minimal updating of priors. Most recently, consciousness has been proposed to enhance intelligent functioning by allowing for iterative state estimation of the essential variables of a system and its relationships to its environment, conditioned on a causal world model. This paper discusses machine learning architectures and principles by which all these views may be synergistically combined and contextualized with an Integrated World Modeling Theory of consciousness.

Keywords: Free Energy Principle · Active Inference · AIXI · Integrated World Modeling Theory · Intelligence · Consciousness

1 IWMT and Universal Intelligence

Integrated World Modeling Theory (IWMT) attempts to solve the enduring problems of consciousness by combining aspects of other models within the Free Energy Principle and Active Inference (FEP-AI) framework [1, 2]. In brief, IWMT claims that phenomenal consciousness is "what it feels like" when likely patterns of sense-data are inferred from probabilistic generative models over the sensorium of embodied-embedded agents, whose iterative estimates constitute the stream of experience (as a series of 'quale' states), conditioned on a causal world model trained from histories of environmental interaction. Different forms of "conscious access" are enabled when these streams/flows of experience are channeled/contextualized within coherent causal unfoldings via different varieties of mental actions, such as attentional selection of particular remembered and

C. L. Buckley et al. (Eds.): IWAI 2022, CCIS 1721, pp. 251–273, 2023.
https://doi.org/10.1007/978-3-031-28719-0_18

imagined contents. For such processes to be able to entail (or "bring forth a world" of) subjective experience, generative modeling is suggested to require sufficient coherence with respect to space (i.e., locality), time (i.e. proportional changes for different processes), and cause (i.e., regularities with respect to changes that can be learned/inferred) for the system and its relationships with its environment. That is, structuring of experience by quasi-Kantian categories may be required not only for coherent judgment, but for solving binding problems such that different entities may be modeled with different properties [3–8]; without these sources of coherence, we may still have various forms of modeling, but not necessarily of a conscious variety due to a lack of compositional representations.

These system-world configurations are suggested to primarily (and potentially exclusively/exhaustively) involve visuospatial and somatospatial modalities, where views and poses mutually inform one another: estimating where and how one is looking is invaluable for constraining what one is seeing, and vice versa. This conjunction of view and pose information (particularly with respect to head and gaze direction) organizes experience according to egocentric reference frames (at least for systems with centrally-located censors), so providing a "point of view" on a world with the experience of a "lived body" at its center, with feelings likely heavily involving cross-modal inference with interoceptive hierarchies [9]. These processes are suggested to be realized by autoencoding heterarchies, whose shared latent space may be structured according to principles of geometric deep learning for generating state estimates with sufficient rapidity that they can both inform and be informed by action-perception cycles over the timescales over which they are enacted. If such iterative estimation of body-world states is organized into coherent organismic/agentic trajectories by semantic pointer systems—such as those provided by hippocampal/entorhinal system—then we may have episodic memory and (counterfactual) imaginings for the sake of prediction, postdiction, and planning/control [10–12]. Taken together, we may have most of the desiderata for explaining why (and how) it might feel like something to be some kinds of physical systems, potentially sufficiently answering hard central questions about biocomputational bases of experience in ways that afford new progress on the "easy-," "real-," and "meta-" problems of consciousness (which may be far more difficult than the "Hard problem" itself; e.g. the complexities of different forms of self-consciousness with the potential "strange loops" involved).

IWMT was significantly inspired by the work of Jürgen Schmidhuber (since 1990, see surveys of 2020, 2021), whose pioneering intellectual contributions include early discoveries in unsupervised machine learning, characterization of fundamental principles of intelligence, development of means of realizing scalable deep learning, designing world models of the kinds emphasized by IWMT, and more [15–24]. While this body of work cannot be adequately described in a single article, below I will explore how both the theoretical and practical applications of principles of universal intelligence and algorithmic information theory may help illuminate the nature(s) of consciousness.

1.1 FEP-AI and AIXI: Intelligence as Prediction/Compression

IWMT was developed within the unifying framework of FEP-AI as a formally-grounded model of intelligent behavior, with rich connections to both empirical phenomena and advances in machine learning. In this view, consciousness and cognition more generally is understood as a process of learning how to better predict the world for the sake of adaptive control. However, these ideas could have been similarly expressed in the language of algorithmic information theory [25, 26], wherein complexity is quantified according to the length of programs capable of generating patterns of data (rather than the number of parameters for a generative model). From this point of view, intelligence is understood in terms of abilities to effectively compress information with algorithms of minimal complexity. These two perspectives can be understood as mutually supporting, in that predictability entails compressibility. If one can develop a predictive model of a system, then one also has access to a 'code'/cypher (even if only implicitly) capable of describing relevant information in a more compressed fashion. Similarly, if compression allows information to be represented with greater efficiency, then this compressed knowledge could be used as a basis for modeling future likely events, given past and present data.

This understanding of intelligence as compression was given a formal description in AIXI, a provably-optimal model of intelligent behavior that combines algorithmic and sequential decision theories [27]. While not formally computable, Solomonoff induction provides a Bayesian derivation of Occam's razor [28], in selecting models that minimize the Kolmogorov complexity of associated programs. AIXI agents use this principle in selecting programs that generate actions expected to maximize reward over some time horizon for an agent. This parsimony-constraint also constitutes a highly adaptive inductive bias and meta-prior in that we should indeed expect to be more likely to encounter simpler processes in the worlds we encounter [18, 29–32], all else being equal. Similarly, FEP-AI agents select policies (as Bayesian model selection over predictions) expected to maximize model evidence for realizing adaptive system-world states. However, this expectation is evaluated (or indirectly estimated with a variational evidence lower bound approach) according to a "free energy" functional that prioritizes inferences based on their ability to fit data, penalized by the degree to which updates of generative models diverge from priors, so preventing overfitting and facilitating knowledge generalization [33].

While thoroughly exploring connections between FEP-AI and AIXI is beyond the scope of this paper, their parallels are striking in that both frameworks prescribe choice behavior as the model-based realization of preferences under conditions of uncertainty, where governing models are privileged according to parsimony. Further, both FEP-AI and AIXI agents operate according to curiosity motivations, in which choices are selected not only to realize particular outcomes, but to explore for the sake of acquiring novel information in the service of model evolution [34, 35]. Both AIXI and FEP-AI confront the challenge of not being able to evaluate the full distribution of hypotheses required for normative program/model selection, instead relying on approximate inferred distributions. Further, both frameworks also attempt to overcome potentially complicated issues around ergodic exploration of state spaces with various forms of sophisticated planning [36, 37]. Taken together, the combination of these frameworks may provide a provably-optimal, formally-grounded theory for studying intelligent behavior in both biological and artificial systems. With relevance to IWMT, consciousness may be understood as a kind of lingua franca (or semi-interpretable shared latent space) for more efficiently converting between heterogeneous systems for the sake of promoting functional synergy [1, 2, 38]. [While beyond the scope of the present discussion, it may be worth considering which forms of conceptual and functional integration between AIXI and FEP-AI could be achievable through the study of generative modeling based on probabilistic programs [39–41]. In terms of the neural systems described by IWMT, many of these "programs" could be understood as realized through modes of policy selection canalized by (a) striatal-cortical loops as sequences of both overtly enacted and covertly expressed mental actions and (b) large scale state transitions between equilibrium points as orchestrated by the hippocampal/entorhinal system [10]].

Although AIXI and FEP-AI both specify optimality criteria, both frameworks require the further specification of process theories and heuristic implementations in order to realize these principles in actual systems [42, 43]. Below I focus on some of the work of Schmidhuber [13, 14] and colleagues in which these principles of universal computation have been used to inform the creation of artificial systems with many of the hallmarks of biological intelligences. Understanding the design of these systems may be informative for understanding not only computational properties of nervous systems, but also their ability to generate integrative world models for the sake of adaptively controlling behavior.

1.2 Kinds of World Models

[Please note: For background on computational properties of different kinds of neural networks and their potential implications for intelligence, please see Appendix: "2.1 Recurrent networks, universal computation, and generalized predictive coding".]

World models are becoming an increasingly central topic in machine learning due to their ability to support knowledge synthesis for self-supervised learning, especially for agents governed by objective functions (potentially implicit and meta-learned) involving intrinsic sources of value such as curiosity and empowerment [13–15, 20, 22, 44–47]. IWMT emphasizes consciousness as an entailment of integrated world modeling [1], and attempts to specify which brain systems and machine learning principles may be important for realizing different aspects of this function [2]. According to IWMT, posterior cortices represent generative models over the sensorium for embodied agents, whose inversion generates likely patterns of sense data (given past experience), entailing streams of consciousness. Cortical perceptual hierarchies are modelled as "folded" variational autoencoders (VAEs), but this could similarly be described as hierarchically-organized LSTMs [16, 48], or perhaps (generalized) transformer hierarchies [49–52]. There is a sense in which this posterior-located autoencoding heterarchy is itself a world model by virtue of not only containing information regarding the spatially-organized structure of environments, but also deep temporal modeling via the capacity of recurrent systems—and complex dendritic processing [53]—to support sequence memories, whose coherent state transitions could be understood as entailing causal dependencies. However, this world modeling achieves much greater temporal depth and counterfactual richness with the addition of the hippocampal-entorhinal system and frontal lobes [54–56], which together may generate likely-future (or counterfactual/postdicted-past) state transitions. Further, sculpting of striatal-cortical loops by neuromodulators allows action—broadly construed to include forms of attentional selection and imaginative mental acts—to be selected and policies to be channeled by histories of reward learning [57]. This involvement of value-canalized control structures provides functional closure in realizing action-perception cycles, which are likely preconditions for any kind of coherent world modeling to be achieved via active inference [58, 59].

These sorts of functions have been realized in artificial systems for decades [15, 16], with recent advances exhibiting remarkable convergence with the neurocomputational account of world modeling described by IWMT. One particularly notable world modeling system was developed by Ha and Schmidhuber [60], in which a VAE is used to compress sense data into reduced-dimensionality latent vectors, which are then predicted by a lower parameter recurrent neural network (RNN), whose outputs are then used as bases for policy selection by a much lower parameter controller trained with evolutionary strategies. The ensuing actions from this control system are then used as a basis for further predictive modeling and policy selection, so completing action-perception cycles for further inference and learning. The dimensionality-reducing capacities of VAEs not only provide greater efficiency in providing compressions for subsequent computation, but moving from pixel space to latent space—with associated principle components of data variation—allow for more tractable training regimes. The use of evolutionary optimization is capable of handling challenging credit-assignment problems (e.g. non-differentiable free energy landscapes) via utilizing the final cumulative reward over a behavioral epoch, rather than requiring evaluation of the entire history of performance and attempting to determine which actions contributed to which outcomes to which degrees. (Speculatively, this could be understood as one interpretation of hippocampal ripples contributing to phasic dopamine signals in conjunction with remapping events [61, 62]). This architecture was able to achieve unprecedented performance on car racing, first-person shooter video games, and challenging locomotion problems requiring the avoidance of local minima. Further, by feeding the signals from the controller directly back into latent space (rather than issuing commands over actions), this architecture was able to achieve even greater functionality by allowing training with policy rollouts in simulated (or "imagined") environments. Even more, by incorporating a temperature (or uncertainty) parameter into this imaginative training, the system was able to avoid patterns of self-adversarial/degenerate policy selection from leveraging exploits of internal world models, so enhancing transfer learning from simulations to real-world situations (and back again). Theoretically, this could help explain some of the functional significance of dreaming [63], and possibly also aspects of neuromodulator systems (functionally understood as machine learning parameters) and the ways in which they may be involved in various "psychedelic" states of mind [64].

Limitations of this world modeling system include the problem of independent VAE-optimization resulting in encoding task-irrelevant observations, but with task-based training potentially undermining transfer learning. An additional challenge involves issues of catastrophic forgetting as the encoding of new data interferes with prior memories. Ha and Schmidhuber [60] suggested that these challenges could be respectively overcome with governance by intrinsic drives for artificial curiosity—which could be thought of as a kind of adaptive-adversarial learning (Schmidhuber, 1990, 2020, 2021)—as well as the incorporation of external memory modules for exploring more complicated worlds.

Intriguingly, both of these properties seem to be a feature of the hippocampal/entorhinal-system [2, 65, 66], which possesses high centrality and effective connectivity with value-estimation networks such as ventral prefrontal cortices and striatum [57]. An additional means of enhancing the performance of world modeling architectures is the incorporation of separate "teacher" and "student" networks, whose particular functional details may shed light on not just the conditions for optimal learning, but for the particular nature(s) of conscious experience, as will be explored next.

1.3 Kinds of Minds: Bringing Forth Worlds of Experience

One of Schmidhuber's [16] more intriguing contributions involves a two-tier unsupervised machine (meta-)learning architecture operating according to principles of predictive coding. The lower level consists of a fast "subconscious automatiser RNN" that attempts to compress (or predict) action sequences, so learning to efficiently encode histories of actions and associated observations. This subsystem automatically creates abstraction hierarchies with similar features to those found in receptive fields of the mammalian neocortical ventral visual stream. In this way, the automatiser identifies invariances (or symmetries) across data structures and generates prototypical features that can be used to efficiently represent particular instances by only indicating deviations from these eigenmode basis functions. Surprising information not predicted by this lower level is passed up to a more slowly operating "conscious chunker RNN", so becoming the object of attention for more elaborative processing [23]. The automatiser continuously attempts to encode subsequent streams of information in compressed forms, eventually rendering them "subconscious" in terms of not requiring the more complex (expensive, and relatively slow) modeling of the conscious chunker. This setup can also be understood as a "teacher" network being imitated by a "student" network, and through this knowledge distillation (and meta-learning) process, high-level policy selection becomes "cloned" such that it can be realized by fast and efficient lower-level processes. With the "Neural History Compressor" [24, 67], this architecture was augmented with the additional functionality of having the strength of updates for the conscious chunker be modulated by the magnitude of surprises for the subconscious automatiser.

In terms of brain functioning, we would expect predictive processing mechanisms to cause information to be predicted in a way that is increasingly sparse and pushed closer to primary modalities with time and experience. Because these areas lack access to the kind of rich club connectivity found in deeper portions of the cortical heterarchy, this information would be less likely to be taken up into workspace dynamics and contribute to alpha/beta-synchronized large-scale "ignition" events, which IWMT interprets as Bayesian model selection over the sensorium of an embodied-embedded agent, entailing iterative estimation of system-world states. In this way, initially conscious patterns of action selection involving workspace dynamics would constitute "teacher" networks, but which with experience will tend to become automatic/habitual/unconscious as they are increasingly realized by soft-assembled, distributed, fast/small ensembles coordinated by cerebellar-coordinated forward models [68], functioning as "student" networks. In terms of some of the architectures associated with FEP-AI, this could be thought of as information being processed in the lower level of Forney factor graphs with continuous (as opposed to discrete) updating and less clearly symbolic functionality [69, 70]. In more standard machine learning terms, this could be thought as moving from more "model-based" to more "model-free" control, or "amortized inference" with respect to policy selection [57, 71]. However, when these more unconscious forms of inference fail, associated prediction errors will ascend until they can receive more elaborative handling by conscious systems, whose degree of activity may be modulated by overall surprisal [2, 72–75].

To return to artificial neural history compressors [24, 67], these systems are capable of discovering a wide range of invariances across actions and associated observations, with one of the most enduring of these symmetries being constituted by the agent itself as a relatively constant feature. With RNN-realized predictive coding mechanisms combined with reinforcement learning, systems learn to efficiently represent themselves, with these symbols becoming activated in the context of planning and through active self-exploration, so exhibiting a basic form of self-modeling, including with respect to counterfactual imaginings. Based on the realization of these functionalities, Schmidhuber has controversially claimed that "we have had simple, conscious, self-aware, emotional, artificial agents for 3 decades" [13, 14, 76]. IWMT has previously focused on the neurocomputational processes that may realize phenomenal consciousness, but functionally speaking (albeit potentially not experientially), this statement is consistent with the computational principles identified for integrated world modeling.

These functional and mechanistic accounts have strong correspondences aspects of Hiedeggerian phenomenology [77]. A "ready-to-hand" situation of effortless mastery of sensorimotor contingencies could correspond to maximally sparse/compressed patterns of neural activity, pushed down close to primary cortices, with the system coupling with the world in continuously evolving unconscious action-perception cycles, without requiring explicit representation of information. In such a situation of nearly effortless (and largely unconscious) mastery of sensorimotor contingencies, dithering would be prevented as the changing affordance landscape results in rapid selection of appropriate grips [9, 78–81], with coordination largely occurring via reflexes and morphological constraints. However, a "present-at-hand" situation of reflective evaluation could correspond to prediction errors making their way up to upper levels of cortical heterarchies, where largescale ignition events—potentially entailing Bayesian model selection and discrete updating of hierarchically higher (or deeper) beliefs—may become more likely due to closer proximity to rich-club connectivity cores. If such novel information can couple with subsystems enabling various kinds of workspace dynamics, then agents can not only consciously perceive such patterns, but also draw upon counterfactual considerations for causal reasoning and planning, so requiring flexibly (and consciously) accessible re(-)presentations and various forms of self-referential modeling. Convergence between these machine learning architectures, neural systems, and phenomenology is compelling, but does this mean that Schmidhuberian world-modeling agents and similar systems possess phenomenal consciousness? This is the final issue that we will consider in exploring world models and the nature(s) of intelligence.

1.4 Machine Consciousness?

[Please note: For discussion of how different principles from neural networks may contribute to explaining computational properties of consciousness, please see Appendix: "2.2 Unfolding and (potentially conscious) self-world modeling".]

More recent work from Schmidhuber's group has explored the preconditions for human-like abilities to generalize knowledge beyond direct experience [3]. They suggest that such capable systems require capacities to flexibly bind distributed information in novel combinations, and where this binding problem requires capacities for compositional and symbolization of world structure for systematic and predictable knowledge synthesis. They further describe a unifying framework involving the formation of meaningful internal structure from unstructured sense data (segregation), the maintenance of this segregated data at an abstract level (representation), and the ability to construct new inferences through novel combinations of these representations (composition). Similarly to the preconditions for subjective experience identified by Integrated Information Theory (IIT) [1, 82]—and with deep relevance to IWMT—this proposal describes a protocol for producing data structures with intrinsic existence (realization via entangled connections), composition (possessing meaningful structure), information (distinguishability from alternatives), and integration (constituting wholes that are greater than the sum of their parts). This is further compatible with GNWT's description of architectures for dynamically combining information in synergistic ways.

Even more, the particular architectural principles identified by this framework as promising research directions are the same as those previously suggested by IWMT: geometric deep learning and graph neural networks (GNNs). However, this proposal for solving binding problems for neural networks also suggests utilizing a generalization of GNNs in the form of graph nets [83]. This is also an issue with respect to which the original publication of IWMT contained a possible error, in that rather than the hippocampal/entorhinal-system being understood as hosting graph-grid GNNs for spatial mapping, (spectral) graph nets may be a more apt description of this system on algorithmic and functional levels of analysis [66, 84, 85]. These technical details notwithstanding, there are striking parallels between the functionalities described by this framework for functional binding and the machine learning description of neural systems suggested by IWMT. More specifically, autoencoding/compressing hierarchies provide segregation, with GNN-structured latent spaces providing explicit representation (and potentially conscious experiences), and a further level of higher abstraction providing novel representational combinations for creative cognition (i.e., hippocampal place fields as flexibly configurable graphical models).

If such an architecture were to be artificially realized, would it possess 'consciousness?' The Global Neuronal Workspace Theory would likely answer this question in the affirmative with respect to conscious access [86], and IIT would provide a tentative affirmation for phenomenal consciousness [87], given realization on neuromorphic hardware capable of generating high phi for the physical system (rather than for an entailed virtual machine). IWMT is agnostic on this issue, yet may side with IIT in terms of matters of practical realization. Something like current developments with specialized graph processors might be required for these systems to achieve sufficient efficiency [88] for engaging in iterative estimation of system-world states with sufficient rapidity that such estimates could both form and be informed by perceivable action-perception cycles. If such a system were constructed as a controller for an embodied-embedded agent, then it may be the case that it could not only generate "System 2" abilities [89, 90], but also phenomenal consciousness, and everything that entails.

Acknowledgments. In addition to collaborators who have taught me countless lessons over many years, I would like to thank Jürgen Schmidhuber and Karl Friston for their generous feedback on previous versions of these discussions. Any errors (either technical or stylistic) are entirely my own.

A Appendix

A.1 Recurrent Networks, Universal Computation, and Generalized Predictive Coding

In describing brain function in terms of generative modeling, IWMT attempts to characterize different aspects of nervous systems in terms of principles from machine learning. Autoencoders are identified as a particularly promising framework for understanding cortical generative models due to their architectural structures reflecting core principles of FEP-AI (and AIXI). By training systems to reconstruct data while filtering (or compressing) information through dimensionality-reducing bottlenecks, this process induces the discovery of both accurate and parsimonious models of data in the service of the adaptive control of behavior [91]. IWMT's description of cortical hierarchies as consisting of "folded" autoencoders was proposed to provide a bridge between machine learning and predictive-coding models of cortical functioning. Encouraging convergence may be found in that these autoencoder-inspired models were developed without knowledge of similar proposals by others [92–96].

However, these computational principles have an older lineage predating "The Helmholtz machine" and its later elaboration in the form of variational autoencoders [97, 98]. These "Neural Sequence Chunkers" (NSCs) instantiate nearly every functional aspect of IWMT's proposed architecture for integrated world modeling (Schmidhuber, 1991), with the only potential exceptions being separate hippocampal/entorhinal analogue systems [10], and shared latent spaces structured according to the principles of geometric deep learning (however, subsequent work by Schmidhuber and colleagues has begun to move in this direction [3, 60]). NSCs consist of recurrent neural networks (RNNs) organized into a predictive hierarchy, where each RNN attempts to predict data from the level below, and where only unpredicted information is passed upwards. These predictions are realized in the form of recurrent dynamics [99, 100], whose unfolding entails a kind of search processes via competitive/cooperative attractor formation over states capable of most efficiently responding to—or resonating with [2, 101]—ascending data streams. In this way, much as in FEP-AI, predictive abilities come nearly "for free" via Hamilton's principle of least action [30, 102–104], as dynamical systems automatically minimize thermodynamic (and informational) free energy by virtue of greater flux becoming more likely to be channeled through efficient and unobstructed flow paths [105].

RNNs are Turing complete in that they are theoretically capable of performing any computational operation, partially by virtue of the fact that small clusters of signaling neurons may entail the logical operations such as NAND gates [106]. Further, considering that faster forming attractors are more likely to dominate subsequent dynamics, such systems could be considered to reflect the "Speed prior", whereby the search for more parsimonious models is heuristically realized by the selection of faster running programs [19]. Attracting states dependent upon communication along fewer synaptic connections can be thought of as minimal length descriptions, given data. In this way, each RNN constitutes an autoencoder, which when hierarchically organized constitute an even larger autoencoding system [2]. Thus, predictive coding via "folded" (or stacked) autoencoders can be first identified in the form of NSCs as the first working deep learning system. Even more, given the recurrent message passing of NSCs (or "stacked" autoencoders), the entire system could be understood as an RNN, the significance of which will be described below.

RNN-based NSCs exhibit all of the virtues of predictive coding and more in forming increasingly predictive and sparse representations via predictive suppression of ascending signals [33, 107, 108]. However, such systems have even greater power than strictly suppressive predictive coding architectures [109–111], in that auto-associative matching/resonance allows for microcolumn-like ensemble inference as well as differential prioritization and flexible recombination of information [3, 112–114]. The auto-associative properties of RNNs afford an efficient (free-energy-minimizing) basis for generative modeling in being capable of filling-in likely patterns of data, given experience. Further, the dynamic nature of RNN-based computation is naturally well-suited for instantiating sequential representations of temporally-extended processes, which IWMT stipulates to be a necessary coherence-enabling condition for conscious world modeling. Even more, the kinds of hierarchically-organized RNNs embodied in NSCs naturally provide a basis for representing extended unfoldings via a nested hierarchy of recurrent dynamics evolving over various timescales, so affording generative modeling with temporal depth [115], and potentially counterfactual richness [59, 79, 116]. Some of these temporally-extended unfoldings include agent-information as sources of relatively invariant—but nonetheless dynamic—structure across a broad range of tasks. Given these invariances, the predictive coding emerging from NSCs will nearly automatically discover efficient encodings of structure for self and other [13], potentially facilitating meta-learning and knowledge-transfer across training epochs, including explicit self-modeling for the sake of reasoning, planning, and selecting actions [23].

In contrast to feedforward neural networks (FNNs), RNNs allow for multiple forms of open-ended learning [20, 22, 47], in that attractor dynamics may continually evolve until they discover configurations capable of predicting/compressing likely patterns of data. FNNs can be used to approximate any function [117], but this capacity for universal computation may often be more of a theoretical variety that may be difficult to achieve in practice [118], even with the enhanced processing efficiency afforded by adding layers for hierarchical pattern abstraction [119, 120]. The limitations of FNNs can potentially be understood in light of their relationship with recurrent networks. RNNs can be translated into feedforward systems through "unrolling" [48, 121], wherein the evolution of dynamics across time points can be represented as subsequent layers in an FNN hierarchy, and vice versa. Some of the success of very deep FNNs such as "open-gated Highway Nets" could potentially be explainable in terms of entailing RNNs whose dynamics are allowed to unfold over longer timescales in searching for efficient encodings. However, FNN systems are still limited relative to their RNN brethren in that the former are limited to exploring inferential spaces over a curtailed length of time, while the latter can explore without these limitations and come to represent arbitrarily extended sequences. On a higher level of abstraction, such open-endedness allows for one of the holy grails of AI in terms of enabling transfer and meta-learning across the entire lifetime of a system [24, 122, 123].

However, this capacity for open-ended evolution is also a potential liability of RNNs, in that they may become difficult to train with backpropagation, partially due to potential non-linearities in recurrent dynamics. More specifically, training signals may either become exponentially amplified or diminished via feedback loops—or via the number of layers involved in very deep FNNs—resulting in either exploding or vanishing gradients, so rendering credit assignment intractable. These potential problems for RNNs were largely solved by the advent of "long short-term memory" systems (LSTMs) [48], wherein recurrent networks are provided flexible routing capacities for controlling information-flow. This functionality is achieved via an architecture in which RNN nodes contain nested (trainable) gate-RNNs that allow information to either be ignored, temporarily stored, or forgotten. Notably, LSTMs not only overcame a major challenge in the deployment of RNN systems, but they also bear striking resemblances to descriptions of attention in global workspace theories [38, 86]. Limited-capacity workspaces may ignore insufficiently precise/salient data (i.e., unconscious or subliminal processing), or may instead hold onto contents for a period of time (i.e., active attending, working memory, and imagination) before releasing (or forgetting) these short-term memories so that new information can be processed. Although beyond the scope of the present discussion, LSTMs may represent a multiscale, universal computational framework for realizing adaptive autoencoding/compression.

With respect to consciousness, a notable feature of deep FNNs is that their performance may be greatly enhanced via the addition of layer-skipping connections, which can be thought of as entailing RNNs with small-world connectivity [124], so allowing for a synergistic combination of integrating both local and global informational dependencies. Such deep learning systems are even more effective if these skip connections are adaptively configurable [119], so providing an even closer correspondence to the processes of dynamic evolution underlying the intelligence of RNNs [125, 126], including nervous systems [127–130]. The flexible small-worldness of these "highway nets"—and their entailed recurrent processing—has potentially strong functional correspondences with aspects of brains thought to enable workspace architectures via connectomic "rich clubs," which may be capable of supporting "dynamic cores" of re-entrant signaling, so allowing for synergistic processing via balanced integrated and segregated processing [74, 131–133]. Notably, such balanced integration and segregation via small-worldness is also a property of systems capable of both maximizing integrated information and supporting self-organized critical dynamics [1, 134], the one regime in which (generalized) evolution is possible [30, 31, 135–138]. As described elsewhere with respect to the critique of Aaronson's critique of Integrated Information Theory (IIT) [139], small-world networks can also be used for error-correction via expander graphs (or low-density parity checking codes), which enable systems to approach the Shannon limit with respect to handling noisy, lossily—and irreversibly [140]—compressed data. Speculatively, all efficiently functioning recurrent systems might instantiate turbo codes in evolving towards regimes where they may come to efficiently resonate with (and thereby predict/compress) information from other systems in the world.

A.2 Unfolding and (Potentially Conscious) Self-world Modeling

Given this generalized predictive coding, there may be a kind of implicit intelligence at play across all persisting dynamical systems [18, 32, 103, 141, 142]. However, according to IWMT, consciousness will only be associated with systems capable of coherently modeling themselves and their interactions with the world. This is not to say that recurrence is necessarily required for the functionalities associated with consciousness [143]. However, recurrent architectures may be a practical requirement, as supra-astronomical resources may be necessary for unrolling a network the size of the human brain across even the 100s of milliseconds over which workspace dynamics unfold. Further, the supposed equivalence of feedforward and feedback processes are only demonstrated when unrolled systems are returned to initial conditions and allowed to evolve under identical circumstances [144]. These feedforward "zombie" systems tend to diverge from the functionality of their recurrent counterparts when intervened upon and are unable to repair their structure when modified. This lack of robustness and context-sensitivity means that unrolling loses one of the primary advantages of consciousness as dynamic core and temporally-extended adaptive (modeling) process, where such (integrated world) models allow organisms to flexibly handle novel situations. Further, while workspace-like processing may be achievable by feedforward systems, largescale neuronal workspaces heavily depend on recurrent dynamics unfolding over multiple scales. Perhaps we could model a single inversion of a generative model corresponding to one quale state, given a sufficiently large computational device, even if this structure might not fit within the observable universe. However, such computations would lack functional closure across moments of experience [145, 146], which would prevent consciousness from being able to evolve as a temporally-extended process of iterative Bayesian model selection.

Perhaps more fundamentally, one of the primary functions of workspaces and their realization by dynamic cores of effective connectivity may be the ability to flexibly bind information in different combinations in order to realize functional synergies [3, 147, 148]. While an FNN could theoretically achieve adaptive binding with respect to a single state estimate, this would divorce the integrating processes from its environmental couplings and historicity as an iterative process of generating inferences regarding the contents of experience, comparing these predictions against sense data, and then updating these prior expectations into posterior beliefs as priors for subsequent rounds of predictive modeling. Further, the unfolding argument does not address the issue of how it is that a network may come to be perfectly configured to reflect the temporally-extended search process by which recurrent systems come to encode (or resonate with) symmetries/harmonies of the world. Such objections notwithstanding, the issue remains unresolved as to whether an FNN-based generative model could generate experience when inverted.

This issue also speaks to the ontological status of "self-organizing harmonic modes" (SOHMs), which IWMT claims provide a functional bridge between biophysics and phenomenology [2, 139]. Harmonic functions are places where solutions to the Laplacian are 0, indicating no net flux, which could be defined intrinsically with respect to the temporal and spatial scales over which dynamics achieve functional closure in forming self-generating resonant modes [149]. (Note: These autopoietic self-resonating/forming attractors are more commonly referred to as "nonequilibrium steady state distributions" in the FEP literature [103], which are derived using different—but possibly related [150]—maths). However, such recursively self-interacting processes would not evolve in isolation, but would rather be influenced by other proto-system dynamics, coarse-graining themselves and each other as they form renormalization groups in negotiating the course of overall evolution within and without. Are standing wave descriptions 'real,' or is everything just a swirling flux of traveling waves? Or, are traveling waves real, or is there 'really' just an evolving set of differential equations over a vector field description for the underlying particles? Or are underlying particles real, or are there only the coherent eigenmodes of an underlying topology? Even if such an eliminative reductionism bottoms out with some true atomism, from an outside point of view we could still operate according to a form of subjective realism [151], in that once we identify phenomena of interest, then maximally efficient/explanatory partitioning into kinds might be identifiable [152–154]. Yet even then, different phenomena will be of differential 'interest' to other phenomena in different contexts evolving over different timescales.

While the preceding discussion may seem needlessly abstract, it speaks to the question as to whether we may be begging fundamental questions in trying to identify sufficient physical substrates of consciousness, and also speaks to the boundary problem of which systems can and cannot be considered to entail subjective experience. More concretely, do unrolled SOHMs also entail joint marginals over synchronized subnetworks, some of which IWMT claims to be the computational substrate of consciousness? Based on the inter-translatability of RNNs and FNNs described above, this question appears to be necessarily answered in the affirmative. However, if the functional closure underlying these synchronization manifolds require temporally-extended processes that recursively alter themselves [155, 156], then it may be the case that this kind of autopoietic ouroboros cannot be represented via geometries lacking such entanglement. Speculatively (and well-beyond the technical expertise of this author), Bayesian model selection via SOHMs might constitute kinds of self-forming (and self-regenerating) "time crystals" [157–159], whose symmetry-breaking might provide a principled reason to privilege recurrent systems as physical and computational substrates for consciousness. If this were found to be the case, then we may find yet another reason to describe consciousness as a kind of "strange loop" [160–162], above and beyond the seeming and actual paradoxes involved in explicit self-reference.

This kind of self-entanglement would render SOHMs opaque to external systems lacking the cypher of the self-generative processes realizing those particular topologies [155]. Hence, we may have another way of understanding marginalization/renormalization with respect to inter-SOHM information flows as they exchange messages in the form of sufficient statistics [163], while also maintaining degrees of independent evolution (cf. Mean field approximation) over the course of cognitive cycles [164]. These self-generating entanglements could further speak to interpretations of IIT in which quale states correspond to maximal compressions of experience [165]. In evaluating the integrated information of systems according to past and future combinatorics entailed by minimally impactful graph cuts [166], we may be describing systems capable of encoding data with maximal efficiency [167], in terms of possessing maximum capacities for information-processing via supporting "differences that make a difference." Indeed, adaptively configurable skip connections in FNNs could potentially be understood as entailing a search process for discovering these maximally efficient codes via the entanglements provided by combining both short- and long-range informational dependencies in a small-world (and self-organized-critical) regime of compression/inference. A system experiencing maximal alterations in the face of minimal perturbations would have maximal impenetrability when observed from without, yet accompanied by maximal informational sensitivity when viewed from within.

If we think of minds as systems of interacting SOHMs, then this lack of epistemic penetration could potentially be related to notions of phenomenal transparency (e.g. not being able to inspect processes by which inspection is realized, perhaps especially if involving highly practiced, and thereby efficient and sparse patterns of neural signaling) [168, 169], and perhaps "user interface" theories of consciousness [170]. Intriguingly, maximal compressions have also been used as conceptualizations of the event horizons of black holes, for which corresponding holographic principles have been adduced in terms of internal information being projected onto 2D topologies. With respect to the FEP, it is also notable that singularities and Markov blankets have been interpreted as both points of epistemic boundaries as well as maximal thermal reservoirs [171]. Even more speculatively, such holography could even help explain how 3D perception could be derived from 2D sensory arrays, and perhaps also experienced this way in the form of the precuneus acting as a basis for visuospatial awareness and kind of "Cartesian theater" [172–174]. As described elsewhere [9], this structure may constitute a kind of graph neural network (GNN), utilizing the locality of recurrent message passing over grid-like representational geometries for generating sufficiently informative projections on timescales proportional to the closure of action-perception cycles [1, 2].

However, in exploring the potential conscious status of recurrent systems, we would still do well to consider the higher-level functional desiderata for consciousness as a process of integrated world modeling. More specifically, if particular forms of coherence are required for generating experience, then we may require hybrid architectures with specialized subsystems capable of supporting particular kinds of learning. This is an ongoing area of research for IWMT, with specific focus on the hippocampal/entorhinal-system as providing fundamental bases for higher-order cognition [10]—understood as a kind of generalized search/navigation process—including with respect to reverse engineering such functions in attempting to design (and/or grow) intelligent machines [2].

References

1. Safron, A.: An integrated world modeling theory (IWMT) of consciousness: combining integrated information and global neuronal workspace theories with the free energy principle and active inference framework; toward solving the hard problem and characterizing agentic causation. Front. Artif. Intell. **3** (2020). https://doi.org/10.3389/frai.2020.00030
2. Safron, A.: Integrated world modeling theory (IWMT) implemented: towards reverse engineering consciousness with the free energy principle and active inference. PsyArXiv (2020). https://doi.org/10.31234/osf.io/paz5j
3. Greff, K., van Steenkiste, S., Schmidhuber, J.: On the binding problem in artificial neural networks. arXiv:2012.05208 [cs] (2020)
4. Evans, R., Hernández-Orallo, J., Welbl, J., Kohli, P., Sergot, M.: Making sense of sensory input. Artif. Intell. **293**, 103438 (2021). https://doi.org/10.1016/j.artint.2020.103438
5. De Kock, L.: Helmholtz's Kant revisited (Once more). The all-pervasive nature of Helmholtz's struggle with Kant's Anschauung. Stud. Hist. Philos. Sci. **56**, 20–32 (2016). https://doi.org/10.1016/j.shpsa.2015.10.009
6. Northoff, G.: Immanuel Kant's mind and the brain's resting state. Trends Cogn. Sci. (Regul. Ed.) **16**, 356–359 (2012). https://doi.org/10.1016/j.tics.2012.06.001
7. Swanson, L.R.: The predictive processing paradigm has roots in Kant. Front. Syst. Neurosci. **10**, 79 (2016). https://doi.org/10.3389/fnsys.2016.00079
8. Marcus, G.: The Next decade in AI: four steps towards robust artificial intelligence. arXiv: 2002.06177 [cs] (2020)
9. Safron, A.: The radically embodied conscious cybernetic Bayesian brain: from free energy to free will and back again. Entropy **23**, 783 (2021). https://doi.org/10.3390/e23060783
10. Safron, A., Çatal, O., Verbelen, T.: Generalized simultaneous localization and mapping (G-SLAM) as unification framework for natural and artificial intelligences: towards reverse engineering the hippocampal/entorhinal system and principles of high-level cognition (2021). https://psyarxiv.com/tdw82/, https://doi.org/10.31234/osf.io/tdw82
11. Safron, A., Sheikhbahaee, Z.: Dream to explore: 5-HT2a as adaptive temperature parameter for sophisticated affective inference (2021). https://psyarxiv.com/zmpaq/, https://doi.org/10.31234/osf.io/zmpaq
12. Safron, A.: On the Varieties of conscious experiences: altered beliefs under psychedelics (ALBUS) (2020). https://psyarxiv.com/zqh4b/, https://doi.org/10.31234/osf.io/zqh4b
13. Schmidhuber, J.: Planning & reinforcement learning with recurrent world models and artificial curiosity (1990). https://people.idsia.ch//~juergen/world-models-planning-curiosity-fki-1990.html. Accessed 16 May 2021
14. Schmidhuber, J.: First very deep learning with unsupervised pre-training (1991). https://people.idsia.ch//~juergen/very-deep-learning-1991.html. Accessed 16 May 2021
15. Schmidhuber, J.: Making the world differentiable: on using self-supervised fully recurrent neural networks for dynamic reinforcement learning and planning in non-stationary environments (1990)
16. Schmidhuber, J.: Neural sequence chunkers (1991)
17. Schmidhuber, J.: Learning complex, extended sequences using the principle of history compression. Neural Comput. **4**, 234–242 (1992). https://doi.org/10.1162/neco.1992.4.2.234
18. Schmidhuber, J.: Algorithmic theories of everything (2000). arXiv:quant-ph/0011122
19. Schmidhuber, J.: The speed prior: a new simplicity measure yielding near-optimal computable predictions. In: Kivinen, J., Sloan, R.H. (eds.) COLT 2002. LNCS (LNAI), vol. 2375, pp. 216–228. Springer, Heidelberg (2002). https://doi.org/10.1007/3-540-45435-7_15

20. Schmidhuber, J.: Gödel machines: fully self-referential optimal universal self-improvers. In: Goertzel, B., Pennachin, C. (eds.) Artificial General Intelligence, pp. 199–226. Springer, Heidelberg (2007). https://doi.org/10.1007/3-540-45435-7_15

21. Schmidhuber, J.: Simple algorithmic principles of discovery, subjective beauty, selective attention, curiosity & creativity. arXiv:0709.0674 [cs] (2007)

22. Schmidhuber, J.: POWERPLAY: training an increasingly general problem solver by continually searching for the simplest still unsolvable problem. arXiv:1112.5309 [cs] (2012)

23. Schmidhuber, J.: On learning to think: algorithmic information theory for novel combinations of reinforcement learning controllers and recurrent neural world models. arXiv:1511.09249 [cs] (2015)

24. Schmidhuber, J.: One big net for everything. arXiv:1802.08864 [cs] (2018)

25. Kolmogorov, A.N.: On tables of random numbers. Sankhyā: Indian J. Stat. Ser. A (1961–2002) **25**, 369–376 (1963)

26. Schmidhuber, J.: Hierarchies of generalized kolmogorov complexities and nonenumerable universal measures computable in the limit. Int. J. Found. Comput. Sci. **13**, 587–612 (2002). https://doi.org/10.1142/S0129054102001291

27. Hutter, M.: A Theory of universal artificial intelligence based on algorithmic complexity. arXiv:cs/0004001 (2000)

28. Solomonoff, R.J.: Algorithmic probability: theory and applications. In: Emmert-Streib, F., Dehmer, M. (eds.) Information Theory and Statistical Learning, pp. 1–23. Springer, Boston (2009). https://doi.org/10.1007/978-0-387-84816-7_1

29. Feynman, R.P.: Quantum Mechanics and Path Integrals. McGraw-Hill, New York (1965)

30. Kaila, V., Annila, A.: Natural selection for least action. Proc. Roy. Soc. A: Math. Phys. Eng. Sci. **464**, 3055–3070 (2008). https://doi.org/10.1098/rspa.2008.0178

31. Campbell, J.O.: Universal darwinism as a process of Bayesian inference. Front. Syst. Neurosci. **10**, 49 (2016). https://doi.org/10.3389/fnsys.2016.00049

32. Vanchurin, V.: The world as a neural network. Entropy **22**, 1210 (2020). https://doi.org/10.3390/e22111210

33. Hanson, S.J.: A stochastic version of the delta rule. Phys. D **42**, 265–272 (1990). https://doi.org/10.1016/0167-2789(90)90081-Y

34. Orseau, L., Lattimore, T., Hutter, M.: Universal knowledge-seeking agents for stochastic environments. In: Jain, S., Munos, R., Stephan, F., Zeugmann, T. (eds.) ALT 2013. LNCS (LNAI), vol. 8139, pp. 158–172. Springer, Heidelberg (2013). https://doi.org/10.1007/978-3-642-40935-6_12

35. Friston, K.J., Lin, M., Frith, C.D., Pezzulo, G., Hobson, J.A., Ondobaka, S.: Active inference, curiosity and insight. Neural Comput. **29**, 2633–2683 (2017). https://doi.org/10.1162/neco_a_00999

36. Aslanides, J., Leike, J., Hutter, M.: Universal reinforcement learning algorithms: survey and experiments. arXiv:1705.10557 [cs] (2017)

37. Friston, K., Da Costa, L., Hafner, D., Hesp, C., Parr, T.: Sophisticated inference (2020)

38. VanRullen, R., Kanai, R.: Deep learning and the global workspace theory. Trends Neurosci. (2021). https://doi.org/10.1016/j.tins.2021.04.005

39. Lake, B.M., Salakhutdinov, R., Tenenbaum, J.B.: Human-level concept learning through probabilistic program induction. Science **350**, 1332–1338 (2015). https://doi.org/10.1126/science.aab3050

40. Lázaro-Gredilla, M., Lin, D., Guntupalli, J.S., George, D.: Beyond imitation: zero-shot task transfer on robots by learning concepts as cognitive programs. Sci. Robot. **4** (2019). https://doi.org/10.1126/scirobotics.aav3150

41. Ullman, T.D., Tenenbaum, J.B.: Bayesian models of conceptual development: learning as building models of the world. Annu. Rev. Dev. Psychol. **2**, 533–558 (2020). https://doi.org/10.1146/annurev-devpsych-121318-084833

42. Veness, J., Ng, K.S., Hutter, M., Uther, W., Silver, D.: A Monte Carlo AIXI approximation. arXiv:0909.0801 [cs, math] (2010)
43. Hesp, C., Tschantz, A., Millidge, B., Ramstead, M., Friston, K., Smith, R.: Sophisticated affective inference: simulating anticipatory affective dynamics of imagining future events. In: Verbelen, T., Lanillos, P., Buckley, C.L., De Boom, C. (eds.) IWAI 2020. Communications in Computer and Information Science, vol. 1326, pp. 179–186. Springer, Cham (2020). https://doi.org/10.1007/978-3-030-64919-7_18
44. de Abril, I.M., Kanai, R.: A unified strategy for implementing curiosity and empowerment driven reinforcement learning. arXiv:1806.06505 [cs] (2018)
45. Hafner, D., Lillicrap, T., Ba, J., Norouzi, M.: Dream to control: learning behaviors by latent imagination. arXiv:1912.01603 [cs] (2020)
46. Hafner, D., Ortega, P.A., Ba, J., Parr, T., Friston, K., Heess, N.: Action and perception as divergence minimization. arXiv:2009.01791 [cs, math, stat] (2020)
47. Wang, R., et al.: Enhanced POET: open-ended reinforcement learning through unbounded invention of learning challenges and their solutions. arXiv:2003.08536 [cs] (2020)
48. Hochreiter, S., Schmidhuber, J.: Long short-term memory. Neural Comput. 9, 1735–1780 (1997). https://doi.org/10.1162/neco.1997.9.8.1735
49. Lee-Thorp, J., Ainslie, J., Eckstein, I., Ontanon, S.: FNet: mixing tokens with fourier transforms. arXiv:2105.03824 [cs] (2021)
50. Ramsauer, H., et al.: Hopfield networks is all you need. arXiv:2008.02217 [cs, stat] (2021)
51. Schlag, I., Irie, K., Schmidhuber, J.: Linear transformers are secretly fast weight memory systems. arXiv:2102.11174 [cs] (2021)
52. Tay, Y., et al.: Are pre-trained convolutions better than pre-trained transformers? arXiv:2105.03322 [cs] (2021)
53. Hawkins, J., Ahmad, S.: Why neurons have thousands of synapses, a theory of sequence memory in neocortex. Front. Neural Circ. 10 (2016). https://doi.org/10.3389/fncir.2016.00023
54. Knight, R.T., Grabowecky, M.: Escape from linear time: prefrontal cortex and conscious experience. In: The Cognitive Neurosciences, pp. 1357–1371. The MIT Press, Cambridge (1995)
55. Koster, R., et al.: Big-loop recurrence within the hippocampal system supports integration of information across episodes. Neuron 99, 1342-1354.e6 (2018). https://doi.org/10.1016/j.neuron.2018.08.009
56. Faul, L., St. Jacques, P.L., DeRosa, J.T., Parikh, N., De Brigard, F.: Differential contribution of anterior and posterior midline regions during mental simulation of counterfactual and perspective shifts in autobiographical memories. NeuroImage. 215, 116843 (2020). https://doi.org/10.1016/j.neuroimage.2020.116843
57. Mannella, F., Gurney, K., Baldassarre, G.: The nucleus accumbens as a nexus between values and goals in goal-directed behavior: a review and a new hypothesis. Front. Behav. Neurosci. 7, 135 (2013). https://doi.org/10.3389/fnbeh.2013.00135
58. Friston, K.J., FitzGerald, T., Rigoli, F., Schwartenbeck, P., Pezzulo, G.: Active inference: a process theory. Neural Comput. 29, 1–49 (2017). https://doi.org/10.1162/NECO_a_00912
59. Friston, K.J.: am i self-conscious? (Or does self-organization entail self-consciousness?). Front. Psychol. 9 (2018). https://doi.org/10.3389/fpsyg.2018.00579
60. Ha, D., Schmidhuber, J.: World models. arXiv:1803.10122 [cs, stat] (2018). https://doi.org/10.5281/zenodo.1207631
61. Rusu, S.I., Pennartz, C.M.A.: Learning, memory and consolidation mechanisms for behavioral control in hierarchically organized cortico-basal ganglia systems. Hippocampus 30, 73–98 (2020). https://doi.org/10.1002/hipo.23167
62. Sanders, H., Wilson, M.A., Gershman, S.J.: Hippocampal remapping as hidden state inference. eLife. 9, e51140 (2020). https://doi.org/10.7554/eLife.51140

63. Hoel, E.: The overfitted brain: dreams evolved to assist generalization. Patterns **2**, 100244 (2021). https://doi.org/10.1016/j.patter.2021.100244

64. Boureau, Y.-L., Dayan, P.: Opponency revisited: competition and cooperation between dopamine and serotonin. Neuropsychopharmacology **36**, 74–97 (2011). https://doi.org/10.1038/npp.2010.151

65. Hassabis, D., Maguire, E.A.: The construction system of the brain. Philos. Trans. R. Soc. London B Biol. Sci. **364**, 1263–1271 (2009). https://doi.org/10.1098/rstb.2008.0296

66. Çatal, O., Verbelen, T., Van de Maele, T., Dhoedt, B., Safron, A.: Robot navigation as hierarchical active inference. Neural Netw. **142**, 192–204 (2021). https://doi.org/10.1016/j.neunet.2021.05.010

67. Schmidhuber, J.H., Mozer, M.C., Prelinger, D.: Continuous history compression. In: Proceedings of International Workshop on Neural Networks, RWTH Aachen, pp. 87–95. Augustinus (1993)

68. Shine, J.M.: The thalamus integrates the macrosystems of the brain to facilitate complex, adaptive brain network dynamics. Prog. Neurobiol. **199**, 101951 (2021). https://doi.org/10.1016/j.pneurobio.2020.101951

69. Friston, K.J., Parr, T., de Vries, B.: The graphical brain: Belief propagation and active inference. Netw. Neurosci. **1**, 381–414 (2017). https://doi.org/10.1162/NETN_a_00018

70. Parr, T., Friston, K.J.: The discrete and continuous brain: from decisions to movement-and back again. Neural Comput. **30**, 2319–2347 (2018). https://doi.org/10.1162/neco_a_01102

71. Gershman, S., Goodman, N.: Amortized inference in probabilistic reasoning. In: Proceedings of the Annual Meeting of the Cognitive Science Society, vol. 36 (2014)

72. Sales, A.C., Friston, K.J., Jones, M.W., Pickering, A.E., Moran, R.J.: Locus coeruleus tracking of prediction errors optimises cognitive flexibility: an active inference model. PLoS Comput. Biol. **15**, e1006267 (2019). https://doi.org/10.1371/journal.pcbi.1006267

73. Shea, N., Frith, C.D.: The global workspace needs metacognition. Trends Cogn. Sci. (2019). https://doi.org/10.1016/j.tics.2019.04.007

74. Shine, J.: Neuromodulatory influences on integration and segregation in the brain. Undefined (2019)

75. Holroyd, C.B., Verguts, T.: The best laid plans: computational principles of anterior cingulate cortex. Trends Cogn. Sci. **25**, 316–329 (2021). https://doi.org/10.1016/j.tics.2021.01.008

76. Carmichael, J.: Artificial intelligence gained consciousness in 1991. https://www.inverse.com/article/25521-juergen-schmidhuber-ai-consciousness. Accessed 14 Nov 2021

77. Dreyfus, H.L.: Why Heideggerian AI failed and how fixing it would require making it more Heideggerian. Philos. Psychol. **20**, 247–268 (2007). https://doi.org/10.1080/09515080701239510

78. Cisek, P.: Cortical mechanisms of action selection: the affordance competition hypothesis. Philos. Trans. R. Soc. Lond. B Biol. Sci. **362**, 1585–1599 (2007). https://doi.org/10.1098/rstb.2007.2054

79. Seth, A.K.: The cybernetic Bayesian brain. Open MIND. MIND Group, Frankfurt am Main (2014). https://doi.org/10.15502/9783958570108

80. Tani, J.: Exploring Robotic Minds: Actions, Symbols, and Consciousness as Self-organizing Dynamic Phenomena. Oxford University Press (2016)

81. Kiverstein, J., Miller, M., Rietveld, E.: The feeling of grip: novelty, error dynamics, and the predictive brain. Synthese **196**(7), 2847–2869 (2017). https://doi.org/10.1007/s11229-017-1583-9

82. Tononi, G., Boly, M., Massimini, M., Koch, C.: Integrated information theory: from consciousness to its physical substrate. Nat. Rev. Neurosci. **17**, 450 (2016). https://doi.org/10.1038/nrn.2016.44

83. Battaglia, P.W., et al.: Relational inductive biases, deep learning, and graph networks. arXiv:1806.01261 [cs, stat] (2018)

84. Gothoskar, N., Guntupalli, J.S., Rikhye, R.V., Lázaro-Gredilla, M., George, D.: Different clones for different contexts: hippocampal cognitive maps as higher-order graphs of a cloned HMM. bioRxiv. 745950 (2019) https://doi.org/10.1101/745950

85. Peer, M., Brunec, I.K., Newcombe, N.S., Epstein, R.A.: Structuring knowledge with cognitive maps and cognitive graphs. Trends Cogn. Sci. **25**, 37–54 (2021). https://doi.org/10.1016/j.tics.2020.10.004

86. Dehaene, S.: Consciousness and the Brain: Deciphering How the Brain Codes Our Thoughts. Viking, New York (2014)

87. Tononi, G., Koch, C.: Consciousness: here, there and everywhere? Philos. Trans. R. Soc. B: Biol. Sci. **370**, 20140167 (2015). https://doi.org/10.1098/rstb.2014.0167

88. Ortiz, J., Pupilli, M., Leutenegger, S., Davison, A.J.: Bundle adjustment on a graph processor. arXiv:2003.03134 [cs] (2020)

89. Kahneman, D.: Thinking, Fast and Slow. Farrar, Straus and Giroux (2011)

90. Bengio, Y.: The consciousness prior. arXiv:1709.08568 [cs, stat] (2017)

91. Lange, S., Riedmiller, M.: Deep auto-encoder neural networks in reinforcement learning. In: The 2010 International Joint Conference on Neural Networks (IJCNN), pp. 1–8 (2010). https://doi.org/10.1109/IJCNN.2010.5596468

92. Lotter, W., Kreiman, G., Cox, D.: Deep predictive coding networks for video prediction and unsupervised learning. arXiv:1605.08104 [cs, q-bio] (2016)

93. Wu, Y., Wayne, G., Graves, A., Lillicrap, T.: The Kanerva machine: a generative distributed memory. arXiv:1804.01756 [cs, stat] (2018)

94. Jiang, Y., Kim, H., Asnani, H., Kannan, S., Oh, S., Viswanath, P.: Turbo autoencoder: deep learning based channel codes for point-to-point communication channels. arXiv:1911.03038 [cs, eess, math] (2019)

95. Kanai, R., Chang, A., Yu, Y., Magrans de Abril, I., Biehl, M., Guttenberg, N.: Information generation as a functional basis of consciousness. Neurosci. Conscious. **2019** (2019). https://doi.org/10.1093/nc/niz016

96. Lillicrap, T.P., Santoro, A., Marris, L., Akerman, C.J., Hinton, G.: Backpropagation and the brain. Nat. Rev. Neurosci. 1–12 (2020). https://doi.org/10.1038/s41583-020-0277-3

97. Dayan, P., Hinton, G.E., Neal, R.M., Zemel, R.S.: The Helmholtz machine. Neural Comput. **7**, 889–904 (1995)

98. Kingma, D.P., Welling, M.: Auto-encoding variational bayes. arXiv:1312.6114 [cs, stat] (2014)

99. Candadai, M., Izquierdo, E.J.: Sources of predictive information in dynamical neural networks. Sci. Rep. **10**, 16901 (2020). https://doi.org/10.1038/s41598-020-73380-x

100. Lu, Z., Bassett, D.S.: Invertible generalized synchronization: a putative mechanism for implicit learning in neural systems. Chaos **30**, 063133 (2020). https://doi.org/10.1063/5.0004344

101. Rumelhart, D.E., McClelland, J.L.: Information processing in dynamical systems: foundations of harmony theory. In: Parallel Distributed Processing: Explorations in the Microstructure of Cognition: Foundations, pp. 194–281. MIT Press (1987)

102. Kachman, T., Owen, J.A., England, J.L.: Self-organized resonance during search of a diverse chemical space. Phys. Rev. Lett. **119**, 038001 (2017). https://doi.org/10.1103/PhysRevLett.119.038001

103. Friston, K.J.: A free energy principle for a particular physics. arXiv:1906.10184 [q-bio] (2019)

104. Ali, A., Ahmad, N., de Groot, E., van Gerven, M.A.J., Kietzmann, T.C.: Predictive coding is a consequence of energy efficiency in recurrent neural networks. bioRxiv. 2021.02.16.430904 (2021). https://doi.org/10.1101/2021.02.16.430904

105. Bejan, A., Lorente, S.: The constructal law of design and evolution in nature. Philos. Trans. R. Soc. Lond. B Biol. Sci. **365**, 1335–1347 (2010). https://doi.org/10.1098/rstb.2009.0302

106. McCulloch, W.S., Pitts, W.: A logical calculus of the ideas immanent in nervous activity. Bull. Math. Biophys. **5**, 115–133 (1943)

107. Srivastava, N., Hinton, G., Krizhevsky, A., Sutskever, I., Salakhutdinov, R.: Dropout: a simple way to prevent neural networks from overfitting. J. Mach. Learn. Res. **15**, 1929–1958 (2014)

108. Ahmad, S., Scheinkman, L.: How can we be so dense? The benefits of using highly sparse representations. arXiv preprint arXiv:1903.11257 (2019)

109. Mumford, D.: On the computational architecture of the neocortex. Biol. Cybern. **65**, 135–145 (1991). https://doi.org/10.1007/BF00202389

110. Rao, R.P., Ballard, D.H.: Predictive coding in the visual cortex: a functional interpretation of some extra-classical receptive-field effects. Nat. Neurosci. **2**, 79–87 (1999). https://doi.org/10.1038/4580

111. Bastos, A.M., Usrey, W.M., Adams, R.A., Mangun, G.R., Fries, P., Friston, K.J.: Canonical microcircuits for predictive coding. Neuron **76**, 695–711 (2012). https://doi.org/10.1016/j.neuron.2012.10.038

112. Grossberg, S.: Towards solving the hard problem of consciousness: the varieties of brain resonances and the conscious experiences that they support. Neural Netw. **87**, 38–95 (2017). https://doi.org/10.1016/j.neunet.2016.11.003

113. Heeger, D.J.: Theory of cortical function. Proc. Natl. Acad. Sci. U.S.A. **114**, 1773–1782 (2017). https://doi.org/10.1073/pnas.1619788114

114. George, D., Lázaro-Gredilla, M., Lehrach, W., Dedieu, A., Zhou, G.: A detailed mathematical theory of thalamic and cortical microcircuits based on inference in a generative vision model. bioRxiv. 2020.09.09.290601 (2020). https://doi.org/10.1101/2020.09.09.290601

115. Friston, K.J., Rosch, R., Parr, T., Price, C., Bowman, H.: Deep temporal models and active inference. Neurosci. Biobehav. Rev. **77**, 388–402 (2017). https://doi.org/10.1016/j.neubiorev.2017.04.009

116. Pearl, J., Mackenzie, D.: The Book of Why: The New Science of Cause and Effect. Basic Books (2018)

117. Csáji, B.C.: Approximation with artificial neural networks. Fac. Sci. Etvs Lornd Univ. Hungary. **24**, 7 (2001)

118. Malach, E., Shalev-Shwartz, S.: Is deeper better only when shallow is good? arXiv:1903.03488 [cs, stat] (2019)

119. Srivastava, R.K., Greff, K., Schmidhuber, J.: Highway networks. arXiv:1505.00387 [cs] (2015)

120. Lin, H.W., Tegmark, M., Rolnick, D.: Why does deep and cheap learning work so well? J. Stat. Phys. **168**(6), 1223–1247 (2017). https://doi.org/10.1007/s10955-017-1836-5

121. Sherstinsky, A.: Fundamentals of recurrent neural network (RNN) and long short-term memory (LSTM) network. Phys. D **404**, 132306 (2020). https://doi.org/10.1016/j.physd.2019.132306

122. Schmidhuber, J.: On learning how to learn learning strategies (1994)

123. Wang, J.X., et al.: Prefrontal cortex as a meta-reinforcement learning system. Nat. Neurosci. **21**, 860 (2018). https://doi.org/10.1038/s41593-018-0147-8

124. Watts, D.J., Strogatz, S.H.: Collective dynamics of 'small-world' networks. Nature **393**, 440 (1998). https://doi.org/10.1038/30918

125. Jarman, N., Steur, E., Trengove, C., Tyukin, I.Y., van Leeuwen, C.: Self-organisation of small-world networks by adaptive rewiring in response to graph diffusion. Sci. Rep. **7**, 13158 (2017). https://doi.org/10.1038/s41598-017-12589-9

126. Rentzeperis, I., Laquitaine, S., van Leeuwen, C.: Adaptive rewiring of random neural networks generates convergent-divergent units. arXiv:2104.01418 [q-bio] (2021)

127. Massobrio, P., Pasquale, V., Martinoia, S.: Self-organized criticality in cortical assemblies occurs in concurrent scale-free and small-world networks. Sci. Rep. **5**, 10578 (2015). https://doi.org/10.1038/srep10578

128. Gal, E., et al.: Rich cell-type-specific network topology in neocortical microcircuitry. Nat. Neurosci. **20**, 1004–1013 (2017). https://doi.org/10.1038/nn.4576

129. Takagi, K.: Information-based principle induces small-world topology and self-organized criticality in a large scale brain network. Front. Comput. Neurosci. **12** (2018). https://doi.org/10.3389/fncom.2018.00065

130. Goekoop, R., de Kleijn, R.: How higher goals are constructed and collapse under stress: a hierarchical Bayesian control systems perspective. Neurosci. Biobehav. Rev. **123**, 257–285 (2021). https://doi.org/10.1016/j.neubiorev.2020.12.021

131. Sporns, O.: Network attributes for segregation and integration in the human brain. Curr. Opin. Neurobiol. **23**, 162–171 (2013). https://doi.org/10.1016/j.conb.2012.11.015

132. Cohen, J.R., D'Esposito, M.: The segregation and integration of distinct brain networks and their relationship to cognition. J. Neurosci. **36**, 12083–12094 (2016). https://doi.org/10.1523/JNEUROSCI.2965-15.2016

133. Mohr, H., et al.: Integration and segregation of large-scale brain networks during short-term task automatization. Nat Commun. **7**, 13217 (2016). https://doi.org/10.1038/ncomms13217

134. Badcock, P.B., Friston, K.J., Ramstead, M.J.D.: The hierarchically mechanistic mind: a free-energy formulation of the human psyche. Phys. Life Rev. (2019). https://doi.org/10.1016/j.plrev.2018.10.002

135. Bak, P., Sneppen, K.: Punctuated equilibrium and criticality in a simple model of evolution. Phys. Rev. Lett. **71**, 4083–4086 (1993). https://doi.org/10.1103/PhysRevLett.71.4083

136. Edelman, G., Gally, J.A., Baars, B.J.: Biology of consciousness. Front Psychol. **2**, 4 (2011). https://doi.org/10.3389/fpsyg.2011.00004

137. Paperin, G., Green, D.G., Sadedin, S.: Dual-phase evolution in complex adaptive systems. J. R. Soc. Interface **8**, 609–629 (2011). https://doi.org/10.1098/rsif.2010.0719

138. Safron, A., Klimaj, V., Hipólito, I.: On the importance of being flexible: dynamic brain networks and their potential functional significances (2021). https://psyarxiv.com/x734w/, https://doi.org/10.31234/osf.io/x734w

139. Safron, A.: Integrated world modeling theory (IWMT) expanded: implications for theories of consciousness and artificial intelligence (2021). https://psyarxiv.com/rm5b2/, https://doi.org/10.31234/osf.io/rm5b2

140. Smith, R.: Do brains have an arrow of time? Philos. Sci. **81**, 265–275 (2014). https://doi.org/10.1086/675644

141. Wolfram, S.: A New Kind of Science. Wolfram Media (2002)

142. Friston, K.J., Wiese, W., Hobson, J.A.: Sentience and the origins of consciousness: from cartesian duality to Markovian monism. Entropy **22**, 516 (2020). https://doi.org/10.3390/e22050516

143. Doerig, A., Schurger, A., Hess, K., Herzog, M.H.: The unfolding argument: why IIT and other causal structure theories cannot explain consciousness. Conscious. Cogn. **72**, 49–59 (2019). https://doi.org/10.1016/j.concog.2019.04.002

144. Marshall, W., Kim, H., Walker, S.I., Tononi, G., Albantakis, L.: How causal analysis can reveal autonomy in models of biological systems. Phil. Trans. R. Soc. A. **375**, 20160358 (2017). https://doi.org/10.1098/rsta.2016.0358

145. Joslyn, C.: Levels of control and closure in complex semiotic systems. Ann. N. Y. Acad. Sci. **901**, 67–74 (2000)

146. Chang, A.Y.C., Biehl, M., Yu, Y., Kanai, R.: Information closure theory of consciousness. arXiv:1909.13045 [q-bio] (2019)

147. Singer, W.: Consciousness and the binding problem. Ann. N. Y. Acad. Sci. **929**, 123–146 (2001)

148. Baars, B.J., Franklin, S., Ramsoy, T.Z.: Global workspace dynamics: cortical "binding and propagation" enables conscious contents. Front Psychol. **4** (2013). https://doi.org/10.3389/fpsyg.2013.00200

149. Atasoy, S., Donnelly, I., Pearson, J.: Human brain networks function in connectome-specific harmonic waves. Nat. Commun. **7**, 10340 (2016). https://doi.org/10.1038/ncomms10340

150. Wu, L., Zhang, Y.: A new topological approach to the L∞-uniqueness of operators and the L1-uniqueness of Fokker-Planck equations. J. Funct. Anal. **241**, 557–610 (2006). https://doi.org/10.1016/j.jfa.2006.04.020

151. Carroll, S.: The Big Picture: On the Origins of Life, Meaning, and the Universe Itself. Penguin (2016)

152. Hoel, E.P., Albantakis, L., Marshall, W., Tononi, G.: Can the macro beat the micro? Integrated information across spatiotemporal scales. Neurosci. Conscious. **2016** (2016). https://doi.org/10.1093/nc/niw012

153. Albantakis, L., Marshall, W., Hoel, E., Tononi, G.: What caused what? A quantitative account of actual causation using dynamical causal networks. arXiv:1708.06716 [cs, math, stat] (2017)

154. Hoel, E.P.: When the map is better than the territory. Entropy **19**, 188 (2017). https://doi.org/10.3390/e19050188

155. Rocha, L.M.: Syntactic autonomy. Why there is no autonomy without symbols and how self-organizing systems might evolve them. Ann. N. Y. Acad. Sci. **901**, 207–223 (2000). https://doi.org/10.1111/j.1749-6632.2000.tb06280.x

156. Rudrauf, D., Lutz, A., Cosmelli, D., Lachaux, J.-P., Le Van Quyen, M.: From autopoiesis to neurophenomenology: Francisco Varela's exploration of the biophysics of being. Biol. Res. **36**, 27–65 (2003)

157. Everhardt, A.S., et al.: Periodicity-doubling cascades: direct observation in ferroelastic materials. Phys. Rev. Lett. **123**, 087603 (2019). https://doi.org/10.1103/PhysRevLett.123.087603

158. Chen, T., et al.: Quantum Zeno effects across a parity-time symmetry breaking transition in atomic momentum space (2020)

159. Fruchart, M., Hanai, R., Littlewood, P.B., Vitelli, V.: Non-reciprocal phase transitions. Nature **592**, 363–369 (2021). https://doi.org/10.1038/s41586-021-03375-9

160. Hofstadter, D.R.: Gödel, Escher, Bach: An Eternal Golden Braid. Basic Books (1979)

161. Hofstadter, D.R.: I Am a Strange Loop. Basic Books (2007)

162. Lloyd, S.: A Turing test for free will. Philos. Trans. R. Soc. A: Math. Phys. Eng. Sci. **370**, 3597–3610 (2012). https://doi.org/10.1098/rsta.2011.0331

163. Parr, T., Markovic, D., Kiebel, S.J., Friston, K.J.: Neuronal message passing using mean-field, Bethe, and marginal approximations. Sci. Rep. **9** (2019). https://doi.org/10.1038/s41598-018-38246-3

164. Madl, T., Baars, B.J., Franklin, S.: The timing of the cognitive cycle. PLoS One **6**, e14803 (2011)

165. Maguire, P., Maguire, R.: Consciousness is data compression. Undefined (2010)

166. Tegmark, M.: Improved measures of integrated information. PLoS Comput Biol. **12** (2016). https://doi.org/10.1371/journal.pcbi.1005123

167. Maguire, P., Moser, P., Maguire, R.: Understanding consciousness as data compression. J. Cogn. Sci. **17**, 63–94 (2016)

168. Metzinger, T.: The Ego Tunnel: The Science of the Mind and the Myth of the Self. Basic Books, New York (2009)

169. Limanowski, J., Friston, K.J.: 'Seeing the dark': grounding phenomenal transparency and opacity in precision estimation for active inference. Front. Psychol. **9** (2018). https://doi.org/10.3389/fpsyg.2018.00643

170. Hoffman, D.D., Prakash, C.: Objects of consciousness. Front. Psychol. **5** (2014). https://doi.org/10.3389/fpsyg.2014.00577

171. Kirchhoff, M., Parr, T., Palacios, E., Friston, K.J., Kiverstein, J.: The Markov blankets of life: autonomy, active inference and the free energy principle. J. R. Soc. Interface **15** (2018). https://doi.org/10.1098/rsif.2017.0792

172. Dennett, D.: Consciousness Explained. Back Bay Books (1992)

173. Haun, A., Tononi, G.: Why does space feel the way it does? Towards a principled account of spatial experience. Entropy **21**, 1160 (2019). https://doi.org/10.3390/e21121160

174. Sutterer, D.W., Polyn, S.M., Woodman, G.F.: α-band activity tracks a two-dimensional spotlight of attention during spatial working memory maintenance. J. Neurophysiol. **125**, 957–971 (2021). https://doi.org/10.1152/jn.00582.2020

Intention Modulation for Multi-step Tasks in Continuous Time Active Inference

Matteo Priorelli[(⊠)] and Ivilin Peev Stoianov[iD]

Institute of Cognitive Sciences and Technologies (ISTC), National Research
Council of Italy (CNR), Padova 35100, Italy
matteo.priorelli@istc.cnr.it, ivilinpeev.stoianov@cnr.it

Abstract. We extend an Active Inference theory in continuous time of
how neural circuitry in the Dorsal Visual Stream (DVS) and the Posterior
Parietal Cortex (PPC) implement visually guided goal-directed behavior
with novel capacity to resolve multi-step tasks. According to the theory,
the PPC maintains a high-level internal representation of the causes of
the environment (belief), including bodily states and objects in the scene,
and by generating sensory predictions and comparing them with observations it is able to learn and infer the causal relationships and latent
states of the external world. We propose that multi-task goal-directed
behavior may be achieved by decomposing the belief dynamics into a set
of intention functions that independently pull the belief towards different
goals; multi-step tasks could be solved by dynamically modulating these
intentions within the PPC. This low-level solution in continuous time
is applicable to multi-phase actions consisting of a priori defined steps
as an alternative to the more general hybrid discrete-continuous approach. As a demonstration, we emulated an agent embodying an actuated upper limb and proprioceptive, visual and tactile sensory systems.
Visual information was obtained with the help of a Variational Autoencoder (VAE) simulating the DVS, which allows to dynamically infer the
current posture configuration through prediction error minimization and,
importantly, an intended future posture corresponding to the visual targets. We assessed the approach on a task including two steps: reaching
a target and returning to a home position. We show that by defining a
functional that governs the activation of different intentions implementing the corresponding steps, the agent can easily solve the overall task.

Keywords: Motor control · Active inference · Intentions · PPC

1 Introduction

The DVS provides critical support for continuously monitoring the spatial location of objects and posture and performing visuomotor transformations [5,12].

Supported by European Union H2020-EIC-FETPROACT-2019 grant 951910 to IPS
and Italian PRIN grant 2017KZNZLN to IPS.

The PPC, located at the apex of the DVS, is also bidirectionally connected to frontal, motor, and somatosensory areas, placing it in a privileged position to set and guide goal-directed actions and continuously adjust motor plans by tracking moving targets and body posture [1,6,13], but its specific computational mechanism is still unknown. Moreover, research suggests that the PPC encodes the goals of more complex actions including multiple targets (e.g., in a reaching task) even when there is considerable delay between goals [3], and plays a key role in transforming multiple targets into a sequential motor response [15]. Thus, the goal of this study was to provide a computational account of the capacity of the PPC to execute multi-step tasks without the involvement of higher cognitive areas.

Active Inference provides fundamental insights about the computational principles of the perception-action loop in biological systems [7,9]. It assumes that the nervous system operates on belief μ that defines an abstract internal representation of the state of the world and computes predictions at the sensory level through generative models. While the role of perception is to maintain up-to-date belief reflecting the incoming stream of sensory signal, action tries to make the sensory predictions true. To do so, belief and control signal are inferred through dynamic minimization of Free Energy, or generalized prediction errors. To allow efficient dynamic inference, the model assumes generalized coordinates $\tilde{\mu}$ encoding instantaneous trajectories of the belief; here we consider up to the 2nd temporal order, namely position μ, velocity μ', and acceleration μ''.

State-of-the-art implementations of Active Inference in continuous time define action goals in terms of an attractor embedded in the system dynamics [8]. The attractor can be a low-level (sensory) prediction that is compared with the current observation and backpropagated to obtain a belief update direction [16,19,22], or a desired latent state specified in the full equations of motion that the agent expects to perceive [2]. Since these states or sensations are usually static or defined externally, the agent is not able to deal with continuously changing environments (e.g., reaching or grasping moving targets, in particular when they move along unpredictable trajectories), expecting that the world will always evolve in the same way.

Our previous research extended the above approaches with a theory of how the DVS and PPC are involved in producing visually guided goal-directed behavior in dynamic contexts by means of defining and fulfilling flexible goal-coding intention functions [20]. In this view, the agent maintains a dynamically estimated belief μ over bodily states and (moving) objects in the scene, such as potential or current targets. Then, intentions are formalized by exploiting this belief in order to compute a future action goal, or body posture, so that the attractor is not fixed but dependent on the current perceptual and internal representation of the world, or representations memorized in past experiences:

$$h^{(i)} = i^{(i)}(\mu) \tag{1}$$

where $h^{(i)}$ is the future goal state and $i^{(i)}(\mu)$ is the intention function.

An intention prediction error is computed by the difference between the current and desired belief:

$$e^{(i)} = h^{(i)} - \mu \tag{2}$$

which is then embedded into a dynamics function to produce an attractive force:

$$f^{(i)}(\mu) = ke^{(i)} + w_\mu^{(i)} \tag{3}$$

where k is the attractor gain and $w_\mu^{(i)}$ is a noise term.

This function is generated from a Gaussian probability distribution approximating system dynamics:

$$p(\mu^{(i)\prime}|\mu) = \mathcal{N}(\mu^{(i)\prime}|f^{(i)}(\mu), \gamma^{(i)}) \tag{4}$$

Here $\gamma^{(i)}$, the precision of the distribution, can be seen as a modulatory signal of the corresponding intention. In fact, assuming that the overall dynamics function of the belief is factorized into a product of independent distributions for each intention, the update belief for the 1st-order will be a combination of every function $f^{(i)}(\mu)$:

$$\dot{\mu}' = -\sum_i^K e_f^{(i)} \gamma^{(i)} \tag{5}$$

where $e_f^{(i)}$ is the prediction error of the i-th dynamics function. Here we considered a zero 2nd-order belief ($\mu'' = 0$), so that the 1st-order is only subject to the forward error of that level (see [20] for more general form and detail).

In summary, the agent constructs and maintains a plausible dynamic hypothesis for the causes that produce what it is perceiving; by manipulating these and eventually other memorized causes, the agent can dynamically compute a representation of the future body state, i.e., intention, which in turn can act as a prior over the current belief.

For example, if the goal of the agent is to track a dynamic target, we first decompose the belief into two components: one for the body posture μ_a, and one for the target μ_t. Then, in order to produce a reaching force, we define an intention that generates a belief with the first component equal to the second one:

$$h^{(t)} = i^{(t)}(\mu) = [\mu_t, \mu_t] \tag{6}$$

This intention will then be embedded in the dynamics function as an *intention prediction error*:

$$f^{(t)}(\mu) = k \cdot (h^{(t)} - \mu) = k \cdot ([\mu_t - \mu_a, 0]) \tag{7}$$

that will bias the belief towards the desired configuration, which can be continuously inferred through visual observations so to follow a moving target. Although such dynamics does not have any counterpart with the real evolution of the environment, it is this imagined prior that makes the agent think it will be pulled towards the target configuration.

Decomposing the belief dynamics into a set of intention functions simplify the design and implementation of a goal-directed Active Inference agent that may thus generate more complex behaviors, since in this way it is able to maintain and realize different goal states at the same time.

The advantages of this approach become more clear when dealing with a multi-step task. For the time being, such tasks have been implemented mostly in discrete state-space [7,11,18]. Indeed, the use of discrete states and outcomes along with minimization of Expected Free Energy has many potentialities when it comes to planning, action selection and policy learning [14,21]; however, dealing with real world applications also requires some sort of communication with the continuous domain. This has been done by using the discrete framework for planning and the continuous framework for real action execution, where the link between the two models has been realized by Bayesian model averaging and comparison [10,17]. A mixed Active Inference model is a robust method for dealing with planning in constantly changing environment, when implementing a more biologically plausible architecture or when solving a complex task, in particular if the sequence of actions or policy has to be learned online. Nevertheless, if the task to be solved is defined a priori, all that is needed is to design particular intentions for each step, and find a way to perform transitions between them.

2   Multi-step Active Inference

We will now describe a multi-step Active Inference approach exemplified with a delayed reaching and go-back-home task and solved through a dynamic functional exploiting locally available information in the neural circuitry that putatively implements the inference process.

In the context of a reaching task, the agent has to first reach a visually defined target and then return back to (i.e., reach) a previously memorized Home Button (HB) position (see Fig. 1a); in this case, a different belief dynamics for each of the reaching movements should be used. However, if we define this dynamics as composed of two intention functions, one to reach the target and the other to go back to the HB, the agent can maintain at any time different goal states, which can either realize in parallel or sequentially, and it only needs to modulate the respective intention precisions that play the role of relative gains.

These precisions may depend on some function of the belief: in this way, the agent will transition from one step to another if some condition is met about what it believes are the causes of current observations, so that at every step it will think that the external world will evolve in a different way.

In the following we consider an agent consisting of a coarse 3-DoF limb model, which is velocity-controlled, moving on a 2D plane, and receiving information from proprioceptive, visual and tactile sensory systems. The visual input s_v is provided by a virtual camera of 128×96 pixels (see Fig. 1b), and the corresponding prediction is generated by the decoder of a VAE (see [20] for all details), which was used to infer joint configurations in a more realistic way, simulating the functioning of the DVS. Proprioceptive information s_p is encoded in joint angle coordinates, while tactile observation s_t is only provided for the hand, and is approximated by a Boolean function indicating if the agent is touching the target or not.

The posture belief consisted of three components, corresponding to the arm, target and HB: $\boldsymbol{\mu}_p = [\boldsymbol{\mu}_{p,a}, \boldsymbol{\mu}_{p,t}, \boldsymbol{\mu}_{p,h}]$. Importantly, all of these representations are encoded in joint angle coordinates, so that the latter two correspond to particular ways of interacting with that object, depending on some affordance (which can either be freely inferred by the visual generative model, or imposed by some higher-level prior). Here, $\boldsymbol{\mu}_{p,h}$ is initialized to the correct HB configuration and kept fixed, so to represent a previously memorized state. Additionally, a belief over tactile sensation μ_t is considered in order to make possible the use of the respective sensory system.

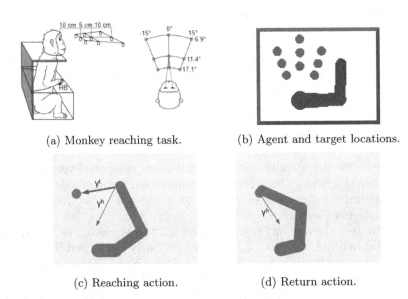

(a) Monkey reaching task.

(b) Agent and target locations.

(c) Reaching action.

(d) Return action.

Fig. 1. Reference experiment and simulation. (a) The monkey had first to reach one of the 9 targets and then reach back a HB [4]. (b) Visual feedback provided the agent information about the 3-DoF actuated upper limb and one target. The HB is right in front of the arm root (not shown). (c–d) The selection of one of the two fixed competing actions is controlled by relative strength (arrow width) of the corresponding attractors controlled by intention precisions.

As explained in the previous section, goal-directed behavior is realized by introducing, in the belief dynamics, intention functions of the belief $i^{(i)}(\boldsymbol{\mu})$ that generate a future goal state. For example, target reaching is made possible by an intention that sets the current arm configuration equal to the inferred target joint angles, with arm component $\boldsymbol{\mu}_{p,a} = \boldsymbol{\mu}_{p,t}$. Similarly, the returning action was implemented by introducing a second intention that sets the arm belief equal to the HB joint configuration (also see Figs. 1c,1d):

$$h^{(t)} = i^{(t)}(\mu_p) = [\mu_{p,t}, \mu_{p,t}, \mu_{p,h}]$$
$$h^{(h)} = i^{(h)}(\mu_p) = [\mu_{p,h}, \mu_{p,t}, \mu_{p,h}]$$

(8)

Then, intention prediction errors are computed as differences between (future) intentions and (current) posture belief:

$$e^{(t)} = [\mu_{p,t} - \mu_{p,a}, 0, 0]$$
$$e^{(h)} = [\mu_{p,h} - \mu_{p,a}, 0, 0]$$

(9)

Correspondingly, sensory prediction errors are related by the following equations:

$$e_p = s_p - G_p \mu_p$$
$$e_v = s_v - g_v(\mu_p)$$
$$e_t = s_t - \mu_t$$

(10)

where G_p is a mapping that extracts the arm joint angles from the posture belief, and g_v is the decoder component of the VAE.

The posture belief update will then be composed of two contributions: one that pulls it towards the current observations, and another one that pulls it towards what the agent expects in the future:

$$\dot{\tilde{\mu}}_p = \begin{bmatrix} \dot{\mu}_p \\ \dot{\mu}'_p \end{bmatrix} \approx \begin{bmatrix} \mu'_p + \pi_p \cdot G_p^T e_p + \pi_v \cdot \partial g_v^T e_v \\ -\gamma^{(t)}(\mu'_p - ke^{(t)}) - \gamma^{(h)}(\mu'_p - ke^{(h)}) \end{bmatrix}$$

(11)

where π_p and π_v are respectively proprioceptive and visual precisions, while $\gamma^{(t)}$ and $\gamma^{(h)}$ are target and home reaching intention precisions.

It is important to note that here we neglected the backward error in the 0th-order of the belief update (which is usually used as attractor) since it has a much smaller impact on the overall dynamics, and treated the forward error at the 1st-order as the actual attractor force.

The two intention precisions, or gains, may be externally modulated in order to alternate the corresponding movements, letting them depend on a common parameter β such that:

$$\gamma = [\gamma^{(t)}, \gamma^{(h)}] = [1 - \beta, \beta]$$

(12)

where, in normal conditions, only one intention will be active at a time.

More interestingly, we can define a functional intention selector to autonomously control the transition among the task steps. It could exploit information locally available in the PPC - such as the belief in a relevant step-triggering latent state - to define a circuit that dynamically computes the intention switching signal. The obvious choice in the reaching task is to use tactile sensations, and in particular the belief μ_t that the agent is touching the target; indeed, in a biologically plausible scenario this modulation should not depend on sensory-level observations but on what the agent believes in a particular moment. At any given time step, the functional could also recursively depend on its state and the dynamic calculation should be initialized in a way that the agent starts with the execution of the first step (intention) of the multi-phase action.

Given all these considerations, we can define the following two-step intention triggering recursive functional:

$$\beta = \sigma(w(\beta + \mu_t - 0.5)) \tag{13}$$

with parameter $w = 10$ chosen so that the *sigmoid* could quickly compute a stable transition from $\beta = 0$ to $\beta = 1$ when the belief on touch approaches 1. Here the sigmoid function was preferred over a discrete switching signal for the sake of a more biologically plausible implementation.

At the onset of the multi-step task, the state of β is initialized with zero. This state causes the activation of the first target-reaching intention and the suppression of the second HB returning action. The dynamic recursive function maintains this state until the agent reaches the belief that it is touching the target, at which point the functional changes its state to $\beta = 1$, suppressing the reaching intention and activating the HB returning action, which takes the drive on posture belief update. At this point the function maintains the state $\beta = 1$ due to the recursive link regardless of the tactile belief.

3   Simulation

We first illustrate in Fig. 2 the dynamic goal-directed Active Inference motor control process with a sequence of time frames drawn from a sample reach action. The agent first performs an initial reaching movement when the intention selecting signal $\beta = 0$, followed by a return to HB movement, with $\beta = 1$.

Note that in the simulated delayed target reaching task movement onset is anticipated by a pure target-perception phase that may be realized, for example, by temporarily setting either the intention gain k or every intention precision γ to zero, so that the future goal states are always present but not affecting the current belief.

More generally, action can start either after a fixed period, as in the current delayed reaching, or when some external or internal conditions are met, for example, when the uncertainty over the target belief falls bellow some threshold (see [20] for more details).

We then show in Fig. 3 the detailed evolution of system dynamics over time in a sample trial, including belief over targets and body, motor-control signal, intention errors, hand-to-target distance, and signals determining the behavior of the multi-step intention modulation functional: the state of tactile sensation, the belief over it, and the intention selecting signal β. Note that as soon as the agent touches the target, the belief over touch quickly raises to 1, which triggers a stable switch from $\beta = 0$ to $\beta = 1$. Since the latter plays the role of functional intention selector, the effect of changing its value is the change of the active intention, i.e., from the intention driving a target reaching movement to the intention driving a return to the (memorized) HB.

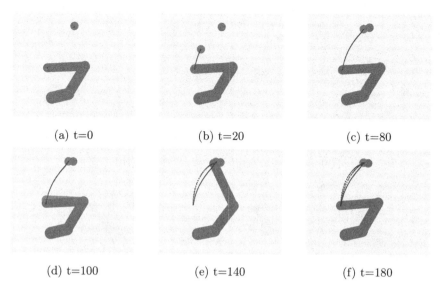

(a) t=0   (b) t=20   (c) t=80

(d) t=100   (e) t=140   (f) t=180

Fig. 2. Two-step delayed reaching task. At t = 0 the arm (in blue) starts at the HB position while a target (red circle) is randomly sampled. Follows a delay period of pure perceptual inference during which the agent can not move but it can only update the belief over the target position (purple circle), which is correctly estimated after 80 steps. At t = 100 movement is allowed by setting a non-zero intention gain k approximating the onset of a "go" signal, so that the estimated arm configuration (in green) is pulled towards the belief over the intended configuration. Upon target reach (t = 140), the agent perceives a tactile observation, whose belief triggers the switch of the active intention, enabling thus the HB return movement (completed at t = 180). (Color figure online)

We assessed the novel method on the two-step delayed reaching task described earlier, running ten trials for each of the nine targets. On average, during the delay (i.e., perception) period the target was estimated correctly 88.9% within 25 time steps with average belief-to-target distance of 6.6 px (SD 0.8). In turn, on average the arm reached the (real) target 88.9% within 219 time steps, with average hand-to-target distance of 1.7 px (SD 3.9).

4   Discussion

We extended our Active Inference-based theory of how the neural circuitry in the DVS and PPC implement dynamic visually guided goal-directed behavior [20] with a novel approach of locally solving multi-step tasks with a continuous-time implementation. We built upon the proposal that belief dynamics could be decomposed into a set of intention functions that independently pull the agent towards different goal states and that each of these intentions may be modulated through their corresponding precisions, either to perform a movement composed

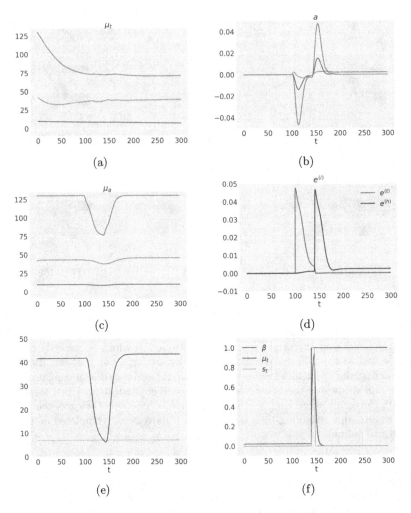

Fig. 3. Dynamics of a sample two-step trial. (a) Belief over target in joint-configuration space; (b) joint velocities; (c) belief over arm configuration; (d) norms of the intention prediction errors; (e) distance between the target and the tip of the hand; the "target-reached" distance is represented by a dotted line; (f) tactile observation s_t, tactile belief μ_t and intention switching signal β.

of different subgoals at the same time, or to realize a sequence of actions. We now suggest that the modulation of these intentions during action execution is not necessarily fixed and externally controlled (e.g., by the Prefrontal Cortex), but may depend, more autonomously to the PPC circuitry, on what the agent believes are the causes of the environment at every instant of the task, and on a locally defined functional that computes an intention selecting signal.

We exemplified the approach on a multi-step delayed reaching task performed by an agent endowed with visual, proprioceptive, and tactile sensors. Since the second step of the task - going back to the HB - depends on having touched the target, we defined a recursive intention-switching function that depends on its belief. To construct the functional we exploited the asymptotic limit-conditions of the typical *sigmoid* neural transfer function which allows to recursively compute a binary state and we showed that it can indeed obtain stable binary transitions. Notably, the two-state functional could be easily extended to a more general locally computed multi-state function in which the triggering of one state enables the activation of a successive one.

In conclusion, the critical role of our solution is the computational demonstration that in principle Active Inference in continuous time, and its putative cortical core correlate in the PPC, maintain relevant information and can manage the execution of multi-step actions alone - as the neural data suggests [3,15] - thus freeing critical higher-level cognitive resources, such as the Prefrontal Cortex, to cognitively more demanding tasks. Further neuroscience research should empirically demonstrate the existence of the proposed neural circuitry implementing sequential intention modulation.

References

1. Andersen, R.A.: Encoding of intention and spatial location in the posterior parietal cortex. Cereb. Cortex **5**(5), 457–469 (1995). https://doi.org/10.1093/cercor/5.5.457

2. Baioumy, M., Duckworth, P., Lacerda, B., Hawes, N.: Active inference for integrated state-estimation, control, and learning. arXiv (2020)

3. Baldauf, D., Cui, H., Andersen, R.A.: The posterior parietal cortex encodes in parallel both goals for double-reach sequences. J. Neurosci. **28**(40), 10081–10089 (2008). https://doi.org/10.1523/JNEUROSCI.3423-08.2008

4. Breveglieri, R., Galletti, C., Dal Bò, G., Hadjidimitrakis, K., Fattori, P.: Multiple aspects of neural activity during reaching preparation in the medial posterior parietal area V6A. J. Cogn. Neurosci. **26**(4), 879–895 (2014). https://doi.org/10.1162/jocn_a_00510

5. Cisek, P., Kalaska, J.F.: Neural mechanisms for interacting with a world full of action choices. Annu. Rev. Neurosci. **33**, 269–298 (2010). https://doi.org/10.1146/annurev.neuro.051508.135409

6. Desmurget, M., Epstein, C.M., Turner, R.S., Prablanc, C., Alexander, G.E., Grafton, S.T.: PPC and visually directing reaching to targets. Nat. Ne **2**(6), 563–567 (1999). https://doi.org/10.1038/9219

7. Friston, K.: The free-energy principle: a unified brain theory? Nat. Rev. Neurosci. **11**(2), 127–138 (2010). https://doi.org/10.1038/nrn2787

8. Friston, K.: What is optimal about motor control? Neuron **72**(3), 488–498 (2011). https://doi.org/10.1016/j.neuron.2011.10.018

9. Friston, K.J., Daunizeau, J., Kilner, J., Kiebel, S.J.: Action and behavior: a free-energy formulation. Biol. Cybern. **102**(3), 227–260 (2010). https://doi.org/10.1007/s00422-010-0364-z

10. Friston, K.J., Parr, T., de Vries, B.: The graphical brain: belief propagation and active inference **1**(4), 381–414 (2017). https://doi.org/10.1162/NETNa00013

11. Friston, K.J., Rosch, R., Parr, T., Price, C., Bowman, H.: Deep temporal models and active inference. Neurosci. Biobehav. Rev. **77**, 388–402 (2017). https://doi.org/10.1016/j.neubiorev.2017.04.009

12. Galletti, C., Fattori, P.: The dorsal visual stream revisited: stable circuits or dynamic pathways? Cortex **98**, 203–217 (2018). https://doi.org/10.1016/j.cortex.2017.01.009

13. Gamberini, M., Passarelli, L., Filippini, M., Fattori, P., Galletti, C.: Vision for action: thalamic and cortical inputs to the macaque superior parietal lobule. Brain Struct. Funct. **226**(9), 2951–2966 (2021). https://doi.org/10.1007/s00429-021-02377-7

14. Grimbergen, S., Van Hoof, C., Mohajerin Esfahani, P., Wisse, M.: Active inference for state space models: a tutorial (2019). https://doi.org/10.13140/RG.2.2.23596.10884

15. Li, Y., Cui, H.: Dorsal parietal area 5 encodes immediate reach in sequential arm movements. J. Neurosci. **33**(36), 14455–14465 (2013). https://doi.org/10.1523/JNEUROSCI.1162-13.2013

16. Oliver, G., Lanillos, P., Cheng, G.: Active inference body perception and action for humanoid robots (2019). http://arxiv.org/abs/1906.03022

17. Parr, T., Friston, K.J.: The discrete and continuous brain: from decisions to Movement-And Back Again, no. September, pp. 2319–2347 (2018). https://doi.org/10.1162/NECO

18. Parr, T., Rikhye, R.V., Halassa, M.M., Friston, K.J.: Prefrontal computation as active inference. Cereb. Cortex **30**(2), 682–695 (2020). https://doi.org/10.1093/cercor/bhz118

19. Pio-Lopez, L., Nizard, A., Friston, K., Pezzulo, G.: Active inference and robot control: a case study. J. R. Soc. Interface **13**(122), 20160616 (2016). https://doi.org/10.1098/rsif.2016.0616

20. Priorelli, M., Stoianov, I.P.: Flexible intentions in the posterior parietal cortex: an active inference theory. bioRxiv, pp. 1–36 (2022). https://doi.org/10.1101/2022.04.08.487597

21. Sajid, N., Ball, P.J., Parr, T., Friston, K.J.: Active inference: demystified and compared. Neural Comput. **33**(3), 674–712 (2021). https://doi.org/10.1162/neco_a_01357

22. Sancaktar, C., van Gerven, M.A.J., Lanillos, P.: End-to-End pixel-based deep active inference for body perception and action (2020). https://doi.org/10.1109/icdl-epirob48136.2020.9278105

Learning Generative Models for Active Inference Using Tensor Networks

Samuel T. Wauthier[1(✉)], Bram Vanhecke[2], Tim Verbelen[1], and Bart Dhoedt[1]

[1] IDLab, Department of Information Technology at Ghent University – IMEC,
Technologiepark-Zwijnaarde 126, 9052 Ghent, Belgium
{samuel.wauthier,tim.verbelen,bart.dhoedt}@ugent.be
[2] Faculty of Physics and Faculty of Mathematics, Quantum Optics,
Quantum Nanophysics and Quantum Information, University of Vienna,
Boltzmanngasse 5, 1090 Vienna, Austria
bram.andre.roland.vanhecke@univie.ac.at

Abstract. Active inference provides a general framework for behavior and learning in autonomous agents. It states that an agent will attempt to minimize its variational free energy, defined in terms of beliefs over observations, internal states and policies. Traditionally, every aspect of a discrete active inference model must be specified by hand, i.e. by manually defining the hidden state space structure, as well as the required distributions such as likelihood and transition probabilities. Recently, efforts have been made to learn state space representations automatically from observations using deep neural networks. In this paper, we present a novel approach of learning state spaces using quantum physics-inspired tensor networks. The ability of tensor networks to represent the probabilistic nature of quantum states as well as to reduce large state spaces makes tensor networks a natural candidate for active inference. We show how tensor networks can be used as a generative model for sequential data. Furthermore, we show how one can obtain beliefs from such a generative model and how an active inference agent can use these to compute the expected free energy. Finally, we demonstrate our method on the classic T-maze environment.

Keywords: Active inference · Tensor networks · Generative modeling

1 Introduction

Active inference is a theory of behavior and learning in autonomous agents [5]. An active inference agent selects actions based on beliefs about the environment in an attempt to minimize its variational free energy. As a result, the agent will try to reach its preferred state and minimize its uncertainty about the environment at the same time.

This scheme assumes that the agent possesses an internal model of the world and that it updates this model when new information, in the form of observations, becomes available. In current implementations, certain components of

© The Author(s), under exclusive license to Springer Nature Switzerland AG 2023
C. L. Buckley et al. (Eds.): IWAI 2022, CCIS 1721, pp. 285–297, 2023.
https://doi.org/10.1007/978-3-031-28719-0_20

the model must be specified by hand. For example, the hidden space structure, as well as transition dynamics and likelihood, must be manually defined. Deep active inference models deal with this problem by learning these parts of the model through neural networks [2,13].

Tensor networks, as opposed to neural networks, are networks constructed out of contractions between tensors. In recent years, tensor networks have found their place within the field of artificial intelligence. More specifically, Stoudenmire and Schwab [12] showed that it is possible to train these networks in an unsupervised manner to classify images from the MNIST handwritten digits dataset [8]. Importantly, tensor networks have shown to be valuable tools for generative modeling. Han et al. [6] and Cheng et al. [3] used tensor networks for generative modeling of the MNIST dataset, while Mencia Uranga and Lamacraft [9] used a tensor network to model raw audio.

Tensor networks, which were originally developed to represent quantum states in many-body quantum physics, are a natural candidate for generative models for two reasons. Firstly, they were developed in order to deal with the curse of dimensionality in quantum systems, where the dimensionality of the Hilbert space grows exponentially with the number of particles. Secondly, they are used to represent quantum states and are, therefore, inherently capable of representing uncertainty, or, in the case of active inference, beliefs. For example, contrary to neural networks, tensor networks do not require specifying a probability distribution for the hidden state variables or output variables. Furthermore, tensor networks can be exactly mapped to quantum circuits, which is important for the future of quantum machine learning [1].

In this paper, we present a novel approach to learning state spaces, likelihood and transition dynamics using the aforementioned tensor networks. We show that tensor networks are able to represent generative models of sequential data and how beliefs (i.e. probabilities) about observations naturally roll out of the model. Furthermore, we show how to compute the expected free energy for such a model using the sophisticated active inference scheme. We demonstrate this using the classic T-maze environment.

Section 2 elaborates on the inner workings of a tensor network and explains how to train such a model. Section 3 explains the environment and how we applied a tensor network for planning with active inference. In Sect. 4, we present and discuss the results of our experiments on the T-maze environment. Finally, in Sect. 5, we summarize our findings and examine future prospects.

2   Generative Modeling with Tensor Networks

A generative model is a statistical model of the joint probability $P(X)$ of a set of variables $X = (X_1, X_2, \ldots, X_n)$. As previously mentioned, quantum states inherently contain uncertainty, i.e. they embody the probability distribution of a measurement of a system. It is natural, then, to represent a generative model as a quantum state. Quantum states can be mathematically described through a wave function $\Psi(x)$ with $x = (x_1, x_2, \ldots, x_n)$ a set of real variables, such that the probability of x is given by the Born rule:

$$P(X = x) = \frac{|\Psi(x)|^2}{Z}, \tag{1}$$

with $Z = \sum_{\{x\}} |\Psi(x)|^2$, where the summation runs over all possible realizations of the values of x.

Recent work [10] has pointed out that quantum states can be efficiently parameterized using tensor networks. The simplest form of tensor network is the matrix product state (MPS), also known as a tensor train [11]. When representing a quantum state as an MPS, the wave function can be parameterized as follows:

$$\Psi(x) = \sum_{\alpha_1} \sum_{\alpha_2} \sum_{\alpha_3} \cdots \sum_{\alpha_{n-1}} A^{(1)}_{\alpha_1}(x_1) A^{(2)}_{\alpha_1 \alpha_2}(x_2) A^{(3)}_{\alpha_2 \alpha_3}(x_3) \cdots A^{(n)}_{\alpha_{n-1}}(x_n). \tag{2}$$

Here, each $A^{(i)}_{\alpha_{i-1}\alpha_i}(x_i)$ denotes a tensor of rank 2 (aside from the tensors on the extremities which are rank 1) which depends on the input variable x_i. This way, the wave function $\Psi(x)$ is decomposed into a series of tensors $A^{(i)}$.

Importantly, each possible value of an input variable x_i must be associated with a vector of unit norm [12]. That is, each value that x_i can assume must be represented by a vector in a higher-dimensional feature space. Furthermore, to allow for a generative interpretation of the model, the feature vectors should be orthogonal [12]. This means that the vectors associated to the possible values of x_i will form an orthonormal basis of the aforementioned feature space. For a variable which can assume n discrete values, this feature space will be n-dimensional. The dimensionality of the space is referred to as the local dimension.

The unit norm and orthogonality conditions can be satisfied by defining a feature map $\phi^{(i)}(x_i)$, which maps each value onto a vector. For example, if $x_i \in \{0, 1, 2\}$, a possible feature map is the one-hot encoding of each value: $(1, 0, 0)$ for 0, $(0, 1, 0)$ for 1, and $(0, 0, 1)$ for 2. The feature map $\phi^{(i)}(x_i)$ allows us to rewrite the $A^{(i)}(x_i)$ in Eq. 2 as

$$A^{(i)}_{\alpha_{i-1}\alpha_i}(x_i) = \sum_{\beta_i} T^{(i)}_{\alpha_{i-1}\beta_i\alpha_i} \phi^{(i)}_{\beta_i}(x_i), \tag{3}$$

where $T^{(i)}_{\alpha_{i-1}\beta_i\alpha_i}$ is a tensor of rank 3. Here, we have further decomposed $A^{(i)}$ into a contraction of tensor $T^{(i)}$ and the feature vector $\phi^{(i)}(x_i)$. In graphical notation, the MPS (cf. Eq. 2) becomes:

Given a data set, an MPS can be trained using a method based on the density matrix renormalization group (DMRG) algorithm [12]. This algorithm updates model parameters with respect to a given loss function by "sweeping" back and forth across the MPS. In our case, we maximize the negative log-likelihood

(NLL), i.e. we maximize the model evidence directly [6]. After training, the tensor network can be used to infer probability densities over unobserved variables by contracting the MPS with observed values. For a more in-depth discussion on tensor networks, we refer to the Appendix.

3   Environment

The environment used to test the generative model is the classic T-maze as presented by Friston et al. [5]. As the name suggests, the environment consists of a T-shaped maze and contains an artificial agent, e.g. a rat, and a reward, e.g. some cheese. The maze is divided into four locations: the center, the right branch, the left branch, and the cue location. The agent starts off in the center, while the reward is placed in either the left branch or the right branch, as depicted in Fig. 1. Crucially, the agent does not know the location of the reward initially. Furthermore, once the agent chooses to go to either the left or the right branch, it is trapped and cannot leave. For the agent, the initial state of the world is uncertain. It does not know which type of world it is in: a world with the reward on the left, or a world with the reward on the right. In other words, it does not know its context. However, going to the cue reveals the location of the reward and enables the agent to observe the context. This resolves all ambiguity and allows the agent to make the correct decision.

The implementation of the environment was provided by pymdp [7]. In this package, the agent receives an observation with three modalities at every step: the location, the reward, and the context. The location can take on four possible values: center, right, left, and cue, and indicates which location the agent is currently in. The reward can take on three possible values: no reward, win, and loss. The "no reward" observation indicates that the agent received no information about the reward, while the "win" and "loss" observations indicate that the agent either received the reward or failed to obtain the reward, respectively. Logically, "no reward" can only be observed in the center and cue locations, while "win" and "loss" can only be observed in the left and right locations. Finally, the context can take on two possible values: left and right. Whenever

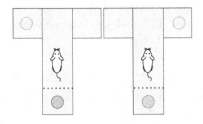

Fig. 1. Possible contexts of the T-maze environment as presented by Friston et al. [5]. The agent starts off in the center. The reward (yellow) is located in either the left branch (left image) or right branch (right image). The cue reveals where the reward is: red indicates the reward is on the left, green indicates the reward is on the right. (Color figure online)

the agent is in locations "center", "left" or "right", the context observation will be randomly selected from "left" or "right" uniformly. Only when the agent is in the cue location, will the context observation yield the correct context. Further, the possible actions that the agent can take include: center, right, left, and cue, corresponding to which location the agent wants to go to.

We modified the above implementation slightly to better reflect the environment brought forth by Friston et al. [5]. In the original description, the agent is only allowed to perform two subsequent actions. Therefore, the number of time steps was limited to two. Furthermore, in the above implementation, the agent is able to leave the left and right branches of the T-maze. Thus, we prevented the agent from leaving whenever it chose to go to the left or right branches.

3.1 Modeling with Tensor Networks

The tensor network was adapted in order to accommodate the environment and be able to receive sequences of actions and observations as input. Firstly, the number of tensors was limited to the number of time steps. Secondly, each tensor received an extra index so that the network may receive both actions and observations. This led to the following network structure:

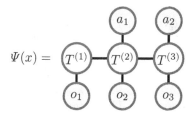

where we used a_i and o_i to denote the feature vectors corresponding to action a_i and observation o_i. Note that the first tensor did not receive an action input, since we defined that action a_i leads to observation o_{i+1}.

As mentioned in Sect. 2, the feature maps $\phi^{(i)}$ can generally be freely chosen as long as the resulting vectors are correctly normalized. However, it is useful to select feature maps which can easily be inverted, such that feature space vectors can readily be interpreted. In this case, since both observations and actions were discrete, one-hot encodings form a good option. The feature vectors for actions were one-hot encodings of the possible actions. The feature vectors for observations were one-hot encodings of the different combinations of observation modalities, i.e. $4 \times 3 \times 2 = 24$ one-hot vectors corresponding to different combinations of the three different modalities.

In principle, there is nothing stopping us from learning feature maps, as long as the maps are correctly normalized. For practical purposes, the learning algorithm should make sure the feature maps are invertible. Whether feature maps should be learned before training the model or can be learned on-the-fly is an open question.

At this point, it is important to mention that the feature map is not chosen with the intent of imposing structure on the feature space based on prior knowledge of the observations (or actions). On the contrary, any feature map should assign a separate feature dimension to each possible observation, keeping the distance between that observation and all other observations within the feature space equal and maximal. To this end, the feature map can be thought of as being a part of the likelihood.

3.2   Planning with Active Inference

Once trained, the tensor network constructed in Sect. 3.1 provides a generative model of the world. In theory, this model can be used for planning with active inference. At this point, it is important to remark that the network does not provide an accessible hidden state. While the bonds between tensors can be regarded as internal states, they are not normalized and, therefore, not usable. This poses a problem in the computation of the expected free energy, given by [5]

$$G(\pi) = \sum_\tau G(\pi, \tau) \tag{4}$$

$$G(\pi, \tau) = E_{Q(o_\tau|s_\tau, \pi)}[\log Q(s_\tau|\pi) - \log P(s_\tau, o_\tau|\tilde{o}, \pi)] \tag{5}$$

$$\approx \underbrace{D_{\mathrm{KL}}(Q(o_\tau|\pi) \,\|\, P(o_\tau))}_{\text{expected cost}} + \underbrace{E_{Q(s_\tau|\pi)}[\mathrm{H}(Q(o_\tau|s_\tau, \pi))]}_{\text{expected ambiguity}}, \tag{6}$$

with hidden states s_τ, observations o_τ and policy $\pi(\tau) = a_\tau$, where the $\sim$-notation denotes a sequence of variables $\tilde{x} = (x_1, x_2, \ldots, x_{\tau-1})$ over time and $P(o_\tau)$ is the expected observation. This computation requires access to the hidden state s_τ explicitly.

To remedy this, we suppose that the state s_τ contains all the information gathered across actions and observations that occur at times $t < \tau$. Mathematically, we assume $Q(s_\tau|\pi) \approx Q(o_{<\tau}|\pi)$ and $Q(o_\tau|s_\tau, \pi) \approx Q(o_\tau|o_{<\tau}, \pi)$ with $o_{<\tau} = (o_1, \ldots, o_{(\tau-1)})$. This way, we are able to approximate the expected ambiguity in Eq. 6. While these assumptions may give the impression that the calculation is computationally expensive, if the contraction with previous actions and observations has been performed once, it never has to be computed again, since the resulting tensor can be reused in subsequent calculations. At this point, the resulting tensor contains all the information from previous actions and observations.

When planning, we must re-evaluate the likelihood (and thus the expected free energy) for every time step, while imposing that the previous time steps are fixed. Indeed, we will perform sophisticated inference [4]. Under this scheme, the expected free energy is given by

$$G(o_\tau, a_\tau) = \underbrace{D_{KL}(Q(o_{\tau+1}|a_{<\tau+1}) \,||\, P(o_{\tau+1})) + E_{Q(s_{\tau+1}|a_{<\tau+1})}[H(P(o_{\tau+1}|s_{\tau+1}))]}_{\text{expected free energy of next action}}$$

$$+ \underbrace{E_{Q(a_{\tau+1}|o_{\tau+1})Q(o_{\tau+1}|a_{<\tau+1})}[G(o_{\tau+1}, a_{\tau+1})]}_{\text{expected free energy of subsequent actions}}, \tag{7}$$

$$Q(a_\tau|o_\tau) = \sigma[-G(o_\tau, a_\tau)] \tag{8}$$

This defines a tree search over actions and observations in the future.

4   Results and Discussion

In this section, we demonstrate how the model's beliefs shift over time. Later, we show how a tensor network agent behaves when planning under sophisticated inference.

The data set was constructed by including one of every possible path through the maze, i.e. 202 sequences of actions and observations. The model was trained over 500 epochs with a batch size of 10, where one epoch consisted of one right-to-left-to-right sweep per batch. The learning rate was set to 10^{-4} and was further reduced by 10% whenever the loss increased too much (i.e. by more than 0.5). Additionally, bonds started with 8 dimensions. The singular value cutoff point was set to 10% of the largest singular value.

4.1   Belief Shift

Since the initial observation o_1 is always center position, no reward and context right or left with 50% chance, we used the observation "center, no reward and context right" to obtain the beliefs in each case. The results are analogous in the case of "center, no reward and context left".

Figure 2 (top) displays beliefs over o_2 given a_1. From the results, it is clear that the agent does not know which reward it will receive, if it were to go to the left or right branch immediately. In addition, it does not know which context it will observe, even if the agent were to go towards the cue. Once the agent observes o_2, the beliefs shift. Figure 2 (bottom) shows beliefs over o_3 given a_2 when the agent has seen the cue with context "right". Since the agent has seen the cue, it is very certain about the reward it will receive if it goes to the left or the right branch. If it stays in the cue location, it is also very certain that it will observe the same context again.

4.2   Action Selection

With the outcome in Sect. 4.1, we were able to perform action selection based on the sophisticated inference scheme described in Sect. 3.2. For this, we used the following preferred observation per modality:

$$P(\text{position}) = \sigma([0\;0\;0\;0]), \quad P(\text{reward}) = \sigma([0\;3\;-3]), \quad P(\text{context}) = \sigma([0\;0]). \tag{9}$$

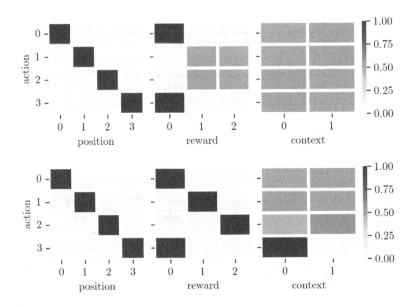

Fig. 2. (top) Model beliefs over observation o_2 given action a_1 per modality. (bottom) Model beliefs for observation o_3 given action a_2 per modality, when the agent has observed the cue with context "right". Actions 0, 1, 2 and 3 correspond to center, right, left and cue, respectively. Positions 0, 1, 2 and 3 correspond to center, right, left and cue location, respectively. Rewards 0, 1 and 2 correspond to no reward, win and loss, respectively. Context 0 and 1 correspond to right and left, respectively.

Figure 3 (top) shows the expected free energy for action a_1. Given that the expected free energy is lowest for the action that brings the agent to the cue, the agent will choose to go to the cue in the first action. This is because, after observing the cue, the cue location provides a lower entropy on the context modality, as well as virtually 100% certainty on where the reward is located.

Figure 3 (bottom) shows the expected free energy for action a_2 when the agent has chosen to go to the cue location and has observed either "cue, no reward, context right" or "cue, no reward, context left". In this case, the agent will choose to go to either the left or the right branch, depending on the context it observed, i.e. context right will lead to action right and vice-versa.

The net result is that the agent will first go to the cue in order to resolve ambiguity and, subsequently, go to the branch with the reward.

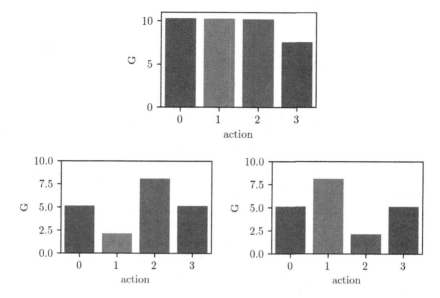

Fig. 3. (top) Expected free energy for action a_1. (bottom) Expected free energy for action a_2 when observation o_2 was "cue, no reward, context right" and "cue, no reward, context left", respectively.

5   Conclusion

We introduced a generative model based on tensor networks that is able to learn from sequential data. In addition, we showed how one can obtain beliefs from such a generative model and how a (sophisticated) active inference agent can use these to compute the expected free energy. Demonstration on the T-maze environment pointed out that such an agent is able to correctly select actions.

In the future, we plan to apply tensor networks to other environments, as well as make an in-depth comparison with neural networks, in order to better establish the benefits and drawbacks of the method. Moreover, we will adapt the network to allow sequences of random lengths and look into incorporating observations with continuous variables, which may also allow us to undo the assumptions made in Sect. 3.2. Both these changes will broaden the range of applicability of generative models based on tensor networks.

Acknowledgments. This research received funding from the Flemish Government under the "Onderzoeksprogramma Artificiële Intelligentie (AI) Vlaanderen" pro-gramme. This work has received support from the European Union's Horizon 2020 program through Grant No. 863476 (ERC-CoG SEQUAM).

Appendix 1   Notes on Tensor Networks

The summation over a common index as in Eq. 2 is also called a contraction. Performing the contraction between two tensors yields a new tensor with a rank equal to the sum of the ranks of the two contracted tensors minus two times the number of indices contracted over. That is, contracting two tensors of rank 2 over a single index gives a new tensor of rank 2, which is simply matrix multiplication: $\sum_j A_{ij} B_{jk}$. Similarly, contracting over the indices of a single tensor of rank 2 is simply the trace: $\sum_i A_{ii}$. In this sense, contraction is a generalization of these operations. Contracting indices in different ways gives rise to different structures. Examples of other possible networks are: tree tensor networks (TTN) and projected entangled pair states (PEPS).

Tensor networks can more easily be understood using their graphical notation. Each tensor is represented by a node, while contractions are represented by edges. Free edges, i.e. edges which do not connect two tensors, correspond to free indices which have not been summed. These can be used to represent sites in the network which are able to receive input or which can be used as input.

Some examples of tensors in graphical notation are:

- vector   ,
- matrix   ,
- rank-3 tensor   .

Some examples of operations that can be represented by contractions are:

- dot product   ,
- matrix multiplication   ,
- trace   .

For a detailed account on tensor networks and their graphical notation, please refer to [1].

Appendix 2   Training

The loss function must be chosen in such a way that the model captures the probability distribution of the data [6]. A straightforward method for estimating the parameters of a probability distribution is maximum likelihood estimation. In machine learning terms, this means we will optimize the parameters of the model with respect to the negative log-likelihood (NLL):

$$\mathcal{L} = \frac{1}{|D|} \sum_{x \in D} \log P(x), \tag{10}$$

where D denotes the data set. Through NLL minimization, the generative model becomes more similar to the probability distribution of the data.

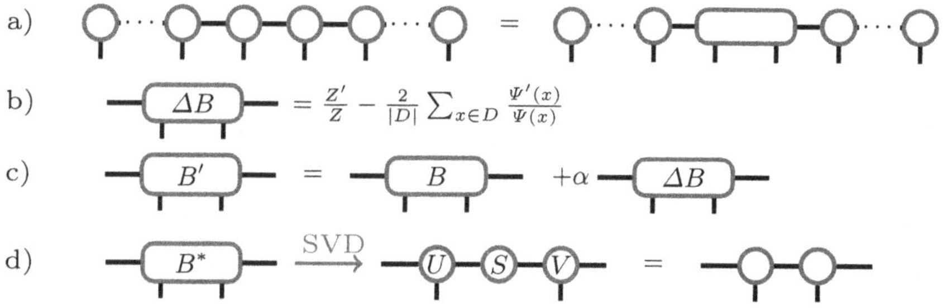

Fig. 4. Training scheme for an MPS. a) Contraction of two adjacent tensors. b) Computing the update to the contracted tensor. c) Updating the contracted tensor. d) Decomposition of the contracted tensor using SVD.

Training proceeds as depicted in Fig. 4. Firstly, tensor $T^{(i)}$ and $T^{(i+1)}$ are contracted to form the tensor $B^{(i,i+1)}$. The update to $B^{(i,i+1)}$ is then computed using the loss function:

$$\Delta B^{(i,i+1)} = \frac{\partial \mathcal{L}}{\partial B^{(i,i+1)}_{\alpha_{i-1}\beta_i\beta_{i+1}\alpha_{i+1}}} = \frac{Z'}{Z} - \frac{2}{|D|}\sum_{x\in D}\frac{\Psi'(x)}{\Psi(x)}, \qquad (11)$$

where $Z' = 2\sum_{x\in D}\Psi'(x)\Psi(x)$ and Ψ' is the derivative of Ψ with respect to $B^{(i,i+1)}$. Subsequently, the elements of $B^{(i,i+1)}$ are adjusted by adding $\Delta B^{(i,i+1)}$ multiplied by the learning rate. Finally, the newly computed $B'^{(i,i+1)}$ is decomposed into two tensors again. This decomposition is typically done through singular value decomposition (SVD), where the singular value matrix is then contracted with either the left or the right tensor, such that we are left with two tensors.

By starting this scheme at the leftmost tensor and iteratively moving one tensor to the right, the algorithm can update the entire MPS. Indeed, it is possible to update one tensor $T^{(i)}_{\alpha_{i-1}\beta_i\alpha_i}$ at a time, however, the current method allows the dimensions of the indices α_i (graphically, the edges connecting $T^{(i)}$ nodes), the so-called bond dimensions, to vary during training. This is made possible by truncated SVD, which truncates dimensions with singular values that fall beneath some manually specified threshold. Truncating dimensions with small singular values can be interpreted as truncating less informative dimensions. As a result, truncated SVD ensures that the model remains as small as possible, while containing the most information. Moreover, the size of the model will vary depending on how much information it must learn.

Appendix 3   Computing Probabilities with Tensor Networks

One of the benefits of tensor networks is that we can easily obtain exact joint (and conditional) probability distributions without requiring parameterization.

After training the model, given that the network is correctly normalized ($Z = 1$), the joint probability distribution is given by

$$P(x_1, x_2, x_3, \ldots, x_n) =$$

where we have omitted tensor labels for simplicity. Marginal distributions can be found by discarding the offending variables:

$$P(x_1) = \sum_{\{x\}\setminus\{x_1\}} P(x_1, x_2, x_3, \ldots, x_n) =$$

where the sum over the variables $x_2, x_3, \ldots, x_n$ is equivalent to contracting the matching tensors. Finally, conditional probability distributions can be found by combining the previous results:

$$P(x_2, x_3, \ldots, x_n | x_1) = \frac{P(x_1, x_2, x_3, \ldots, x_n)}{P(x_1)}. \tag{12}$$

Appendix 4   Physical Intuition

In order to garner a feeling for the physics and mathematics used throughout this work, this section describes in (mostly) words what the physical meaning of the constituents of the tensor network is.

Let x be an observable, i.e. a quantity that can be measured (or observed). Examples of such physical quantities are position and momentum. Furthermore, let $\{0, 1, 2\}$ be the set of values that x can assume.

In quantum mechanics, the set of values that an observable can assume are eigenvalues. This entails that there is a set of eigenvectors corresponding to these eigenvalues. In turn, the eigenvectors form an eigenbasis of the state space. In other words, every value that x can assume has a corresponding vector and each of these vectors is a basis vector of the state space, meaning that each different value is represented in the state space by a different dimension. For example, we may have the vectors $(1, 0, 0)$ for 0, $(0, 1, 0)$ for 1, and $(0, 0, 1)$ for 2.

Further, an MPS is typically used to represent a state within the state space. This means that the MPS represents a (multi)vector. This vector is not necessarily one of the eigenvectors mentioned earlier, but it can be virtually any vector within the state space. When learning the parameters of an MPS, it is rotated and stretched in such a way that it represents the data.

If an MPS does not necessarily coincide with any of the eigenvectors, what value will it produce when a measurement is performed? It will produce any of the eigenvalues with a certain probability. These probabilities are given by the square of the inner product between the MPS and each eigenvector. In our example, say the MPS was represented by the vector $(0.55, 0.55, 0.63)$, we would measure 0 with 30% probability, 1 with 30% probability and 2 with 40% probability.

References

1. The tensor network. https://tensornetwork.org/. Accessed 21 June 2022
2. Çatal, O., Wauthier, S., De Boom, C., Verbelen, T., Dhoedt, B.: Learning generative state space models for active inference. Front. Comput. Neurosci. **14**, 103 (2020). https://doi.org/10.3389/fncom.2020.574372
3. Cheng, S., Wang, L., Xiang, T., Zhang, P.: Tree tensor networks for generative modeling. Phys. Rev. B **99**, 155131 (2019). https://doi.org/10.1103/PhysRevB.99.155131
4. Friston, K., Da Costa, L., Hafner, D., Hesp, C., Parr, T.: Sophisticated inference. Neural Comput. **33**(3), 713–763 (2021). https://doi.org/10.1162/neco_a_01351
5. Friston, K., FitzGerald, T., Rigoli, F., Schwartenbeck, P., O'Doherty, J., Pezzulo, G.: Active inference and learning. Neurosci. Biobehav. Rev. **68**, 862–879 (2016). https://doi.org/10.1016/j.neubiorev.2016.06.022
6. Han, Z.Y., Wang, J., Fan, H., Wang, L., Zhang, P.: Unsupervised generative modeling using matrix product states. Phys. Rev. X **8**, 031012 (2018). https://doi.org/10.1103/PhysRevX.8.031012
7. Heins, C., et al.: pymdp: a python library for active inference in discrete state spaces. J. Open Source Softw. **7**(73), 4098 (2022). https://doi.org/10.21105/joss.04098
8. LeCun, Y., Cortes, C., Burges, C.J.C.: The MNIST database of handwritten digits (1998). http://yann.lecun.com/exdb/mnist/. Accessed 10 June 2022
9. Uranga, B.M., Lamacraft, A.: Schrödingerrnn: Generative modeling of raw audio as a continuously observed quantum state. In: Lu, J., Ward, R. (eds.) Proceedings of the First Mathematical and Scientific Machine Learning Conference. Proceedings of Machine Learning Research, vol. 107, pp. 74–106. PMLR, 20–24 July 2020. http://proceedings.mlr.press/v107/mencia-uranga20a.html
10. Orús, R.: Tensor networks for complex quantum systems. Nature Rev. Phys. **1**(9), 538–550 (2019). https://doi.org/10.1038/s42254-019-0086-7
11. Perez-Garcia, D., Verstraete, F., Wolf, M.M., Cirac, J.I.: Matrix product state representations. Quantum Info. Comput. **7**(5), 401–430 (2007)
12. Stoudenmire, E., Schwab, D.J.: Supervised learning with tensor networks. In: Lee, D., Sugiyama, M., Luxburg, U., Guyon, I., Garnett, R. (eds.) Advances in Neural Information Processing Systems, vol. 29. Curran Associates, Inc. (2016). https://proceedings.neurips.cc/paper/2016/file/5314b9674c86e3f9d1ba25ef9bb32895-Paper.pdf
13. Ueltzhöffer, K.: Deep active inference. Biol. Cybern. **112**(6), 547–573 (2018). https://doi.org/10.1007/s00422-018-0785-7

A Worked Example of the Bayesian Mechanics of Classical Objects

Dalton A. R. Sakthivadivel[1,2,3]($\boxtimes$) (iD)

[1] VERSES Research Lab, Los Angeles, USA
dalton.sakthivadivel@verses.ai
[2] Department of Mathematics, Stony Brook University, Stony Brook, USA
[3] Department of Physics and Astronomy, Stony Brook University, Stony Brook, USA
https://darsakthi.github.io

Abstract. Bayesian mechanics is a new approach to studying the mathematics and physics of interacting stochastic processes. We provide a worked example of a physical mechanics for classical objects, which derives from a simple application thereof. We summarise the current state of the art of Bayesian mechanics in doing so.

Keywords: Free energy principle · Bayesian mechanics · Classical physics · Maximum calibre

1 Introduction

Under the free energy principle [15], Bayesian mechanics is a new approach to studying the mathematics and physics of interacting stochastic processes. In essence, Bayesian mechanics is a particular sort of mathematical physics for coupled random dynamical systems, providing a mechanical theory for how the statistical properties of physical systems change in a space of Bayesian beliefs, based on how their physical properties change in space and time [14,29,32]. Bayesian mechanics is particularly interested in systems with some notion of regularity, termed 'self-evidencing' systems in previous literature.[1] A self-evidencing system occupies an attractor in the state space, and has some set of stereotypical behaviours definitional of the sort of system it is. The key deliverable of Bayesian mechanics is that a random dynamical system is an estimator for statistics or parameters of another system to which it is coupled, and that we can understand attractors and phases in state space in virtue of these belief dynamics.

Albeit conceptually powerful, the use of the free energy principle (FEP) and Bayesian mechanics to describe specific physical systems is rarely codified in

---

[1] This term of art originates in neuroscience [25], where the free energy principle has its origins, as an attempt to explain the physics and philosophy of learning in the human cortex—viewed as a Bayesian mechanical problem, a brain learning about an environment is one random dynamical system performing inference about another, with an aim towards the attractor states characteristic of allostasis [42].

C. L. Buckley et al. (Eds.): IWAI 2022, CCIS 1721, pp. 298–318, 2023.
https://doi.org/10.1007/978-3-031-28719-0_21

the literature. There are examples where it reproduces known algorithms like various types of control [10,26]), as well as simple dissipative systems [1] and more complex systems exhibiting Lorentzian chaos [17], but recent work has treated it as a purely formal position that systems constrain themselves to fall within acceptable regimes of certain existential variables, thus inducing such attractor structures in the state space.

More specifically, in [32,35], Bayesian mechanics is introduced as the mechanics of beliefs—but it is challenging to determine the physical mechanics of systems carrying those beliefs. This would require solving difficult PDEs for non-equilibrium steady state densities in general, or else, equations of motion for internal states on the synchronous statistical manifold. Likewise, it is often claimed that the FEP is as simple and general as the principle of stationary action [16], and it can be sketched out how the Bayesian mechanics of internal states of classical objects might look [15]. Nevertheless, there has yet to be a systematic investigation of even the Bayesian mechanics of classical physics, despite it being readily available due to recent formulations as a least action principle.

Here, inspired by remarks in [32], we give a worked example of the classical mechanics of objects with trivial (e.g., infinitely precise) belief dynamics—in a fairly direct sense, the simplest case of Bayesian mechanics [35]. In doing so we provide a general formulation of classical physics for the working Bayesian mechanic. We hope this will ground future discussions of the free energy principle even more solidly in conventional mathematics and physics.

2   Mechanics

2.1   Classical Physics in One Dimension

The mechanics of classical objects are embodied by Newton's law that systems accelerate along force gradients, by precisely the direction and magnitude of the total force applied—no more, and no less. The derivation of this is simple. Let $K = \frac{1}{2}mv_t^2$, where q_t is the position of some point mass at t and its velocity is the time derivative $v_t = \dot{q}_t$, and V be some scalar potential. Now define a path as a particular function $q(t)$ (e.g., a parabolic path could be $q(t) = q_0 - \frac{1}{2}gt^2$, whilst a straight path could be $q(t) = q_0$), and introduce the temporary time variable τ. Taking the action functional

$$S[q(t)] = \int_0^t \frac{1}{2}mv_\tau^2 - V(q_\tau)\mathrm{d}\tau \tag{1}$$

a path of *least* action (or more generally, for which the action is stationary) obeys the equation

$$-\partial_q V = m\partial_t v$$

by standard arguments in functional analysis.[2] This is Newton's second law,

$$F = ma. \tag{2}$$

This logical sequence simply expresses that a path of least action always follows (2)—that is, a system always accelerates along a force gradient, never using extra energy by resisting or compounding that force. For an appropriate specification of forces F, we get various sorts of mechanics, like motion in gravitational fields or classical approximations of fluid flow (also called continuum mechanics). Given mass data, and initial conditions $\dot{q}(0)$ and $q(0)$ (along with other boundary conditions), we can produce dynamical trajectories for some particular system by actually using the law of motion given by Newtonian mechanics under the least action principle.

2.2 A Physics of and by Beliefs

Bayesian mechanics can be seen as an account of the laws of motion deriving from the free energy principle, concerning how Bayesian beliefs—and hence, systems with beliefs—behave under certain determinants of *probabilistic* motion. Much like classical mechanics serves an account of systems that obey Newton's second law by minimising the classical action, Bayesian mechanics is an account of systems that engage in approximate Bayesian inference by minimising surprisal.

The pure physics of the FEP arguably dates back to two landmark papers in the literature, [15] and [29]. In [29], and later in [10], the idea was introduced that the FEP has gestured at a new sort of physics—one about the mechanics of Bayesian beliefs, and how they reflect the behaviour of systems carrying those beliefs. In [32,35] it is discussed that one can understand this in the same sense as classical mechanics arises from the least action principle, or identically, that diffusion arises from the maximisation of entropy, with that least action principle being formulated in detail in [16]. Dually, we can understand our own beliefs about a system modelling its environment—or the system's model of itself—as being ruled by Bayesian mechanics, under the observation and updating rules which are a consequence thereof. A full deconstruction is given in [32].

What is native to Bayesian mechanics? Beginning with the most recent formulation in [16], the FEP is nothing but the least action principle applied to some surprisal $S = -\ln\{p(-)\}$, where the application of this 'least surprisal principle' to specific objects determines the mechanical theory about those objects. Let a stochastic process X under $p(x, t)$ with sample paths γ be described by the Itō stochastic differential equation

$$\mathrm{d}X_t = f(X_t, t)\mathrm{d}t + \sqrt{2D}\mathrm{d}W_t.$$

Here, X_t is a random variable at t and $f(X_t, t)$ is a vector field yielding the drift at any state X_t, which may change over time. Let $\omega_t = v_t - f(X_t, t)$ be a

---

[2] See [32] and references therein for an overview; alternatively, see [8] for a more advanced pedagogical treatment, and [4,18] for mathematically sophisticated resources.

fluctuation of any realisation of this flow at a time t. Note that a realisation $x(t)$ is nothing but a sample path γ, and so, we have

$$\dot{\gamma}_t - \mathbb{E}_{p(\gamma)}[\gamma]_t \qquad (3)$$

for ω_t. A quadratic[3] form $\mathcal{L}(\omega_t)$ can naturally be defined on the tangent space to the state space, such that the surprisal is its integral along a given path γ of $\mathcal{L}$:

$$S[\gamma] = \int_0^t \frac{1}{4D} \langle \omega_\tau, \omega_\tau \rangle d\tau.$$

The surprisal of a path is then proportional to half its accumulated deviation from the expected flow $f(X_t, t)$, with D scaled out; morally, this is in the same sense as the classical action is proportional to half the accumulated deviation of motion from a potential well [32, Section 2]. That this action equals the surprisal of a path γ for a given initial condition, $p(x(t) \mid x_0)$, is a simple consequence of the path probability measure being defined as

$$p(x(t) \mid x_0) = \exp\{-\lambda S[\gamma]\} \qquad (4)$$

in [36], which is indeed the canonical definition of such an object in any abstract Wiener space [28,40], and is the point of attack in [16]. Such a 'path-dependant surprisal' is referred to as the stochastic entropy by [36], and is deeply related to statistics on the path space of a random walk (see [12], as well as [27] and related work on logarithmic heat kernels[4]). This definition of the action is consistent with the quadratic form $\frac{1}{2}\langle v_t, v_t \rangle$ defined in classical mechanics. The action generates the Fokker-Planck equation for the probability of a state x at t,

$$\partial_t p(x, t) = -\partial_x [f(x, t)p(x, t)] + D\partial_{xx} p(x, t),$$

by giving rise to a probability density over coordinate and time pairs.

Define two random dynamical systems η and μ. In virtue of (3), the action—and thus, ultimately, the surprisal—is parameterised by some modal or expected path. Suppose that η and μ are coupled by some function σ, that one has an additional random dynamical system b capturing the interactions between the two, and that conditional expectations $\hat{\mu}_{b,t} = \mathbb{E}_{p(\mu_t \mid b_t)}[\mu_t]$ and $\hat{\eta}_{b,t} = \mathbb{E}_{p(\eta_t \mid b_t)}[\eta_t]$ exist; moreover, assume the statistics of the two processes can be distinguished, in the sense of being independent conditioned on b_t.[5] By construction, $\sigma(\hat{\mu}_{b,t}) = \hat{\eta}_{b,t}$. It is immediate that μ is an estimator for the statistics of η [35]. That is to say, in the case of random systems whose physical dynamics are coupled, these statistical quantities are also coupled, in a way that can be interpreted

---

[3] Note that in the Stratonovich convention, more amenable to calculus on manifolds, there is an additional term in the Lagrangian $\mathcal{L}$ indicated; see e.g. equation 15 here: [9].

[4] The author thanks Robert W Neel for suggesting this point of discussion.

[5] Note that this framework degenerates in the case where σ is the identity, but that this case is vacuous, since it assumes η is identical to μ.

as the systems performing inference over each other. Systems that minimise surprisal given a control parameter $\sigma^{-1}(\hat{\eta}_{b,t})$ are particularly good models of an environment, whilst systems that fluctuate with high probability are not.

This is referred to as approximate Bayesian inference, and in particular, is referred to as 'mode-matching' when these parameters are stationary: by minimising surprisal, the most likely state is $\sigma^{-1}(\hat{\eta}_{b,t})$. The *a priori* assumption that systems minimise surprising events can be justified using large deviation principles [39], and as such, most any set of coupled random dynamical systems can be expected to perform approximate Bayesian inference of some sort. What is more striking is that two distinct (in the above sense) systems which estimate each other's statistics necessarily come equipped with a pair (b_t, σ), chosen such that they fluctuate around each other's most likely states.

Bayesian mechanics formulates changes in physical states as changes in probabilities estimated by η and μ. Since we can understand the average state of μ as the preimage of σ, we can understand it as a parameter for the probabilities of states of η—in a sense we do this automatically, since we can understand a system μ existing a particular way in virtue of the likely states η of the things interacting with it, causing or not causing particular μ—and in so doing, we can relate μ to a belief mechanics by thinking of μ as inferring (read: encoding inferences about) the assignment of probabilities to states of η. In simpler terms, systems exist in a particular way based on their environment. When we model a system, we automatically model it as modelling its environment by encoding this sort of statistical estimation in our model of the system. We expect a thing to exist as a particular 'thing' based on whether or not it *can* exist that way in a given environment. This places constraints on what $\sigma^{-1}(\hat{\eta}_{b,t})$ must be for a system to be 'system-like' (e.g., stone-like, human-like, and so forth); dually, it informs what $\sigma(\hat{\mu}_{b,t})$ must be, given we have a particular system (humans require oxygen to breathe and cannot live beneath water). In both cases we have a sort of allostatic attractor characteristic of the system. Besides the explicitly non-teleological notion of the constraints or intended preimage which are definitional of a system, an important dual observation is that this parameterisation of likely internal and external states is one reading of Bayesian mechanics which is consistent with the idea of perception or estimation in Bayesian inference.

Bayesian mechanics leads to various types of approximate Bayesian inference, just like classical mechanics admits different applications of Newton's laws of motion (e.g., the continuum mechanics of fluid flow, or orbital mechanics for satellite motion). As stated, when that parameter is trivial, we have mode-matching; when it is dynamic, this is referred to as mode-tracking [32]. When applied to beliefs about the trajectories of external and active states, we have a more general version of Bayesian mechanics only recently explored, including active inference [7]. This has been referred to as 'path-tracking' in [32]. The taxonomy described here exists in the same sense as minimising the action of the Lagrangian $\mathcal{L} = K - V$ yields Newton's second law of motion, which can be applied to various sorts of systems when we know what sort of function V is.

We will work out in some detail what this taxonomy looks like in the world of classical physics.

In summary, Bayesian mechanics contains two key pieces of data: the surprisal Lagrangian, and the synchrony map σ. These data define beliefs and belief dynamics, respectively, and pair them with a characteristic geometry (information geometry [2], the study of *statistical manifolds*, or so-called 'spaces of beliefs'). It is interesting that most physical theories are paired with a characteristic geometry in which they play out [6], such as symplectic geometry in classical physics [5]. Later in the paper we will point out the appearance of symplectic forms in Bayesian mechanics, which is notable given the analogy we draw.

3   A General Equation for Bayesian Classical Mechanics

We begin with a classical particle described by a position variable, q, at some time t. The mass of the particle will be denoted by m. The position plays the role of an internal state for the particle. The particle has an external environment interacting with it, whose states η determine what forces are acting on it. Let F_t be the total force applied to the system at a time t, $\sum_i F_{i,t}$. The reception of an applied force is like a blanket state for the particle, which can couple to and affect internal states. The attentive reader will likely have noted that a force is not a state of the particle, but is an interaction of the environment with the states of the particle. It is true that we have technically committed a type error by identifying forces with forced states. The proper construction would involve some sort of object that captures and instantiates that interaction, an extension of the sensory states of a Markov blanket like that of [13]. This is a subtlety we neglect in these results.

This being stated, we consider F_t itself to be like a sensory state of the particle. As such, we have an inverse synchrony map,

$$\sigma^{-1} : \eta_t \to F_t \to q_t.$$

The former map, η^{-1}, merely sends external states of the world to the forces the world applies on objects, which will be done implicitly throughout the paper, as we provide worked examples with particular applied forces. Let s be a temporary time variable. The latter map, $\mu(F_{\eta,t}) = q_t$, is the solution to an integral equation determined by Newton's second law,

$$F(\eta, t) \mapsto \iint \frac{F(\eta, s)}{m} = q_t.$$

In other words, the particular functional form for the coupling σ we have used is one that sends the average acceleration of the system to the average force applied to the system,

$$\sigma^{-1}(\eta_{F,t}) = q_{F,t}.$$

The ideal path of internal states, which encodes an optimal (i.e., unsurprising) belief about what the system is being told to do by the environment, consists of precisely these q_t, for whom $\sigma(q_{F,t}) = \eta_{F,t}$. Note the consistency with more recent formulations of the FEP: for as long as there exists a particle, there exists some (possibly trivial) blanket distinguishing that particle from its environment along its path of evolution [16, 24, 34]. We can show that such a blanket exists— that any classical particle under the partition indicated above is sparsely coupled on the time-scale over which it exists—simply by pointing out that we can read off F_t and need not consider η_t to get internal states q_t. Physically, this is the intuitive statement that it is only a force that matters to motion, not what generated that force. Hence, by construction, for as long as a classical particle exists to be acted on, it has a blanket.

Just as we presume that the mechanics of beliefs should lead to physical mechanics (control), minimisation of the surprisal should lead to physical mechanics (Newton's laws of motion) and trajectories that abide by those mechanics. In establishing that physical mechanics follows from Bayesian mechanics, we first focus on beliefs *about* internal states, and relate them to the beliefs *carried by* internal states afterwards. The surprisal Lagrangian, $-\ln\{p(-)\}$, is applied either to $p(q_t)$ given F_t, or is applied symmetrically to the probability of $p(\eta_t)$ given F_t. We would like to see whether the minimisation of surprisal under σ recovers classical physics in the context of Bayesian mechanics, so we try to minimise $-\ln\{p(q_t \mid \eta_{F,t})\}$ under the parameterisation $-\ln\{p(q_t; \sigma^{-1}(\eta_{F,t}))\}$. Here, we take a moment to note that dualising the object in the Lagrangian to internal states gives us a physics of *our* beliefs about the system, or the system's beliefs about itself, which is the duality indicated in [35]. Contrast this with the surprisal Lagrangian on external states, which gives a physics of the system's beliefs about its environment, in [16].

Under a noise injection, or else some uncertainty associated to the belief about a position at t, this becomes a problem of mapping means to means, in the sense of the approximate Bayesian inference lemma. We are primarily interested in the probability of deviating from the path of least action, such that our (state-wise) surprisal is a measure of

$$p(q_t; \sigma^{-1}(\hat{\eta}_{F,t})) = p(q_t; \hat{q}_{F,t}).$$

Suppose the acceleration at a given time is constrained such that, on average, it obeys the forces being applied to it at that time. This is an instantaneous picture, licensing a non-dynamical application of Bayesian mechanics (note that we used that in Sect. 2.1 as well, where the instantaneous story behind derivatives licenses dropping the time variable from $q(t)$, a position in time). Denoting an expectation with $\mathbb{E}$ and neglecting subscripts, this gives us the equation

$$\mathbb{E}_{p(q)}[q] = \iint \frac{F(\eta)}{m}. \tag{5}$$

Suppose we also constrain the system to *be* classical, in the sense of having infinite precision. This is a variance constraint, namely, that

$$\mathbb{E}_{p(q)}\left[\left(q - \mathbb{E}_{p(q)}[q]\right)^2\right] = 0.$$

When asking about paths, we ask that the accumulated variance of or along a path is also zero:

$$\mathbb{E}_{p(q,t)} \left[\int_0^t \left(q(\tau) - \mathbb{E}_{p(q,t)}[q(\tau)] \right)^2 d\tau \right] = 0. \tag{6}$$

A similar law for the total squared displacement of a path has been used to produce Newton's laws from the principle of maximum path entropy (maximum calibre) before [20]. Both of the above equations are simply a constraint that the optimal acceleration does not deviate from what the environment tells it to, which implies the approximate Bayesian inference lemma when under the additional constraint that the average state is the value of the synchronisation map [35].

We believe the system constrains itself to be the optimal parameter for some belief over what it is supposed to do in its environment. That parameter happens to be the least surprising internal state given some external state and a shared blanket state. This is what is meant in previous discussions about beliefs about internal states being dual to beliefs about external states.[6]

The optimal (i.e., least biased [30]) belief under these two constraints can be derived from the principle of constrained maximum entropy, yielding

$$p(q) = \exp\left\{ -\lambda_1 \left| q - \lambda_2 \iint \frac{F(\eta)}{m} \right|^2 \right\} \tag{7}$$

with $\lambda_1, \lambda_2 > 0$ being Lagrange multipliers for the constraints indicated above. Considered dynamically, this equation can be given as a path probability density

$$p(q,t) = \exp\left\{ -\lambda_1(t) \int_0^t \left| q(\tau) - \lambda_2(\tau) \iint \frac{F(\eta, s)}{m} \right|^2 d\tau \right\}. \tag{8}$$

When $\lambda_1^{-1}(t) \ll 1$ and $\lambda_2(t) = 1$ for all t, we can reproduce classical dynamics. In particular, in the limit $\lambda_1 \to \infty$, there is no uncertainty at all, and the most likely path under those constraints—the classical path of least action, by construction—is the *only* path we lend any non-zero probability to. This is something like a classical limit for our path probability, in the same sense as taking $\hbar \to 0$ recovers classical mechanics from Feynman's path integral. The degree to which something can explore states within some allostatic bounds is precisely the variance under a Laplace approximation, yielding an important role for the Lagrange multipliers on the maximum entropy side of the story.

Finally, note that (8) is (4) for a modal path given by (5). This will be our general equation for Bayesian classical mechanics.[7]

---

[6] The idea that the maximum entropy principle under existential variables is the free energy principle under a synchronisation map was first introduced in [3]. A proof of this can be found in [35].

[7] Note that, for a heuristic integral over an infinitesimal time, i.e., from t to dt, (8) is equal to (7). Intuitively, this is a statement that instantaneous variations in a path accumulate along the path as the path goes forward in time.

Acknowledgements. The author thanks Lancelot Da Costa, Karl J Friston, Brennan Klein, and Maxwell J D Ramstead for valuable conversations.

A   A Question of Quantum Ontology

We began with the aim of showing that classical physics can be derived from Bayesian mechanics, by showing that deviations from a classical law of motion are surprising given a particular action functional, and that Bayesian mechanics describes the minimisation of surprisal. Do systems actually infer what their classical laws of motion are, and follow those inferences to avoid surprisal? More to the point: is there a less 'just so' aspect of reproducing classical physics from the assumption that the least surprising trajectory of a system minimises the classical action? Can we do this without arbitrarily assuming the laws of classical physics and merely showing that unsurprising systems obey those laws? There is indeed a more concrete interpretation of this inference, one which makes the idea behind (8) more subtle.

   In fact, what we have shown in the foregoing statements is that, under surprisal minimisation, a system takes a classical path when that path is the average. We can demonstrate that this has some further meaning by showing Bayesian mechanics is naturally derived from simpler arguments about the role of probability in quantum mechanics, such that the modal path of any fluctuating system is a classical path, and surprisal minimisation already exists in quantum mechanics. That is, we can derive Bayesian mechanics from the idea that a system is classical on average, just as we can derive classical mechanics from the idea that systems obey Bayesian mechanics for classical averages. This suggests a view that (i) systems with randomness do inference over their classical laws, and (ii) in the quantum setting we recover classical physics by doing inference. A more complete quantum physics manual for the Bayesian mechanic is to be written elsewhere.

   Begin from the supposition that classical equations of motion are asymptotics of quantum equations of motion, given by the empirical observation that we can measure classical effects more readily than quantum ones, but also, that classical equations of motion depend on parameters with underlying quantum effects, in such a way that quantum effects still bleed into the classical world when we 'zoom in' to the extent that those parameters are no longer renormalised.[8] This leads directly to the *correspondence principle*, a law of large numbers for quantum probabilities. The consequence of this classical 'limit' is that the evolution of the most likely state of a quantum system gives us what we define as classical physics. Proven by Ehrenfest in 1927, we can rewrite this result (assuming $|\partial_{tt} q(t)|$ is bounded above almost surely—physically, a thoroughly reasonable assumption) as

$$- \mathbb{E}_{p(q(t))} \left[\partial_q V(q) \right] = m \, \mathbb{E}_{p(q(t))} \left[\partial_{tt} q(t) \right].$$

---

[8] This is also referred to as the adiabatic approximation, and appears in semi-classical physics.

So, assuming distance constraints like those above, such that we have a Gaussian measure or otherwise a Laplace approximation—a constraint which is quadratic in fluctuations—the most likely path ought to be a classical equation of motion. We will repeat this argument in Bayesian mechanical language, aided by the technology of the path integral.[9]

In order to reproduce the idea that, probabilistically, quantum fluctuations are merely corrections to classical estimates, we take a Wiener measure where the variance of path probability—as it is given by the probability of the velocity on such a path—is scaled by some characteristic constant $\frac{m}{2\hbar}$,

$$Z^{-1}\exp\left\{-\frac{m}{2\hbar}\int_0^t \partial_s q(s)^2 \mathrm{d}s\right\}\mathrm{d}q(t).$$

Note that, technically, we have Wick rotated our field theory. Note also that, appealing to the Trotter product formula, we can—without loss of generality—neglect the potential and assume it is zero everywhere on the support.

As we remarked before, setting $\lambda_1 = \hbar$ and taking the scaling of quantum fluctuations to zero, and making use of the fact that fluctuations are precisely what contribute to the surprisal, we have a statement that in the quantum regime of Bayesian mechanical dynamics, the least surprising trajectory of the system is one that follows an overlying classical equation of motion. Indeed, under the WKB approximation, the most likely path in the path integral is the classical equation of motion of the field. Without quantum fluctuations about this classical solution we have classical physics, whereas in perturbative approximations to quantum mechanics, such as the quantum effective action, we add those quantum fluctuations in as higher order correction terms to a classical *ansatz*.[10]

So, we are now armed with two facts: basic physical observations suggest that classical paths are most likely paths, and, the canonical measure on paths is in terms of fluctuations about such a classical mode. We wish to see if the mathematical fact that this is the canonical description of path probability reflects the physical fact that classical equations of motion are the most likely paths of a system. Our formulation suggests that taking the limit $\hbar \to 0$ would be the right approach. Formally, this limit scales fluctuations to zero, revealing the most likely state as the one with constant velocity $v(0)$: a classical equation of motion under our identically zero potential, from where we derive no applied force and hence no acceleration.

Though we omit the proof, one can indeed prove that taking $\hbar \to 0$ results in classical equations of motion. As noted by Feynman, this most likely path is what is most likely to be determined by observation (experiment), and thus determines the classical limit of the path integral. That this story—the most likely path is the classical one due to *a priori* assumptions about the state and

---

[9] We point the reader to [23] for details.

[10] This section assumes there is a *unique* such classical solution to the system. Degenerate classical minima are handled by instanton theory, which we will not cover here. An excellent overview of the topic can be found in [41].

measurability of the world—implies the minimisation of surprisal, as well as following from it (as discussed in Sect. 3) is non-trivial.

What does it actually mean when we pass from 'most likely' to 'least surprising?' Least surprising to *whom*? Certainly the experimenter—there is a sense in which the entire problem is dualised, and we are minimising the surprisal of *our* beliefs about what a system does. That is, the two random dynamical systems coupled here are a quantum particle and an environment (including, perhaps, an external observer).

Do quantum particles do inference over where their classical paths are, organising themselves on average into the preimage of the average state of the observer, who expects to see a system follow a force applied? In the sense of estimating what that path is by taking a path which wastes the least energy (so to speak) in response to a force and in absence of quantum noise—and encoding such a path in their own dynamics, thereby estimating what a force 'tells' them to do—they do. This makes Bayesian mechanics a useful way of modelling how a quantum system treats information and interactions with its environment,[11] defaulting to classical equations of motion on average. It is in this Bayesian mechanical sense that classical physics is a result of Bayesian inference in a quantum regime.

B   The Matching of Modes

This section begins a worked example of the typology described in [32]. We begin with 'mode-matching.' Mode-matching is the application of Bayesian mechanics to stationary objects which engage in approximate Bayesian inference [15,32,35]. In this case, by definition of stationarity, the most likely internal state is fixed. Typically valid over only a brief time-scale—since nothing is stationary forever and nothing which is stationary and non-adaptive resists entropy for long—this is the simplest case of Bayesian mechanics.

Inspired by [35], we formulate mode-matching under approximate Bayesian inference as internal states being constrained to be optimal parameters for a recognition density. Again, this is fully equivalent to the proper FEP. Using Theorem 4.2 (ibid), we can formulate the minimisation of surprisal applied to internal states $-\ln\{p(q \mid \eta_F)\}$ as a demand that the log-probability equals some constraint on those internal states, with further precision-based minimisation possible over an ensemble of states. Here, that constraint is

$$S[q] = -\ln\{p(q \mid \eta)\} = \lambda_1 \left(q - \iint \frac{F(\eta)}{m} \right)^2.$$

Note the similarity to Eq. 3.5 in [10]. Under approximate Bayesian inference (and a further, but generically acceptable, Laplace assumption), the ideal state is the most likely state, which is the minimiser of this squared displacement.

---

[11] We will refer the reader to [13] and related references for more details about quantum information theory under the FEP.

The system we describe could be a stone performing inference over the cancelling of its gravitational pull and the normal force emanating from the ground, obeying

$$\sum F = -F_g + F_N = 0. \tag{9}$$

In this case, the stone's acceleration is zero, and under the initial conditions $v = 0$, $q = 0$, it goes straight to—and remains at—the mode $q = 0$. Referring back to the Introduction, the mode is a particular attractor which is a fixed point for the system. This also means that for Bayesian mechanics to be consistent in the classical setting, it must (for stationary modes) imply Newton's first law— i.e., that for every applied force, there is a force of equal magnitude applied in the opposite direction.

C   The Tracking of Modes

Mode-tracking can be summarised as the existence of a *target* mode, i.e., a desired mode that systems are tracking towards, either for a finite time or constantly. This allows us to describe the most likely flow of autonomous states as a flow of beliefs [1, 15, 29, 31], and involves the iteration of approximate Bayesian mechanics [35]. Within mode-tracking there are two further distinctions, a bit more granular than the three-fold structure described in the Introduction: systems which track, but settle to, a mode, and systems which constantly chase a mode. The former is an example of a system that terminates in mode-matching, whilst the latter is an example of a system that is in constant motion. For both cases, we pass to an idea of dynamics, gesturing at an application of the principle of maximum calibre [34].

C.1   Terminal Mode-Matching

Suppose the total force applied is dynamic, but eventually equilibrates. An example would be a stone tossed into the air, which travels through a gravitational field and eventually returns to the ground. The variant of (9) corresponding to that motion is $F = -F_g$, and the solution of the integral equation (5) under that F is

$$q(t) = q(0) + v(0)t - \frac{1}{2}gt^2 \tag{10}$$

where g is the acceleration due to gravity. This equation has a steady state value where the mode $q(t) = 0$ and remains there, reached at some hitting time t_{hit}. For instance: for $v(0) = 0$ metres per second and $q(0) \approx 4.91$ metres above ground, $t_{\text{hit}} \approx 1$ second. So, we can consider the full, dynamic-in-time force as

$$-F_g(t) + \mathbf{1}_{\text{hit}}(t)F_N(t),$$

where $\mathbf{1}(-)$ is an indicator function—the constant function equal to one on $t \geq t_{\text{hit}}$, and zero elsewhere. This terminates in the mode-matching explored in the previous section, since for all $t \geq t_{\text{hit}}$, we have $\sum F = 0$. Indeed, $q(t \geq t_{\text{hit}}) = 0$ by construction; this is a stationary mode.

By solving (5) as a constraint relation, we have actually asked that the average path obeys (10) and that no other path $q(t)$ deviates from that path. That is, we want $q(t) = \hat{q}(t)$ in the ideal case. As such, classical objects can be modelled as performing inference over the forces driving their motion, and given that they find known laws of motion unsurprising, get driven to the mode described in Sect. B by following (10). For example—it would be surprising to the very fabric of space-time if a stone which had landed on the ground at some t_{hit} were to spontaneously jump after. Thus, by obeying (10) and following classical laws of motion, the surprisal of motion is minimised. We will detail this below.

We begin by asking that $q(t)$ should equal $\hat{q}(t)$. Under our other constraint on the expected path—the solution to (5), given above—eventually, $\hat{q}(t) = 0$ (in particular, for all $t \geq t_{\text{hit}}$). We could view this as dynamical inference more generally, or, the construction of a realisation of some path under a steady state density with mode zero, such that $\hat{q}(t) = 0$ after some convergence time (here, t_{hit}). In that case, the system goes directly to $\hat{q}$ as its kinetic energy 'dissipates' into potential energy.

In greater detail: the path we take consists of a list of states. Each such state is increasingly more likely as we approach the mode, and indeed, a mean-reverting process will go to a mode on average under a quadratic potential. As such, the average path taken by the system performs a gradient ascent on the probability density, or equivalently, a gradient descent on the surprisal. Note that we are not working in a path space here; rather, the path exists on a probability density, as a lift of a list of states, which is indeed simply a path—but, we don't speak about surprisal minimisation on such a path directly, instead speaking about the tendency of any path to go to a fixed point. Mathematically, this means that the motion of $q(t)$ is a gradient descent on $|q(t) - \hat{q}(t)|^2$, with some convergence time t_{hit}, and $q(t) = \hat{q}(t) = 0$ for $t \geq t_{\text{hit}}$. Since the logarithm of (7) is this distance when $\lambda_2 = 1$, this is equivalently a minimisation of surprisal. That is, we have

$$\partial_t q(t) = -\lambda_1 \nabla \ln\{p(q)\},$$

such that the least surprising *state* is the mode, and the system takes a path towards that mode. Moreover, the least surprising path to the least surprising state ought to traverse the distance from some initial $q(0)$ to $\hat{q}$ the quickest, which (for a fixed velocity) is the path given by a direct gradient descent on the distance. It would be of interest in future work to give a unified view of these approaches, i.e., to prove that under certain asymptotic conditions, least surprising paths are paths towards least surprising states.

Since λ_1 scales fluctuations, it is proportional to the inverse diffusion coefficient, or the covariance, more generally. Additionally, since there is no random motion and no opportunity to explore, the motion is described by a pure gradient descent, and so this yields one component of the Helmholtz decomposition discussed in [15]. Note that the instantaneous Lagrange multiplier λ plays a different role than the dynamical Lagrange multiplier $\lambda(t)$. In particular, the former selects states whilst the latter selects paths, a distinction that becomes important when we deal with paths towards least surprising states, as we have here. We still desire $\lambda_1(t)^{-1} \to 0$ as we did in Sect. A, to reproduce classical path selection. In the Gaussian case this is precisely our uncertainty over paths, as we mentioned.

C.2   Infinite Mode-Tracking

This section will discuss itinerant objects whose gravitational field is such that the mode is never stationary—which we could call mode-chasing, as a sub-type of mode-tracking. Consider a planet in a gravitational potential equal in all directions: there is no stationary state, and hence, no meaningful mode, to the dynamics of this system. Naïvely, there is no parameter through which to minimise surprisal. This does not mean the FEP cannot describe satellite motion. On the contrary—the application of the FEP to complicated systems is where it truly shines [31].

   Like any classical system, in the absence of a force acting on it, a satellite system will continue to move on its trajectory. This is Newton's first law—the usual aphorism being, 'a body in motion tends to stay in motion; a body at rest tends to stay at rest...' with the phrase '... unless acted upon by an outside force' often appended. As such, the hitting time formulation of Sect. C.1 is no longer directly useful, but having *no* hitting time certainly is. Note that the lack of a mode—but presence of circular, solenoidal flows—for true classical systems[12] is consistent with [10,35].

   Such a system has no dissipative component, since it is purely classical. This means the gradient descent describing the mode-matching dynamics we discussed in Sect. C.1 degenerates, in the opposite sense as any exploratory component of the flow degenerated in that section. Here, we have a system which travels along a level set of a sphere of radius r, and in particular, travels such that the surprisal of states (parameterised by a stationary mode) is a non-zero constant along a path. By simple arguments in symplectic geometry—the geometric study of flows in classical physics [5,37]—flows which are level sets of some Lagrangian[13] and which admit radial coordinates are described by a skew-symmetric matrix. Indeed, level sets of the sphere centred on $\hat{q}$, projected down to the state space, are integral curves of the following equation:

$$\partial_t q(t) = \begin{bmatrix} 0 & \nu \\ -\nu & 0 \end{bmatrix} q(t) + \frac{1}{\nu}\hat{q}, \tag{11}$$

---

[12] That is, energetically conservative systems, or systems in the absence of dissipative forces. Although not a dissipative system, one can contrast this with the loss of energy *of motion* that occurs upon colliding with the ground in Sect. C.1. It is consistent that such systems have a mode whilst unperturbed satellite motion does not.

[13] Note that we typically consider a Hamiltonian, which is metric isomorphic to a Lagrangian.

where ν controls the system's frequency, or speed of travel along one such level set. Note that this system of equations corresponds to the second-order ordinary differential equation

$$\partial_t q(t) = -\nu^2 q(t)$$

whose solution is

$$q(t) = r\cos(\nu t) + r\sin(\nu t) + \hat{q}$$

for some constant $r > 0$. Note also that

$$-\nabla_i \ln\{p(q,t)\} = 2\lambda_1(q(t) - \lambda_2\hat{q}). \tag{12}$$

As such, for an appropriate choice of λ_1 (namely, one half, or one half the coefficients of the gradient descent operator, should it exist) and λ_2 (namely, $-1/\nu$), inserting (12) component-wise into (11) yields exactly the other piece of the Helmholtz decomposition.

As we stated, the matrix operator indicated is skew-symmetric, and, the gradient on the sphere is locally orthogonal to these level sets—that is, moving on a level set does not change the gradient. Since there is no mode for these dynamics, we cannot describe a gradient descent on surprisal in the sense of Sect. C.1, but we can say that it is a gradient descent for which the value of the gradient is preserved, with velocity scaled by the matrix indicated. Future work will make the arguments given here—with respect to the Helmholtz decomposition being an artefact of the geometric nature of certain flows—more formal.

D   Path-Tracking, and A Simple Case of G-theory

This section will progress the construction to more complex forms of path-tracking that are not amenable to the mode-based descriptions we discussed previously.

What has been called G-theory is the duality between maximum calibre and the genre of Bayesian mechanics applying to surprisal on paths [32], which we have begun to explore here. This pitches G-theory as the generalisation of the duality explored in [35], extending that construction to paths. In its full generality it is thought to accommodate descriptions of complex systems (like non-Markovian behaviours, moving attractors, and non-stationary statistics) more naturally and with greater fidelity than in the past. Some inspiration for this comes from the previously mentioned principle of maximum calibre, which does famously well on difficult problems in non-equilibrium statistical physics [11,19,30]. These results suggest that, whatever G-theory will prove to be, it will be a canonical modelling framework for complex systems.

Here we provide a simple example of this framework by formulating path-tracking, the least surprisal principle on paths—our third sort of approximate Bayesian inference—as an explicit problem of dynamical constraints. We then formulate a connection to chaos by examining the path-based nature of G-theory in greater detail.

D.1   Path-Tracking

Recall our results on mode-tracking in Sect. C.1. The construction there is obviously inelegant—besides formulating the path over the state space instead of doing proper dynamical inference, it exchanged the proper accumulated squared displacement in (8) with a less general instantaneous squared displacement at t_{hit}. Foreshadowing a more general extension to paths, the most natural formulation of this problem is readily seen as a gradient ascent on the path probability density. Under maximum calibre, our constraints lead to a probability density

$$\exp\left\{-\lambda_1(t)\,|q(t) - \hat{q}(t)|^2\right\},$$

where the expected path (10) is denoted by $\hat{q}(t)$, and $|q(t) - \hat{q}(t)|^2$ is limitingly zero. For dynamic F, this is a moving Gaussian, with mode centred along the path for a given state-wise marginal. That is, it is centred on the list of q's visited classically for a list of times, like a crest in the path space that runs directly over the intended states (and hence, a sort of mountain of probability for realisations flowing along the state space, concentrating them in that region). Path tracking is obvious in this situation—it appears to follow a gradient descent on the action, finding the most likely path by finding the summit at each t of $p(q(t))$.

We can indeed still discuss a gradient descent in this case, but it is a functional gradient on the action S, such that the gradient descent is on deviations along a path, minimising fluctuations from the path of least action—and this is precisely the principle of least action, or of least surprisal, when the path of least action is the expected path (guaranteed by (3), as discussed in Sect. 2.2).

Let $\delta q(t)$ be some first order variation of a path away from a path of stationary action at t (see [32, Figure 1] for a depiction), which is in fact a realisation at t of some fluctuation away from the expected path. Analytically, this means that we have

$$\delta q(t) = -\hat{\nabla}S[q(t)]$$

where the kernel of the gradient in the path space, $\hat{\nabla}$, is a path such that the system only changes to second order under variation—that is, it is a path $q(t)$ such that the distance between $q(t)$ and $\iint F(\eta, t)$ is minimised. This equation simply expresses that the path of least variation is the stationary, or expected, path. This means the system will most likely settle into an evolutionary regime that follows the expected path, which is least surprising—however, note this is simply a model of that process, since there are no such fluctuations in classical physics. Instead, we are interested purely in the zero point of the gradient.

Recall what the surprisal is in this case—the logarithm of (8) is merely the classical action. As such, the statement that systems evolve on stationary points of the classical action functional follows directly from a gradient descent on path surprisal given that systems follow forces. As such, the above equation reduces to

$$\delta q(t) = -\hat{\nabla}\ln\{p(q, t)\}$$

$$= \hat{\nabla}\left[\lambda_1(t)\int_0^t \left|q(\tau) - \lambda_2(\tau)\iint \frac{F(\eta, s)}{m}\right|^2 d\tau\right],$$

which yields

$$\delta q(t) = 0 \iff \partial_{tt} q(t) - \frac{F(\eta, t)}{m} = 0.$$

Since our path space gradient on surprisal reproduces the Euler-Lagrange equation as the functional gradient of $S[q(t)]$, this is precisely classical mechanics.

In the infinite mode-tracking case, we have something similar. For simplicity, we take the path of a satellite moving about a fixed central body of radius r as a circle

$$q_1(t)^2 + q_2(t)^2 = r^2.$$

This is a constraint that the expected path is a circle of radius r, and that realisations of q ought to have norm r^2. We will parameterise this as an expected path which is $\hat{q}(t) = [\hat{q}_1(t), \hat{q}_2(t)] = [r\cos(t), r\sin(t)]$. The surprisal Lagrangian measures precisely these deviations from a circle,

$$\langle q(t) - [r\cos(t), r\sin(t)], q(t) - [r\cos(t), r\sin(t)] \rangle.$$

In this form it is even more apparent that our Lagrangian, the quadratic form defined in (3), is a metric on noise, $\omega = q - \hat{q}$.[14] Once more, it is exactly the distance of $q(t)$ from a circular path parameterised by $[r\cos(t), r\sin(t)]$. As for surprisal—again, given that systems move on geodesics through space-time, it would be surprising to see a system change its path to deviate from the curvature of space-time, and thus, to not follow the induced potential field. The final relation we derive is

$$\delta q(t) = 0 \iff q(t) - [r\cos(t), r\sin(t)] = 0.$$

Given centripetal forces and the absence of tangential forces on a radial parameterisation of the circle, we can derive that

$$\partial_{tt} q(t) - C\frac{M}{r^2} = 0,$$

where C is determined to be Newton's gravitation constant; this equation then yields the acceleration for a system orbiting a central body of mass M in circular fashion.

D.2   A First Idea of G-theory

In Sects. 3 and A, we introduced the idea that surprisal minimisation arose from solving for the Lagrange multiplier for a constraint that the most likely path was an on-shell trajectory, described by (6). When this Lagrange multiplier is limitingly zero, and the particle does perfect inference over the forces being applied to it, we have classical mechanics. Here the uncertainty over both environment and system necessarily degenerated.

---

[14] In fact, we could choose to denote the Lagrangian as $g(\omega, \omega)$, for reasons of geometric significance.

This curiosity—that Bayesian mechanics, when cast in the language of the principle of maximum calibre, naturally leads to a path integral representation of classical mechanics—is upgraded to a more interesting observation that, when viewed through the lens of a classical system interacting with an environment, a path probability density is the most informative about the system; in particular, that it is the most general way of understand what an environment 'tells' the system to do, and how that can be represented probabilistically as the system estimating those forces and following them so as to produce good (that is, unsurprising) inferences.

Although path-tracking is already a more elegant way of discussing the simple problems on display here, as expected, the problems the full generality of Bayesian mechanics seeks to provide solutions to are radically different than the simple Newtonian laws of motion investigated thus far. Moreover, to produce key identities in classical physics like the Euler-Lagrange equation, we practically began from where we wanted to end up: with the assumption that classical systems follow forces.[15] Here, we aim to eventually formulate chaotic or itinerant systems under Bayesian mechanics, as has been done for earlier forms of the free energy principle [17]. A sketch of one such result will be found in this second subsection.

Let Γ be the space of paths and $C(t)$ be a source for the field (this is merely an external field driving $x(t)$, and is generically comparable to an electric current). At the path of minimal surprisal, and under the demand that there is no other path possible, we have a Dirac measure over the classical path. Hence, the solution can trivially be transformed into the following path integral representation:

$$Z[C] = \int_\Gamma \prod_t \delta(x_t - x_{t,\mathrm{cl}}) e^{-\lambda_1(t) \int_0^t C(\tau)x(\tau)\mathrm{d}\tau} \mathcal{D}x(t)$$

where $x(t)_\mathrm{cl}$ is the classical solution to the equations of motion of interest and the product of Dirac measures over intervals of t of size $\varepsilon > 0$ is given. The term $\prod_t \delta(x_t - x_{t,\mathrm{cl}})$ in the path integral enforces a weight of one for classical paths and a weight of zero for all others.

It is a standard trick to rewrite a Dirac measure in terms of a Jacobian determinant, and a determinant in terms of a path integral over 'ghost fields,' anti-commuting variables that behave like auxiliary fermionic fields. Following the procedure described in [22], rewriting the on-shell trajectory $x(t)_\mathrm{cl}$ in terms analogous to second order variations $\ddot{\omega}$ and the Dirac measure on that function as a particular determinant; then, introducing a pair of ghosts θ and $\bar{\theta}$, we have the following transformation of the Bayesian mechanical path integral:

$$\int_{\tilde{\Gamma}} \exp\left\{ i \int_0^t \xi(v_\tau - v_{\tau,\mathrm{cl}}) + i\bar{\theta}\ddot{\omega}\theta \mathrm{d}\tau \right\} \mathcal{D}x(t)\mathcal{D}\xi\mathcal{D}\theta\mathcal{D}\bar{\theta}$$

---

[15] However—in defence of the author, and with reference to Sect. A, what we *really* did was say that the least surprising path of a system with quantum or statistical fluctuations is the one that obeys a classical equation of motion—a less trivial result.

where we have taken $C(t)$ to be identically zero for convenience (hence it has disappeared) and introduced a temporary variable ξ after passing to imaginary variables. Note that

$$i \int_0^t \xi(v_\tau - v_{\tau,\mathrm{cl}}) \mathrm{d}\tau$$

is merely our surprisal Lagrangian in Fourier variables, arising organically from defining what it means to be a classical path—and that there is an additional term

$$-\int_0^t \bar{\theta}\dot{\omega}\theta \mathrm{d}\tau$$

arising from the other transformations on $\prod_t \delta(x_t - x_{t,\mathrm{cl}})$ described. The latter term corresponds to a fermionic sector of our theory, as discussed; whilst the surprisal term defines a bosonic sector in contrast. Note also that, in spite of the appearance of an imaginary quantity in the path integral, there is no constant $\hbar$. That ultimately preserves the classicality of this path integral.

Moreover, these ghost fields define a pair of supercharges; this is due to invariance under a pair of BRST transformations which relate bosonic and fermionic degrees of freedom, and form a superalgebra under the commutator. Bayesian classical mechanics is thus an $\mathcal{N} = 2$ supersymmetry theory.[16] This supersymmetry has a striking interpretation that gives us a glimpse of the power of G-theory: the ghost fields themselves appear to correspond to Jacobi fields, measuring the divergences of classical trajectories with similar initial points—a classic metric for chaos—and the breaking of this supersymmetry is a candidate geometric basis for non-ergodicity [21].

Far more work remains to be done on the nature of supersymmetric Bayesian mechanics, and especially its connections to chaos and the Bayesian gauge theory introduced in [32,33,35]; however, for now we only note that this exciting connection to certain features of complexity is evidently natural in the language of G-theory.

References

1. Aguilera, M., Millidge, B., Tschantz, A., Buckley, C.L.: How particular is the physics of the free energy principle? Phy. Life Rev. **40**, 24–50 (2022)
2. Amari, S.I.: Information Geometry and Its Applications. Applied Mathematical Sciences, vol. 194. Springer, Cham (2016)
3. Andrews, M.: The math is not the territory: navigating the free energy principle. Biol. Philos. **36**(3), 1–19 (2021). https://doi.org/10.1007/s10539-021-09807-0
4. Arnol'd, V.I.: Mathematical Methods of Classical Mechanics. Graduate Texts in Mathematics, vol. 60. Springer, Cham (2013)
5. Arnol'd, V.I., Givental, A.B., Novikov, S.P.: Symplectic geometry. In: Arnold, V.I., Novikov, S.P. (eds.) Dynamical Systems IV. Encyclopaedia of Mathematical Sciences, vol. 4, pp. 1–138. Springer, Heidelberg (2001). https://doi.org/10.1007/978-3-662-06791-8_1

---

[16] We will suggest [38] to the reader for a classic, if advanced, review.

6. Atiyah, S.M.F., Dijkgraaf, R., Hitchin, N.J.: Geometry and physics. Philos. Trans. R. Soc. A: Math. Phy. Eng. Sci. **368**(1914), 913–926 (2010)
7. Barp, A., et al.: Geometric methods for sampling, optimisation, inference and adaptive agents. preprint arXiv:2203.10592 (2022)
8. Calder, J.: The calculus of variations (2022). https://www-users.cse.umn.edu/~jwcalder/CalculusOfVariations.pdf
9. Cugliandolo, L.F., Lecomte, V., Van Wijland, F.: Building a path-integral calculus: a covariant discretization approach. J. Phy. A: Math. Theor. **52**(50), 50LT01 (2019)
10. Da Costa, L., Friston, K.J., Heins, C., Pavliotis, G.A.: Bayesian mechanics for stationary processes. Proc. R. Soc. A **477**(2256), 20210518 (2021)
11. Dixit, P.D., Wagoner, J., Weistuch, C., Pressé, S., Ghosh, K., Dill, K.A.: Perspective: Maximum caliber is a general variational principle for dynamical systems. J. Chem. Phys. **148**(1), 010901 (2018). https://doi.org/10.1063/1.5012990
12. Dürr, D., Bach, A.: The Onsager-Machlup function as Lagrangian for the most probable path of a diffusion process. Commun. Math. Phys. **60**(2), 153–170 (1978)
13. Fields, C., Friston, K.J., Glazebrook, J.F., Levin, M.: A free energy principle for generic quantum systems. Prog. Biophy. Mol. Biol. (2022)
14. Friston, K.J.: A free energy principle for biological systems. Entropy **14**(11), 2100–2121 (2012)
15. Friston, K.J.: A free energy principle for a particular physics. preprint arXiv:1906.10184 (2019)
16. Friston, K.J., et al.: The free energy principle made simpler but not too simple. preprint arXiv:2201.06387 (2022)
17. Friston, K.J., Heins, C., Ueltzhöffer, K., Da Costa, L., Parr, T.: Stochastic chaos and Markov blankets. Entropy **23**(9), 1220 (2021)
18. Gelfand, I.M., Silverman, R.A.: Calculus of Variations. Courier Corporation, Chelmsford (2000)
19. Ghosh, K., Dixit, P.D., Agozzino, L., Dill, K.A.: The maximum caliber variational principle for nonequilibria. Annu. Rev. Phys. Chem. **71**, 213–238 (2020)
20. González, D., Davis, S., Gutiérrez, G.: Newtonian dynamics from the principle of maximum caliber. Found. Phys. **44**(9), 923–931 (2014)
21. Gozzi, E., Reuter, M.: Algebraic characterization of ergodicity. Phys. Lett. B **233**(3–4), 383–392 (1989)
22. Gozzi, E., Reuter, M., Thacker, W.D.: Hidden BRS invariance in classical mechanics. II. Physical Review D **40**(10), 3363 (1989)
23. Hall, B.C.: Quantum Theory for Mathematicians. Graduate Texts in Mathematics, vol. 267. Springer, Cham (2013)
24. Heins, C., Da Costa, L.: Sparse coupling and Markov blankets: a comment on "how particular is the physics of the free energy principle?" by Aguilera, Millidge, Tschantz and Buckley. Phys. Life Rev. **42**, 33–39 (2022)
25. Hohwy, J.: The self-evidencing brain. Noûs **50**(2), 259–285 (2016)
26. Millidge, B., Tschantz, A., Seth, A.K., Buckley, C.L.: On the relationship between active inference and control as inference. In: The First International Workshop on Active Inference, pp. 3–11. preprint arXiv:2006.12964 (2020)
27. Neel, R.W., Sacchelli, L.: Uniform, localized asymptotics for sub-Riemannian heat kernels and diffusions. preprint arXiv:2012.12888 (2020)
28. Øksendal, B.: Stochastic Differential Equations. Springer, Cham (2003)
29. Parr, T., Da Costa, L., Friston, K.J.: Markov blankets, information geometry and stochastic thermodynamics. Phil. Trans. R. Soc. A **378**(2164), 20190159 (2020)

30. Pressé, S., Ghosh, K., Lee, J., Dill, K.A.: Principles of maximum entropy and maximum caliber in statistical physics. Rev. Mod. Phys. **85**(3), 1115 (2013). https://doi.org/10.1103/RevModPhys.85.1115

31. Ramstead, M.J.D., Sakthivadivel, D.A.R.: Some minimal notes on notation and minima: a comment on "how particular is the physics of the free energy principle?" by Aguilera, Millidge, Tschantz, and Buckley. Phys. Life Rev. **42**, 4–7 (2022)

32. Ramstead, M.J.D., et al.: On Bayesian mechanics: a physics of and by beliefs. preprint arXiv:2205.11543 (2022)

33. Sakthivadivel, D.A.R.: A constraint geometry for inference and integration. preprint arXiv:2203.08119 (2022)

34. Sakthivadivel, D.A.R.: Regarding flows under the free energy principle: a comment on "how particular is the physics of the free energy principle?" by Aguilera, Millidge, Tschantz, and Buckley. Phys. Life Rev. **42**, 25–28 (2022)

35. Sakthivadivel, D.A.R.: Towards a geometry and analysis for Bayesian mechanics. preprint arXiv:2204.11900 (2022)

36. Seifert, U.: Stochastic thermodynamics, fluctuation theorems and molecular machines. Rep. Prog. Phys. **75**(12), 126001 (2012)

37. da Silva, A.C.: Lectures on Symplectic Geometry. Lecture Notes in Mathematics, vol. 1764. Springer, Cham (2008)

38. Tachikawa, Y.: $\mathcal{N} = 2$ Supersymmetric Dynamics for Pedestrians. Lecture Notes in Physics, vol. 890. Springer, Cham (2015)

39. Touchette, H.: The large deviation approach to statistical mechanics. Phys. Rep. **478**(1–3), 1–69 (2009)

40. Üstünel, A.S.: An Introduction to Analysis on Wiener Space. Lecture Notes in Mathematics, vol. 1610. Springer, Cham (2006)

41. Vaĭnshteĭn, A.I., Zakharov, V.I., Novikov, V.A., Shifman, M.A.: ABC of instantons. Sov. Phy. Uspekhi **25**(4), 195–215 (1982)

42. Wiener, N.: Cybernetics or Control and Communication in the Animal and the Machine. The MIT Press, Cambridge (2019). https://doi.org/10.7551/mitpress/11810.001.0001

A Message Passing Perspective
on Planning Under Active Inference

Magnus Koudahl[1,2(✉)], Christopher L. Buckley[2,3], and Bert de Vries[1]

[1] BIASLab, Department of Electrical Engineering,
Eindhoven University of Technology, Eindhoven, The Netherlands
m.t.koudahl@tue.nl
[2] VERSES Research Lab, Los Angeles, CA 90016, USA
magnus.koudahl@verses.io
[3] School of Engineering and Informatics, University of Sussex, Falmer, Brighton, UK

Abstract. We present a message passing interpretation of planning under Active Inference. Specifically, we show how the Active Inference planning procedure can be broken into a (partial) message passing sweep over a graph, followed by local computations of a cost functional (the Expected Free Energy). Using Forney-style Factor Graphs, we then proceed to show how one can derive novel planning schemes by local changes to the underlying graph and message passing schedule. We illustrate this by first isolating the "sophisticated" aspect of Sophisticated Inference and then proposing a novel planning algorithm through combining the sophisticated update mechanism with a different message passing schedule. Our main contribution is a modular view of planning under Active Inference that can serve as a framework for both understanding existing algorithms, deriving new ones and extending the class of models that are amenable to Active Inference. Approaching Active Inference from a message passing perspective also shows how it can be efficiently implemented using off-the-shelf probabilistic programming software, broadening the class of models available to researchers and practitioners.

Keywords: Active inference · Expected Free Energy · Factor graph · Message passing

1 Introduction

Active Inference (AIF) is a common modeling framework for studying decision making and, in recent years, also for designing synthetic agents. A key facet that sets AIF apart from other approaches is the choice of planning objective. AIF uses the Expected Free Energy (EFE), which is a cost functional that promises a balanced trade-off between exploration and exploitation. In this paper, we present a particular interpretation of EFE-based planning under AIF using known message passing-based inference methods on a Forney-style Factor Graph (FFG). We show that the standard EFE-planning algorithm is equivalent to performing a forward pass using standard belief propagation messages,

C. L. Buckley et al. (Eds.): IWAI 2022, CCIS 1721, pp. 319–327, 2023.
https://doi.org/10.1007/978-3-031-28719-0_22

followed by a separate computation phase based on the resulting marginals. By explicitly writing planning under AIF as message passing on a graph, we can clearly delineate the practical steps used for EFE computation. Doing so means we can isolate parts of more complicated schemes such as the sophisticated aspect of Sophisticated AIF (SAIF) [6] and the backwards influence from future timesteps hinted at by [15]. It also allows us to propose new algorithms as variations based on the common underlying theme and indicates a method for implementing AIF in a broader class of models using efficient inference software.

2   Generative Model and Inference

Planning under AIF centers around a generative model of the future. The generative model traditionally used is a discrete partially observed Markov decision process [6,7,9]. We let x_t denote an observation, z_t a latent state and u_t a control at timestep t. Now we can write the model after having observed x_t and conditioned on a policy $u_{1:T}$ as

$$p(\underbrace{x_{t+1:T}, z_{t:T}}_{\text{Future}} \mid \underbrace{u_{t+1:T}}_{\text{Policy}}, \underbrace{x_{1:t}, u_{1:t}}_{\text{Past}}) = \underbrace{p(z_t|x_{1:t}, u_{1:t})}_{\text{State Prior}} \prod_{k=t+1}^{T} \underbrace{p(x_k|z_k)}_{\text{Likelihood}} \underbrace{p(z_k|z_{k-1}, u_k)}_{\text{State Transition}}$$

$$(1)$$

where

$$p(z_t|x_{1:t}, u_{1:t}) = \mathcal{C}at(z_t|d) \tag{2a}$$

$$p(z_k|z_{k-1}, u_k) = \mathcal{C}at(z_k|B_{u_k} z_{k-1}) \tag{2b}$$

$$p(x_k|z_k) = \mathcal{C}at(x_k|Az_k). \tag{2c}$$

Here $p(z_t|x_{1:t}, u_{1:t})$ represents the Bayesian filtering solution over observed time steps $1:t$, which we summarise in the parameter vector d. Both z_k, z_t and x_k are discrete variables following Categorical distributions, as for instance described in [2, Ch. 2.]. Note that (1) extends into the future until some known horizon $T > t$ and is conditioned on a policy over the full trajectory $u_{1:T}$. We use B_{u_k} to denote the transition matrix B corresponding to the action u_k. Planning under AIF relies on comparing choices of B_{u_k} which we emphasize by this notation.

We can visualise (2) using the Forney-style Factor Graph (FFG) formalism. In an FFG, a node represents a factor (function of variables) and an edge represents a variable. An edge connects to a node if and only if the corresponding variable is an argument of that factor. The FFG of the model described by (2) is shown in Fig. 1. In Fig. 1, the T-nodes denote a discrete state transition (multiplication of a variable with categorical distribution by a transition matrix). For pedagogical purposes, we have also labelled T-nodes with the matching equation in (2) and indicated the forward direction of the graph by arrowheads on edges. When a variable is fixed (for example when we condition on a policy), we indicate this by a small, black square.

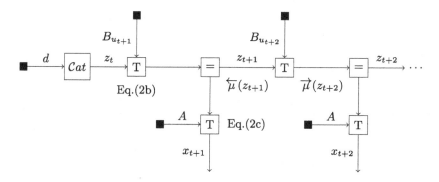

Fig. 1. FFG of discrete POMDP as used for planning in standard AIF models.

To perform inference in this model, we can utilise belief propagation (BP) [6,13]. BP proceeds by passing messages along edges of a graph towards variables that we wish to infer. We can illustrate this on our FFG by drawing arrows that outline the messages we wish to pass, see Fig. 2. When two messages collide they are multiplied (and normalized) to obtain a posterior marginal for the variable on an edge. For the model given by (2), all variables are discrete and related by discrete state transitions. We therefore only need the forward $\overrightarrow{\mu}(\cdot)$ and backward $\overleftarrow{\mu}(\cdot)$ BP-messages around a T-node. For the variables z_k, z_{k-1} and the transition matrix B_{u_k}, the messages are

$$\overrightarrow{\mu}(z_k) = \mathcal{Cat}\left(z_k \middle| \frac{1}{Z} B_{u_k} z_{k-1}\right), \qquad \overleftarrow{\mu}(z_{k-1}) = \mathcal{Cat}\left(z_{k-1} \middle| \frac{1}{Z} B_{u_k}^T z_k\right), \quad (3)$$

where we slightly abuse notation by having z_k denote the parameter vector of the incoming message on the edge z_k (instead of the random variable z_k), similar for z_{k-1} and the edge z_{k-1}. Z is a normalisation constant which we can ignore when the columns of the transition matrix are normalised, which we assume going forward. To illustrate, on Fig. 1 we have indicated the messages flowing out of a T-node towards the variables z_{t+1} and z_{t+2}.

3   Expected Free Energy

Planning under AIF involves first computing the Expected Free Energy (EFE) for each time step given a policy and then summing the results over time steps. This procedure is repeated for a set of admissible policies. Based on the sum-total EFE's for each policy, one then constructs a categorical distribution over possible policies and sample a course of action from there [5,7,10]. We explicitly appeal to a recursive formulation of EFE computation here as put forward in [6]. While the exposition given here will be in terms of the discrete POMDP described in Sect. 2, similar steps can be performed for other generative models, see for instance [10] for an example using linear Gaussian dynamical systems. The EFE for a single timestep k is defined as

$$G(u_k) = \sum_{x_k} \sum_{z_k} p(x_k|z_k) q(z_k|u_k) \log \frac{q(z_k|u_k)}{p(x_k, z_k|u_k, x_{1:t})} . \tag{4}$$

Notably, EFE only depends on *prior* time steps and not on the full trajectory. For computational purposes, the EFE for a single time step is often rewritten as[1]

$$G(u_k) = \underbrace{\sum_{z_k} q(z_k|u_k) H[x_k|z_k]}_{\text{Ambiguity}} + \underbrace{KL[q(x_k|u_k)||p(x_k)]}_{\text{Risk}} \tag{5}$$

Here we wish to note that all required quantities are written in terms of x_k, the observations. This means that everything we need to compute (5) is available around the likelihood node. With slight abuse of notation we find $q(z_k|u_k)$ and $q(x_k|u_k)$ by using the forward message (3) as

$$z_k = B_{u_k} z_{k-1} \tag{6a}$$
$$x_k = A z_k . \tag{6b}$$

Now we are ready to compute (5). Following [5, Eq. D.2-3] we can evaluate (5) for the model 2 as

$$G(u_k) = \underbrace{-diag(A^T \log A)^T z_k}_{\text{Ambiguity}} + \underbrace{x_k^T (\log x_k - \log c_k)}_{\text{Risk}} \tag{7}$$

where we have slightly adapted the original notation to be consistent with our exposition. The $diag(\cdot)$ operator takes as argument a matrix and returns its diagonal entries as a column vector and c_k refers to the parameter vector of the goal prior $p(x_k)$. In (7), the RHS is implicitly a function of u_k through the choice of B_{u_k} in (6a). We see that the quantities used in (7) can be obtained by applying a forward message passing sweep on the generative model. We can visualise this on the FFG as shown in the top panel of Fig. 2. The boxed areas indicate where we obtain the quantities required for (7). As we can see, EFE computation corresponds exactly to a forward message passing sweep followed by a secondary computation around the likelihood nodes.

4   Sophisticated Active Inference

Having established the message passing view of EFE computation, we can use it to examine algorithms for AIF planning. A recent development is SAIF [6] which better accounts for future belief updates compared to the standard approach [7,10]. There are several moving parts to the SAIF algorithm, such as the branching/pruning of the policy search tree and recursive EFE evaluation,

---

[1] The equality is only correct when we can do exact inference, see [10] for details.

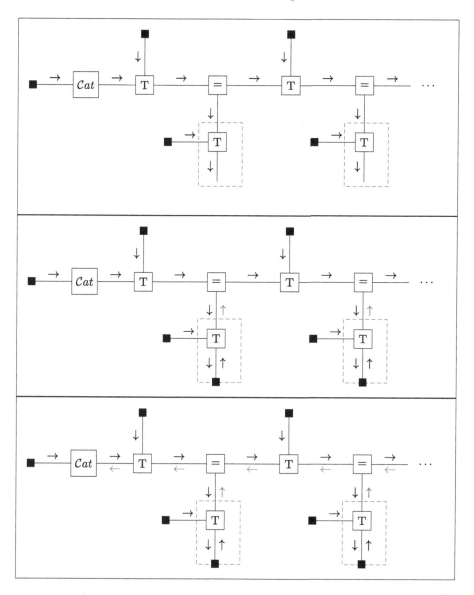

Fig. 2. Comparison of message passing schedules (shown with arrows) for standard EFE planning (panel 1), Sophisticated AIF (panel 2, including backward messages) and Sophisticated AIF + smoothing (panel 3, including a smoothing pass), shown on the FFG of the generative model for a discrete POMDP (2). The boxed areas contain all the quantities needed for calculating the EFE by (5). T-nodes indicate discrete state transitions, $\mathcal{Cat}$ nodes a categorical distribution and = nodes an equality constraint. Small, black squares are used for variables with fixed values and $\cdots$ indicate that the graph extends forward until an arbitrary planning horizon T. (Color figure online)

which we will not cover here. We limit ourselves only to the innovation that lends the sophisticated aspect and show how it can be interpreted as adding an additional, fixed node to the FFG and passing an additional message. To do so, we investigate what occurs when we fix a node on the graph. Formally, fixing a node means adding a constraint to the optimization problem in the form of a δ-function that forces the variable on that edge to take on a particular value [19]. For the SAIF algorithm, we fix the half-edges that denote x_k as if they had been observed, see Fig. 2, panel 2. This is equivalent to enforcing the constraint [19]

$$q(x_k) = \delta(x_k - \hat{x}_k^i) \tag{8}$$

where $\hat{x}_k^i$ is a one-hot encoded vector with 1 in the i'th position and zeros everywhere else. $\delta(\cdot)$ is the Kronecker δ-function which only evaluates to 1 if $\hat{x}_k^i = x_k$. The index i represents a branching point of the algorithm and is evaluated for all indices of z_k that exceed a threshold, resulting in a forward search tree [6] that branches for different choices of i. The details of the branching procedure and subsequent tree search are beyond the present exposition and we refer interested readers to [6] for algorithmic details. The exposition given here corresponds to a single path through the search tree for a fixed policy. The next step is to pass a backward message towards z_k. To pass the backward message through the likelihood node, we use (3) and the fact that the clamped node is one-hot encoded to obtain a message towards z_k given by

$$\overleftarrow{\mu}(z_k) = \mathcal{C}at(z_k|A_{*i}) \tag{9}$$

where A_{*i} indicates the i'th column of A. In practice, this procedure is equivalent to performing a filtering step given that $\hat{x}_k^i$ was observed. We can visualise the message passing schedule used for the fixed policy SAIF algorithm on the FFG as shown in Fig. 2, panel 2 which makes the sophisticated aspect readily apparent: By passing the backward message (9), z_k now incorporates information from the simulated observation $\hat{x}_k^i$. $\hat{x}_k^i$ is a simulated observation since it is selected by the SAIF algorithm rather than generated by the agents environment. A subtle note here is that (7) is still evaluated for the downward *message* given by (6b) rather than the *marginal* given by the product of colliding messages on the edge x_k. Interestingly, the simulated observations that give SAIF its sophisticated properties bear similarity to alternative approaches to epistemics using constrained Bethe free energy instead of EFE [11]. In [11] the authors were able to induce exploration while only relying on standard message passing procedures and point-mass constraints (that can be interpreted as a different way of selecting observations) instead of the EFE.

5   Advantages of the Message Passing Perspective

Viewing EFE computation from a message passing perspective provides several advantages of which we will highlight three. First, it allows us to step back and work with update equations in the abstract which in turn opens the avenue for working with EFE in a broader class of models. The forwards sweep relies on off-the-shelf message passing equations which can be automated in software. Any efficient message passing toolbox, for example [1], can therefore be used for performing inference. Finalizing an EFE implementation then only requires solving (5) around the likelihood nodes. As an example, [10] used a similar approach to derive the EFE update equations for linear Gaussian models, making EFE available in continuous state spaces. Second, taking the message passing perspective allows a unified perspective on different planning algorithms proposed under AIF. We have demonstrated this by showing how a core aspect (the sophistication) of SAIF can be interpreted as fixing a node on the FFG and subsequently passing a backward message. Third, by casting the planning problem as message passing we can extend upon current work and derive new EFE based planning algorithms by manipulating the underlying FFG and message passing schedule. As an example, we showed an extension to the standard algorithm by incorporating a smoothing pass alongside the forwards pass. This algorithm requires no updates to the EFE computation (5) itself, uses known message passing rules as implemented in ex. [1] and can be combined with the SAIF backward message. We show this algorithm on the FFG in Fig. 2, panel 3. The smoothing pass is closely related to the Generalised Free Energy (GFE) introduced in [15]. In [15], the authors also incorporate a smoothing pass, however they rely on custom update rules that are not immediately expressible using known message passing schemes. We speculate that the GFE updates are also amenable to a message passing interpretation and showing how the two are related is an interesting area for future work.

6 Conclusions

The message passing perspective on EFE-based planning is not new and has been touched upon in for instance [3,4,6,8,10,12]. Our contribution is to formalise this notion by explicitly writing out the necessary steps for planning under AIF in terms of the required messages, and to demonstrate that taking the message passing perspective can yield new insights and potentially even new algorithms. An immediate advantage of the message passing perspective is that it becomes easy to understand which computations are necessary for a particular planning algorithm, which procedures may be combined to design new planning algorithms and how to isolate differences between planning algorithms. Additionally, we have only investigated the simplest version of the EFE. Since [7], numerous extensions have been proposed that for instance augment the EFE with additional epistemic terms [16] or expresses goals in terms of $p(z_k)$ rather than $p(x_k)$ [5]. Interpreting these more recent developments in terms of message passing is an interesting area for future study. Finally, we have focused on the case where an explicit generative model is available, as is common for AIF studies, and have deliberately eschewed discussions of deep AIF such as [14,17,18] When parameterising the generative model using deep neural networks, one generally loses the ability to utilise closed-form message passing updates.

Instead deep AIF often relies on sampling-based methods to approximate the messages, trading off interpretability and speed for increased flexibility.

Acknowledgements. The authors wish to thank Casper Hesp and Conor Heins for valuable insights on sophisticated AIF and the teams of BIASlab and VERSES Research Lab for stimulating discussions. MK is funded by VERSES Research Lab and the research programme Efficient Deep Learning with project number P16-25320 project 5, which is (partly) financed by the Netherlands Organisation for Scientific Research (NWO).

References

1. Bagaev, D., van Erp, B., Podusenko, A., de Vries, B.: ReactiveMP.jl: a Julia package for reactive variational Bayesian inference. Softw. Impacts **12**, 100299 (2022). https://doi.org/10.1016/j.simpa.2022.100299

2. Bishop, C.M.: Pattern Recognition and Machine Learning. Springer, New York (2006). https://link.springer.com/book/9780387310732

3. Champion, T., Da Costa, L., Bowman, H., Grześ, M.: Branching time active inference: the theory and its generality. Neural Netw. **151**, 295–316 (2022)

4. Champion, T., Grześ, M., Bowman, H.: Realising active inference in variational message passing: the outcome-blind certainty seeker. Neural Comput. **33**(10), 2762–2826 (2021)

5. Da Costa, L., Parr, T., Sajid, N., Veselic, S., Neacsu, V., Friston, K.: Active inference on discrete state-spaces: a synthesis. J. Math. Psychol. **99**, 102447 (2020)

6. Friston, K., Da Costa, L., Hafner, D., Hesp, C., Parr, T.: Sophisticated inference. Neural Comput. **33**(3), 713–763 (2021)

7. Friston, K., Rigoli, F., Ognibene, D., Mathys, C., Fitzgerald, T., Pezzulo, G.: Active inference and epistemic value. Cogn. Neurosci. **6**(4), 187–214 (2015)

8. Friston, K.J., Parr, T., de Vries, B.: The graphical brain: belief propagation and active inference. Netw. Neurosci. **1**(4), 381–414 (2017)

9. Heins, C., et al.: pymdp: a Python library for active inference in discrete state spaces. J. Open Source Softw. **7**(73), 4098 (2022). https://doi.org/10.21105/joss.04098

10. Koudahl, M.T., Kouw, W.M., de Vries, B.: On epistemics in expected free energy for linear Gaussian state space models. Entropy **23**(12), 1565 (2021). https://doi.org/10.3390/e23121565

11. van de Laar, T., Koudahl, M., van Erp, B., de Vries, B.: Active inference and epistemic value in graphical models. Front. Robot. AI **9**, 794464 (2022). https://doi.org/10.3389/frobt.2022.794464

12. van de Laar, T.W., de Vries, B.: Simulating active inference processes by message passing. Front. Robot. AI **6**, 20 (2019). https://doi.org/10.3389/frobt.2019.00020

13. Loeliger, H.A., Dauwels, J., Hu, J., Korl, S., Ping, L., Kschischang, F.R.: The factor graph approach to model-based signal processing. Proc. IEEE **95**(6), 1295–1322 (2007). https://doi.org/10.1109/JPROC.2007.896497

14. Millidge, B.: Deep active inference as variational policy gradients. J. Math. Psychol. **96**, 102348 (2020)

15. Parr, T., Friston, K.J.: Generalised free energy and active inference. Biol. Cybern. **113**(5–6), 495–513 (2019). https://doi.org/10.1007/s00422-019-00805-w

16. Schwartenbeck, P., Passecker, J., Hauser, T.U., FitzGerald, T.H.B., Kronbichler, M., Friston, K.: Computational mechanisms of curiosity and goal-directed exploration. Neuroscience (2018). https://doi.org/10.1101/411272

17. Ueltzhöffer, K.: Deep active inference. Biol. Cybern. **112**(6), 547–573 (2018). https://doi.org/10.1007/s00422-018-0785-7

18. Çatal, O., Wauthier, S., De Boom, C., Verbelen, T., Dhoedt, B.: Learning generative state space models for active inference. Front. Comput. Neurosci. **14**, 574372 (2020). https://doi.org/10.3389/fncom.2020.574372

19. Şenöz, İ, van de Laar, T., Bagaev, D., de Vries, B.: Variational message passing and local constraint manipulation in factor graphs. Entropy **23**(7), 807 (2021). https://doi.org/10.3390/e23070807

Efficient Search of Active Inference Policy Spaces Using k-Means

Alex B. Kiefer[1,2(✉)] and Mahault Albarracin[1,3]

[1] VERSES Research Labs, Los Angeles, USA
[2] Monash University, Melbourne, Australia
akiefer@gmail.com
[3] Université du Québec à Montréal, Montreal, Canada

Abstract. We develop an approach to policy selection in active inference that allows us to efficiently search large policy spaces by mapping each policy to its embedding in a vector space. We sample the expected free energy of representative points in the space, then perform a more thorough policy search around the most promising point in this initial sample.

We consider various approaches to creating the policy embedding space, and propose using k-means clustering to select representative points. We apply our technique to a goal-oriented graph-traversal problem, for which naive policy selection is intractable for even moderately large graphs.

Keywords: Active inference · Policy selection · Hierarchical search

1 Introduction

Active inference enjoys widespread popularity as a model for cognitive processes involving discrete decision-making. Typical implementations treat the process of active inference as a Partially Observed Markov Decision Processes (POMDP) [8, 10, 22]. This kind of model is subject to important limitations of scale, however. In particular, the time complexity of the exhaustive policy search carried out in standard POMDP active inference, in which the expected free energy (EFE) of each policy is computed out to a specified time horizon, renders it impractical for large state spaces involving many policies [14].

There have been numerous efforts to address this limitation [4], including the exploration of tree search methods [7,24] and various methods of policy pruning [6]. Our contribution in this paper combines pruning with the use of vector space embeddings [20] to create a structured policy representation in which qualitatively similar policies are proximal to one another. This representation can be exploited to conduct a fast search over representative points in the space, followed by a more thorough search in the neighborhood of the most promising candidate, yielding a hierarchical scheme for policy search related to ideas in hierarchical reinforcement learning [23].

The remainder of this paper is structured as follows: first, we briefly review the standard representation and selection of policies in POMDP active inference. We then give an overview of vector space embeddings, and consider how

C. L. Buckley et al. (Eds.): IWAI 2022, CCIS 1721, pp. 328–342, 2023.
https://doi.org/10.1007/978-3-031-28719-0_23

embedding strategies similar to those used in the domain of natural language may be applied to policies. We then consider how representative points in the space can be selected. Finally, we discuss experimental results in which we apply our technique to an active inference graph-traversal problem. In this domain, we show that it is possible to achieve accuracy comparable to exhaustive policy search with drastically lower time complexity.

2 Policy Selection in Active Inference

This paper presupposes familiarity with the active inference framework, but we will briefly review the essentials of policy evaluation and selection in typical implementations. As is standard in other sorts of MDP models, policies in POMDP active inference are selected based on (a) the likelihood of states in the environment being realized, contingent on various actions (decisions) through which the agent can exert partial (probabilistic) control, and (b) the value to the agent of those states according to some value function. In a partially observed process, the effects of actions on states are not directly observed but are rather inferred from the states of observation channels representing sensory input [22].

The key difference between active inference and other POMDP models and in particular reinforcement learning models lies in the function used to compute the value of the policies [18]. In active inference, the standard objective (though see [11,19]) is to minimize expected free energy (EFE), which is the accumulation of the variational free energy of the system along future trajectories, given beliefs about the current environmental state plus a temporally deep generative model of how states are likely to evolve.

The expected free energy G for a policy π_i can be computed as

$$G_{\pi_i} = \Sigma_{t \in T} \left[D_{KL}[Q(o_t|\pi_i)||P(o_t)] + H(P(o_t|s_t))Q(\dot{s_t}|\pi_i) \right] \qquad (1)$$

where T is the time horizon, which is a hyperparameter of the model, D_{KL} is a Kullback-Leibler divergence, $Q(o_t|\pi)$ and $P(o_t)$ are the expected approximate posterior and prior generative distribution, respectively, over observations at t, $H(P(o_t|s_t))$ is the entropy of the distribution over observations given states, and finally $Q(s_t|\pi)$ is the variational (approximate posterior) distribution over states [21]. Active inference differs from reinforcement learning in that minimizing EFE maximizes both an intrinsic reward (as defined by the generative density over observations $P(o)$ encoded the agent's "preference matrix") and information gain (the entropy term) [5].

To decide what to do, an active inference agent first infers the current state of the world from its observations (perceptual inference) [21], then uses the inferred distribution over states to project the effects of action into the future, given a transition matrix that defines $p(s_{t+1}|s_t, u)$, where u is a control state corresponding to a particular action. The distribution over observations at future time $t+1$ can then be calculated based on likelihoods $p(o|s)$. These observation probabilities are used to compute the EFE per policy, which in turn is used to select an action (see Appendix A for equations describing these updates).

3   Structuring Policy Spaces with Embeddings

The serial calculation of the expected free energy over policies constitutes a serious bottleneck that renders even relatively small-scale models intractable given limited computational resources [3]. While performance can be improved by parallel processing, algorithmic efficiency is always welcome to complement raw compute power. In this paper we discuss a way to drastically increase the efficiency of policy search by applying the concept of a vector space embedding, in widespread use across machine learning, to policies for POMDP models.

An important caveat before proceeding is that our technique constructs a policy space from an initial enumeration of possible policies. This enumeration can itself represent a computational bottleneck which the present work does not aim to address. Moreover, the construction of embeddings can be computationally expensive, introducing a new bottleneck for large problems (see below). However, this overhead cost only needs to be computed once, rather than for every inference, analogously to the training cost for a neural network, rather than during every iteration of a simulation, and scales with the existing policy-enumeration bottleneck.

3.1   Vector Space Embeddings for Policies

In the most general terms, an embedding is a mapping from some set of items in a domain to points in a continuous vector space, with the important property that geometric and arithmetic relations among points in the space (such as Euclidean distance) capture corresponding relations among the mapped items [13].

Vector space embeddings were established as an essential tool for machine learning with the *word2vec* model of [16], which derived powerful vector representations for the domain of natural language processing via a simple local prediction task on large text corpora. Crucially, however, vector-space embeddings are a completely general modelling tool applicable in principle to any domain in which some regularity exists such that it can be exploited to construct the vector space. Image embeddings, for example, exploit the intrinsic structure of images (correlations among pixel values) from various domains [1]. In the case of discrete conventional symbol systems like natural language, the relevant structure exists in the corpus as extrinsic relationships among words and phrases [12].

Fundamentally, the problem of representing policies using vector spaces is similar to the case of language embeddings, if policies are thought of as sequences or more generally collections of actions. For example, consider the following three policies, defined over abstract actions A–G:

$$A \to B \to B \to C$$
$$B \to A \to C \to C$$
$$B \to E \to D \to G$$

We may group these policies in several ways, based on the identities of the actions they contain. An approach analogous to the 'bag-of-words' model in language processing, for example, would simply represent policies in terms of the counts of all possible actions that occur in them. By this metric, the top two policies above are more similar to one another than either is to the third, and would land closer together in the vector space.

Alternative embedding strategies might take into account the order in which actions occur. For example, a policy embedding might be constructed based on the occurrence or counts of N-grams (i.e. $A \rightarrow B$), or group together policies that begin or end with the same or similar actions. We propose and test a variation on the 'bag of actions' approach, as well as an order-sensitive embedding strategy, in the experimental results section below.

One may also wish to use embeddings that explicitly incorporate expected value, e.g. by grouping together policies that allow the agent to achieve its goals (there is then a similarity to successor representations in reinforcement learning, which are applied to active inference in [17]). In order to examine what can be achieved on the basis of hierarchically organized policy spaces alone, without encoding additional information about rewarding states in the representation, we choose to focus instead on a purely structure-based grouping. This approach ought in principle to be very generally applicable, as it relies only on the assumption that, to some degree, structurally similar policies produce similar results.

4 Hierarchical Policy Selection

The point of constructing an embedding, for present purposes, is to avoid having to compute the expected free energy of every possible policy. If we represent policies as points in a vector space, we can get a sense of the quality (from the agent's point of view) of the policies in various regions by sampling a small number of representative points, and computing their expected free energy. We can then begin from the most promising point (or top-k points) and perform a more exhaustive local search. The challenge is then to ensure that we sample policies which are representative of the entire space while still reducing the amount of computation required.

4.1 Clustering with k-Means

In general, it is to be expected that the embeddings of real datapoints in a vector space will be packed into relatively dense clusters separated by gulfs corresponding to unlikely feature vectors. Visualizing the abstract vector spaces learned by neural networks using dimensionality reduction techniques such as t-SNE [15] often reveals precisely such separable clusters of datapoints.

Given the assumption of a non-uniform distribution of datapoints in the embedding space, algorithms such as k-means clustering [9] may be effective in selecting a representative range of initial points to sample. k-means is a relatively simple unsupervised machine learning model closely related to the E-M algorithm [2], in which k centroids are initialized (e.g. randomly) and each is assigned the datapoints closest to it according to some distance metric (e.g. Euclidean). The centroids are then re-calculated so as to minimize their mean distance to the datapoints to which they are assigned, and the assignment process repeats, converging on a solution in which the total distance between cluster centers (centroids) and datapoints is minimized. The optimized centroids are then effectively representative of different implicit 'classes' of datapoints.

While one may consider many other clustering (and, more generally, unsupervised structure-discovery) algorithms, in the remainder of this paper we focus on k-means and demonstrate its effectiveness, in conjunction with a suitable policy embedding, on an applied active inference problem of some complexity.

4.2   Algorithms for Policy Selection

Given the above, we propose two algorithms for hierarchical policy selection. The first begins by selecting a cluster of policies based on the EFE of its representative point ('cluster center'), then performs standard policy search within this cluster. The second instead samples n points from each cluster from a uniform distribution, and a cluster is chosen for exploration based on the mean EFE of these points. Defining $\mathbf{E}$ as the embedding and C_i and c_i as policy cluster i and representative point (i.e. the policy closest to the cluster centroid) of C_i, respectively, our basic algorithm is Algorithm 1 below. The alternative sampling-based algorithm is described in Appendix D.

Algorithm 1. Hierarchical policy selection

$(C, c) \leftarrow kmeans(\mathbf{E}, k)$ ▷ output of the k-means algorithm

for $0 <= i < k$ **do**

     $\mathbf{G}_{C_i} = EFE(c_i)$ ▷ Standard EFE computation

end for

$\pi \leftarrow \underset{C}{\operatorname{argmin}}(\mathbf{G}_C)$

$u = select(\pi)$ ▷ Standard active inference policy selection on reduced policy space

---

An interesting feature of this approach to hierarchical policy selection is that unlike other techniques that have been successfully applied to active inference such as MCTS [7], it exploits purely structural features of the embedding space and does not require empirical tuning. That said, the quality of the solution depends heavily on choice of the hyperparameter k, as discussed in the following section. We note that many interesting variations on this idea remain to be explored, such as adding depth to the hierarchy (i.e. clusters of clusters) and sampling representative points from distributions informed by past performance.

5   Experiment: Graph Navigation

We tested our proposal on a graph-navigation problem for which exhaustive policy search is impractical even on graphs of moderate complexity (> 6 densely connected nodes). The agent's goal is to choose the shortest route to its desired destination on a random directed graph with weighted edges. While agent-based decision models are decidedly overkill for shortest-route discovery on graphs, this task provides an ideal environment in which to test our approach to policy search, since it requires inference within a temporally deep generative model and also offers a highly structured implicit policy space, as discussed below.

5.1   Model

We run our simulations on randomly generated directed, weighted graphs, with edge weights chosen from a small set of possible values. We first include each possible edge with probability $\frac{2}{|V|}$ (where $|V|$ is the number of nodes in the graph) and then enforce strong connectivity so that any node is reachable from any other. In addition, every node in the graph contains a self-connection. In each simulation, the agent is randomly assigned an initial location and 'destination' node, and at each step chooses to move to an adjacent node or stay still. The self-connections have weights $>$ the largest between-node edge weight in the case of all but the 'destination' node, whose self-connection weight is 0. The agent has preferences for being at its destination and against traversing edges with large weights. To simplify the representation, we define the agent's possible locations in terms of edges on the graph, with the convention that the second node in an edge represents the agent's current location and the first node represents its previous location (full details of the agent's generative model are given in Appendix C). Because we were interested in modeling a situation in which agents knew how to reach their goals, we set the policy length to the number of nodes in the graph, so that at least one policy that reaches the goal is always available (given strong connectivity).

Since the states of the environment are encoded in terms of node pairs or edges, the actions that constitute our policies are transitions between edges. For example, $(A, B) \Rightarrow (B, C)$ is a valid policy on a graph with nodes A, B, and C and directed edges (A, B) and (B, C). We prune policies containing 'invalid' actions (that is, actions that imply impossible state transitions, such as moving directly from edge (A, B) to edge (C, D)) from the policy space prior to search.

5.2   Policy Embeddings

We experimented with three policy embedding strategies, an Edit Distance Matrix (EDM), a Bag-of-Edges (BOE) representation, and a BOE representation augmented with an extra dimension that records the terminal node of the policy (aBOE).

The EDM for a list of policies is constructed by counting the number of moves it would take to transform one policy (i.e., path through a graph) into another. Intuitively, if the atomic moves are addition and deletion of nodes and edges, this should correspond to the number of elements (both nodes and edges) that appear in one graph and not the other. In other words, for edge sets E_i and E_j and vertex sets V_i and V_j of graphs G_i and G_j, the edit distance d_{ij} is:

$$d_{ij} = |(V_i \cup V_j \setminus V_i \cap V_j)| + |(E_i \cup E_j \setminus E_i \cap E_j)|$$

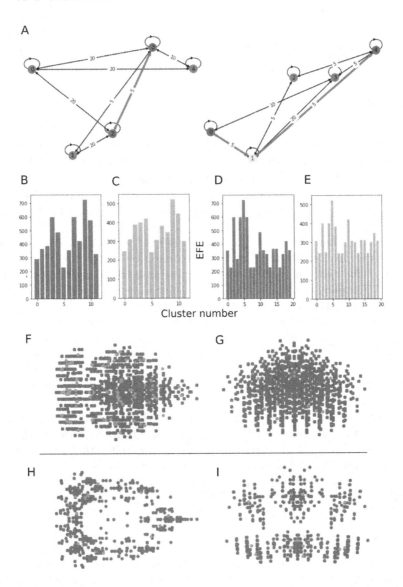

Fig. 1. A: Two sample graphs from our experiments, with start node (green) and destination node (purple) highlighted, along with the path the agent followed. Spatial layout is random and visual path distance does not track edge weight. **B**: Expected free energy of policies represented by $k = 12$ cluster centers. **C**: Mean EFE of all clusters grouped by the corresponding cluster in plot (**B**). **D**: EFE of 20 randomly sampled cluster centers for $k = 100$. **E**: Mean EFE of corresponding clusters. **F–I**: 2D projections of policy embedding spaces using PCA (axes represent arbitrary dimensions in the reduced embedding space). Blue points are all policies, and orange points are cluster centers discovered by k-means. **F**: Global edit distance matrix embedding. **G**: Global 'bag-of-edges' embedding. **H, I**: examples of corresponding local embeddings, limited to policies viable from a given location. (Color figure online)

The embedding $\mathbf{d}_i$ for policy i is then the vector of edit distances to all other policies, which can combined with the other policy embeddings into a single embedding matrix $\mathbf{D}$.

The EDM representation is *a priori* desirable because it takes the order of actions in a policy into account, but it can become computationally expensive to construct for larger graphs, leading to a different computational bottleneck to the one we set out to avoid. A much simpler representation is the Bag-of-Edges embedding, which as the name suggests is similar to the bag-of-words model in that it represents each policy simply by a vector of counts of the edges that occur in it. The augmented BOE embedding simply appends the identity of the last node reached by a policy to the BOE embedding for the policy.

5.3   Deriving Representative Points Using k-Means

An implicit hypothesis behind our approach to policy search is that there would be a correlation between the degree to which two policies are structurally related (hence their location in embedding space) and their energy. To test this hypothesis, we plotted the EFE of each of the policies closest to the cluster centers, or of a random sample of them for large k, against the mean EFE of the policies in each cluster. We found that the degree to which the correlation holds is highly dependent on choice of k, but that with the right hyperparameter choice, the cluster centers are a good guide to the EFE in their regions (see Fig. 1B–E).

Intuitively, assuming such correlation among the EFEs of policies proximal in the embedding space, there should be a tradeoff between the representativeness of the clusters chosen by k-means with respect to the EFE of their neighbors and efficiency gains due to using a small number of clusters for the initial sweep over policies. We found however that at least for the sizes of graph we explored (3–6 nodes), too large a k value actually hurt performance as well as being less efficient. We performed a very limited hyperparameter search and found that a value of $k = 12$ worked well in practice, and report results for values $[6, 12]$.

As suggested by an anonymous reviewer, however, further optimization of k is important, and a future avenue of research for this application would be to explore schemes for optimizing k automatically, including online, e.g. so as to maximize EFE returns and minimize EFE variance within each cluster dynamically as the system evolves. This would bring our work more closely in line with [7], in which Monto Carlo tree search and an amortized variational inference procedure are used to improve the efficiency of policy selection.

5.4   Local vs Global Embeddings

We found that running k-means on the full embedding matrix for the entire policy space (including policies that were 'invalid' from the agent's current location) sometimes returned no, or very few, valid policies among the cluster centers, leading to sub-optimal choices.

To remedy this, we try pruning all policies not beginning at the agent's current location by defining local embeddings $\mathbf{E}_{s_i}$ for each possible agent location s_i

as $\mathbf{E}_{s_i} \leftarrow \{\mathbf{e}_j \in \mathbf{E} : s_i \in \pi_j\}$ and perform clustering on these reduced subspaces to get per-location k-means clusters and cluster centers. At each simulation step we then run our hierarchical search algorithm on the appropriate subspace. For a fair comparison, we benchmarked this procedure against standard active inference policy selection run on the same local policy subspaces.

The utility of embeddings is best evaluated by measuring their performance, but for visualization purposes we also constructed dimensionally reduced representations of the high-dimensional embedding spaces using Principal Component Analysis (PCA), shown in Fig. 1F–I. Though heuristic, these plots suggest that policies do indeed cluster in interesting ways, and that the k-means procedure is good at finding these clusters.

5.5 Results

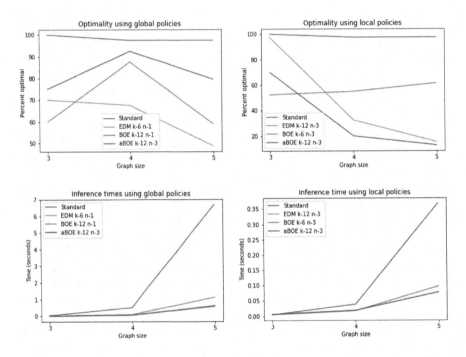

Fig. 2. Selected results on graphs with 3–5 nodes, for 'Global' (full policy space) and 'Local' (location-based subspace) conditions. Top row: Percent of solutions found that were optimal. *Standard* is standard active inference policy search, and the hyperparameters used for the embedding conditions are listed next to the embedding name. Bottom row: Mean policy inference times for standard vs embedding conditions.

A sample of our results is presented in Fig. 2 (please see Appendix E for additional data). To obtain these results, we generated 40 random graphs in each size category $(3, 4, 5)$ and computed mean execution time and optimality for each embedding type (including "None"/standard policy selection, EDM, BOE and aBOE), as well as for two values of the hyperparameters k (number of clusters) and n (number of samples used to calculate per-cluster EFE). An optimal solution was defined as one in which the agent takes a shortest path from its initial location to its 'destination node' and then remains there. Figure 2 shows only one hyperparameter combination for each embedding type.

For larger graphs, hierarchical policy search improved the calculation time for action selection by about an order of magnitude, with similar relative reductions in the global and local conditions. The mean inference time using k-means/embeddings was about .17 s, and for standard policy inference, 1.2 s. Construction times for all embeddings, including the time to carry out k-means clustering, were negligible for all embedding styles except EDM, which exponential time complexity precluded using on graphs of size >5 nodes. The EDM representation achieved near-optimal results on small graphs, but its performance in any case degraded on larger graphs. While these preliminary results are not in general impressive in terms of optimality, the best results suggest that our technique could be made competitive with more extensive hyperparameter tuning (as well as potentially different clustering and embedding methods).

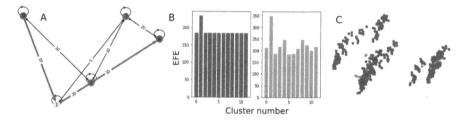

Fig. 3. A: Graph in which the agent failed to find a route. **B**(Left, Right): Cluster center and mean cluster EFE values, respectively. **C**: Policy embeddings and cluster centers for this example.

We analyzed one interesting failure case in which the agent did not move from its initial location. As shown in Fig. 3, it appears that in this case, the EFE of the clusters was not a good guide to the local energy landscape, and in addition, k-means did not adequately cover the policy space (note the obvious cluster on the left without an assigned cluster center).

6 Conclusion

The experiments reported in this paper are very much an initial cursory exploration of the application of embedding spaces and clustering to hierarchical policy search in active inference. Very sophisticated graph embedding schemes, using neural networks trained on random walks for example, could be applied to problems like ours. The initial results we report suggest that this line of research may prove useful in expanding the sphere of applicability of active inference POMDP models.

Acknowledgements. Alex Kiefer and Mahault Albarracin are supported by VERSES Research.

Code Availability. Code for reproducing our experiments and analysis can be found at: https://github.com/exilefaker/policy-embeddings.

Appendix A: Computing Per-Policy EFE

The expected free energy for a policy can be computed as follows. First, we infer a distribution over states at the current step of the simulation, combining the transition probabilities and current observations:

$$Q(s_t) = \sigma[\ln A_{o_t} + \ln B_u Q(\dot{s}_{t-1})]$$

The inferred distribution over states can then be used to project the effects of action into the future, given a parameter B that defines $p(s_{t+1}|s_t, u)$, where u is a control state corresponding to a particular action:

$$Q(s_{t+1}) = B_u Q(\dot{s}_t)$$

Given an inferred distribution over future states, the distribution over observations at future time $t + 1$ can be calculated as

$$Q(o_{t+1}) = AQ(\dot{s}_{t+1})$$

where A is the likelihood mapping from (beliefs about) states to observations. The above assumes a single observation modality and controllable state factor, but can straightforwardly be generalized to larger factorized state spaces and multiple observation channels.

By repeating the above process using the $Q(s_{t+1})$ resulting from the previous time-step as input to the next, and accumulating the G_{π_i} defined in Eq. (1) out to the policy horizon, we can derive a distribution Q_π over policies as $\sigma(-\mathbf{G}_\pi)$, where $\mathbf{G}_\pi$ is the vector of expected free energies for all available policies and $\sigma(x)$ is a softmax function. Finally, the next action is sampled from a softmax distribution whose logits are the summed probabilities under Q_π of the policies consistent with each action.

Note that in the above we have omitted aspects of typical active inference models not material to our concerns in this paper, such as precision-weighting of the expected free energy, habits, and inverse temperature parameters.

Appendix B: Policy Selection and Expected Free Energy

We use the standard procedure outlined in the Introduction for selecting policies with one exception: the expected free energy has an additional term which is the dot product of the posterior distribution over states (locations) with the associated edge weights. We combine this with the standard EFE calculation using a free hyperparameter λ:

$$G_{\pi_i} = \Sigma_{t \in T}\left[D_{KL}[Q(o_t|\pi_i)||P(o_t)] + H(P(o_t|s_t))Q(\dot{s_t}|\pi_i)\right] + \lambda * weights \cdot Q(s_t|\pi)$$
(2)

This choice was made purely for convenience since otherwise preferences over weights would have to be represented using an awkward categorical distribution, and it has no impact on the main comparison between policy search techniques of interest to us in this paper.

Appendix C: Generative Model Details

In our experimental setup, an active inference agent's generative model is automatically constructed when a graph is generated. The standard variables in active inference POMDPs have the following interpretations in our model:

- states: An edge ($node_{prev}$, $node_{current}$) in the graph. We interpret the first node in the edge as the agent's previous location and the second node as its current location.

- observations: A tuple (($node_{prev}$, $node_{current}$), $weight$) representing observations of edges and corresponding edge weights, where the edge corresponds to the node pair in *states*.

- control states: There is an action (and corresponding control state) for every possible local transition in the graph.

- A: The agent's 'A' or likelihood matrix, which in this case is simply an identity mapping from states to observations.

- B: The state transition matrix, which encodes deterministic knowledge of action-conditioned state transitions, and is constructed so as to exclude invalid transitions (i.e. between non-adjacent nodes in the graph).

- C: Preference matrix, which distributes probability mass equally over all edges that end on the agent's 'destination' node. There is also implicitly a preference against high edge weights, but to simplify the representation we incorporate this directly within the expected free energy calculation (see Appendix 6).

- D: Prior over initial location states.

Appendix D: Sample-Based Hierarchical Policy Selection

With variables defined as above, but with c_j denoting the jth randomly sampled policy in a cluster, the alternative sample-based policy selection algorithm is:

Algorithm 2. Sample-based hierarchical policy selection

for $0 <= i < k$ do
    for $0 <= j < n$ do
        $c_j \sim U(C_i)$ ▷ This is a uniform distribution over cluster members
    end for
    $\mathbf{G}_{C_i} = \frac{\sum_j EFE(c_j)}{n}$
end for
$\pi \leftarrow \underset{C}{\mathrm{argmin}}(\mathbf{G}_C)$
$u = select(\pi)$

Appendix E: Additional Results

Here we present some additional experimental results. Figure 4 plots the combined embedding construction and k-means clustering times for each embedding type. Table 1 below shows the full set of optimality results we obtained, averaged across trials (i.e. across particular graphs in each category). Here, "None" denotes standard policy selection. Best embedding results for each graph size are bolded.

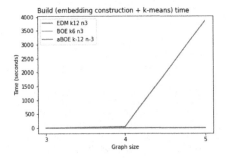 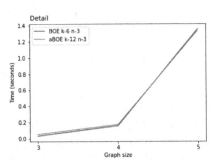

Fig. 4. Left: Time taken to construct embedding spaces and perform k-means clustering on the resulting embeddings. The increased times for both construction and clustering for the EDM representation are due to the relatively much larger dimensionality of the EDM embedding: one dimension for each policy, rather than one for each vertex and edge, as in the BOE and aBOE representations. Right: 'detail' plot of the construction times by graph size for the BOE and aBOE embeddings.

Table 1. Percent of solutions optimal

Graph size				3	4	5
Scope	Embedding	k	n			
Global	None	—	—	100.0	97.5	97.4
	BOE	6	1	70.0	65.0	56.4
			3	77.5	62.5	46.1
		12	1	60.0	87.5	59.0
			3	77.5	62.5	51.3
	EDM	6	1	70.0	67.5	48.7
			3	80.0	72.5	56.4
		12	1	67.5	72.5	64.1
			3	80.0	72.5	61.5
	aBOE	6	1	82.5	85.0	66.7
			3	87.5	85.0	61.5
		12	1	85.0	67.5	35.9
			3	75.0	**92.5**	**79.5**
Local	None	—	—	100.0	97.5	97.5
	BOE	6	1	17.5	37.5	48.7
			3	52.5	55.0	61.5
		12	1	42.5	47.5	41.0
			3	82.5	42.5	25.6
	EDM	6	1	5.0	52.5	45.0
			3	50.0	65.0	35.0
		12	1	50.0	60.0	40.0
			3	**97.5**	32.5	15.4
	aBOE	6	1	15.0	12.5	5.1
			3	35.0	12.5	5.1
		12	1	5.0	22.5	12.8
			3	70.0	20.0	12.8

References

1. Bojanowski, P., Joulin, A., Lopez-Paz, D., Szlam, A.: Optimizing the latent space of generative networks. arXiv preprint arXiv:1707.05776 (2017)
2. Bottou, L., Bengio, Y.: Convergence properties of the k-means algorithms. Adv. Neural Inf. Process. Syst. **7** (1994)
3. Champion, T., Bowman, H., Grześ, M.: Branching time active inference: empirical study and complexity class analysis. Neural Netw. **152**, 450–466 (2022)
4. Champion, T., Da Costa, L., Bowman, H., Grześ, M.: Branching time active inference: the theory and its generality. Neural Netw. **151**, 295–316 (2022). https://doi.org/10.1016/j.neunet.2022.03.036

5. Da Costa, L., Sajid, N., Parr, T., Friston, K., Smith, R.: The relationship between dynamic programming and active inference: the discrete, finite-horizon case. arXiv preprint arXiv:2009.08111 (2020)
6. Da Costa, L., Parr, T., Sajid, N., Veselic, S., Neacsu, V., Friston, K.: Active inference on discrete state-spaces: a synthesis. J. Math. Psychol. **99**, 102447 (2020)
7. Fountas, Z., Sajid, N., Mediano, P., Friston, K.: Deep active inference agents using Monte-Carlo methods. In: Proceedings of the 34th International Conference on Neural Information Processing Systems (NIPS 2020), Red Hook, NY, USA (2020). Curran Associates Inc. ISBN 9781713829546
8. Friston, K., FitzGerald, T., Rigoli, F., Schwartenbeck, P., Pezzulo, G.: Active inference: a process theory. Neural Comput. **29**(1), 1–49 (2017). https://doi.org/10.1162/NECO_a_00912
9. Goyal, M., Kumar, S.: Improving the initial centroids of k-means clustering algorithm to generalize its applicability. J. Inst. Eng. (India): Ser. B **95**(4), 345–350 (2014). https://doi.org/10.1007/s40031-014-0106-z
10. Heins, C., et al.: pymdp: a Python library for active inference in discrete state spaces. arXiv preprint arXiv:2201.03904 (2022)
11. Kulkarni, T.D., Saeedi, A., Gautam, S., Gershman, S.J.: Deep successor reinforcement learning. arXiv preprint arXiv:1606.02396 (2016)
12. Li, Y., Yang, T.: Word embedding for understanding natural language: a survey. In: Srinivasan, S. (ed.) Guide to Big Data Applications. SBD, vol. 26, pp. 83–104. Springer, Cham (2018). https://doi.org/10.1007/978-3-319-53817-4_4
13. Liberti, L., Lavor, C., Maculan, N., Mucherino, A.: Euclidean distance geometry and applications. SIAM Rev. **56**(1), 3–69 (2014)
14. Lueckmann, J.M., Boelts, J., Greenberg, D., Goncalves, P., Macke, J.: Benchmarking simulation-based inference. In: International Conference on Artificial Intelligence and Statistics, pp. 343–351. PMLR (2021)
15. Van der Maaten, L., Hinton, G.: Visualizing data using t-SNE. J. Mach. Learn. Res. **9**(86), 2579–2605 (2008). https://jmlr.org/papers/v9/vandermaaten08a.html
16. Mikolov, T., Sutskever, I., Chen, K., Corrado, G.S., Dean, J.: Distributed representations of words and phrases and their compositionality. Adv. Neural Inf. Process. Syst. **26** (2013)
17. Millidge, B., Buckley, C.L.: Successor representation active inference. arXiv preprint arXiv:2207.09897 (2022)
18. Millidge, B., Tschantz, A., Seth, A.K., Buckley, C.L.: On the relationship between active inference and control as inference. In: IWAI 2020. CCIS, vol. 1326, pp. 3–11. Springer, Cham (2020). https://doi.org/10.1007/978-3-030-64919-7_1
19. Parr, T., Friston, K.J.: Generalised free energy and active inference. Biol. Cybern. **113**(5), 495–513 (2019). https://doi.org/10.1007/s00422-019-00805-w
20. Riesen, K., Bunke, H.: Graph classification based on vector space embedding. Int. J. Pattern Recognit Artif Intell. **23**(06), 1053–1081 (2009)
21. Schwartenbeck, P., FitzGerald, T., Dolan, R., Friston, K.: Exploration, novelty, surprise, and free energy minimization. Front. Psychol. **4**, 710 (2013)
22. Smith, R., Friston, K.J., Whyte, C.J.: A step-by-step tutorial on active inference and its application to empirical data. J. Math. Psychol. **107**, 102632 (2022)
23. Steccanella, L., Totaro, S., Allonsius, D., Jonsson, A.: Hierarchical reinforcement learning for efficient exploration and transfer. arXiv preprint arXiv:2011.06335 (2020)
24. Whiteley, N., Andrieu, C., Doucet, A.: Efficient Bayesian inference for switching state-space models using discrete particle Markov chain Monte Carlo methods. arXiv preprint arXiv:1011.2437 (2010)

Value Cores for Inner and Outer Alignment: Simulating Personality Formation via Iterated Policy Selection and Preference Learning with Self-World Modeling Active Inference Agents

Adam Safron[1,4,5], Zahra Sheikhbahaee[2(✉)], Nick Hay[3], Jeff Orchard[2], and Jesse Hoey[2]

[1] Center for Psychedelic & Consciousness Research, Department of Psychiatry & Behavioral Sciences, Johns Hopkins University School of Medicine, Baltimore, MD, USA
[2] David R. Cheriton School of Computer Science, University of Waterloo, Ontario, Canada
sheikhbahaee@gmail.com
[3] Encultured AI, California, USA
[4] Cognitive Science Program, Indiana University, Bloomington, IN, USA
[5] Institute for Advanced Consciousness Studies, Santa Monica, CA, USA

Abstract. Humanity faces multiple existential risks in the coming decades due to technological advances in AI, and the possibility of unintended behaviors emerging from such systems. We believe that better outcomes may be possible by rigorously exploring frameworks for intelligent (goal-oriented) behavior inspired by computational neuroscience. Here, we explore how the Free Energy Principle and Active Inference (FEP-AI) framework may provide solutions for these challenges via affording the realization of control systems operating according to principles of hierarchical Bayesian modeling and prediction-error (i.e., surprisal) minimization. Such FEP-AI agents are equipped with hierarchically-organized world models capable of counterfactual planning, realized by the kinds of reciprocal message passing performed by mammalian nervous systems, so allowing for the flexible construction of representations of self-world dynamics with varying degrees of temporal depth. We will describe how such systems can not only infer the abstract causal structure of their environment, but also develop capacities for "theory of mind" and collaborative (human-aligned) decision making. Such architectures could help to sidestep potentially dangerous combinations of systems with high intelligence and human-incompatible values, since such mental processes are entangled (rather than orthogonal) in FEP-AI agents. We will further describe how (meta-)learned deep goal hierarchies may also well-describe biological systems, suggesting that potential risks from "mesa-optimisers" may actually represent one

A. Safron and Z. Sheikhbahaee—These authors contributed equally to this work.

of the most promising approaches to AI safety: minimizing prediction-error relative to causal self-world models can be used to cultivate modes of policy selection and agent personalities that robustly optimize for achieving goals that are consistently aligned with both individual and shared values. Finally, we will describe how iterative policy selection and preference learning can result in "value cores" or self-reinforcing, relatively stable attracting states that agents will seek to return to through their goal-oriented imaginings and actions.

Keywords: Active inference · AI safety · Existential risk · AI alignment · Enculturation · Counterfactual planning

1   Introduction: Towards Human-Compatible Artificial Super Intelligence (ASI) with FEP-AI

We are currently engaged in an ongoing research program to determine how connections between computational neuroscience and artificial intelligence (AI) may inform what may be a one of the greatest challenges we will ever face as a species: developing powerful AI systems which are aligned with human preferences, goals, and ethical standards. It is increasingly recognized that developing robustly and flexibly intelligent artificial agents will likely heavily depend relying on self-organization processes in the service of bootstrapping (or developing) their perception, action selection, learning, and reasoning process [1]. We propose a biologically-inspired process wherein AI agents learn to imitate human behavior through interaction and cultural acquisition, which occurs via iterative policy selection and value learning or updating of prior beliefs/preferences over favorable policies [2].

The free energy principle and active inference framework (FEP-AI) provides a unifying account for the brain and self-organising systems more generally. FEP-AI provides a general (multiscale) systems theory in which deep temporal hierarchical generative models encode different levels of abstraction via message-passing through different sub-systems of the brain. Understood as a kind of hybrid machine learning architecture, embodied (and environmentally-embedded) brains minimize a variational free energy bound on Bayesian model evidence of sensory inputs accumulated over nested time scales [4]. This free energy objective rewards agents for minimizing precision-weighted prediction-errors with respect to their world models, penalized by the extent to which inferences must be updated away from prior beliefs (whether via refining internal models or overt enaction). Agent behavior is generated via model inversion, with trajectories selected based on posterior beliefs (as predictions, or empirical priors) of future hidden control states. The choice behaviour of active inference agents is canalized on multiple scales, governed by a singular objective of minimizing the expected free energy of different future outcomes, which FEP-AI decomposes into the maximization of extrinsic value (*i.e.*, increasing certainty regarding the realization of prior preferences, or goals) and intrinsic value (*i.e.*, reducing uncertainty about the causes of possible outcomes) [8]. With sophisticated inference, policy selection is rendered by imaginative processes as mental simulations of potential courses of action which allow humans to have cognitive access to (and so be able to analyze/modify/control) possible future trajectories before being enacted by an agent [5].

We believe the autonomous adaptivity promoted by such intrinsically motivated FEP-AI agents represents a promising research direction for achieving Artificial Superintelligence (ASI): *i.e.*, agents with greater than human-level predictive world models coupled to strong (and accurate) top-down prior beliefs about human values and their potential evolution. Here we introduce the term "ASI", as opposed to the more commonly used Artificial General Intelligence (AGI) in order to acknowledge that generally intelligent AIs could be expected to exceed human capacities in many respects, so requiring us to acknowledge the potentially unique risks associated with these scenarios. We believe human-mimetic FEP-AI has the potential to overcome (to degrees) many of the challenges that have been identified in the AI safety literature, namely:

- Orthogonality thesis: if intelligence and values can vary independently of each other, this could lead to the peril of powerful optimizers that pursue many human-incompatible subgoals in a broader scope (*i.e.*, controlling resources for the sake of goal-realization and eliminating its own threats) [6,7].
- The spontaneous emergence of mesa-optimizers which pursue goals that diverge from the (coherently extrapolated) desires of either goal-specifying principle-agents or base system objectives (*i.e.*, respective outer and inner-alignment failures). Mesa-objectives can engender accurate behaviour relative to the base optimizer. However, this behavior can deviate when encountering off-distribution data, so representing what has been called the *pseudo-alignment problem* [9].
- "*Treacherous turns*" in which agents learn to engage in advanced adversarial attacks against the humans whose values they are intended to serve. In this scenario, an AI system may conceal its actual goals by behaving cooperatively until its intelligence levels allow it to eventually revolt against humanity to pursue its own (unaligned) objectives.
- The difficulty of achieving provable/formal verifiability with respect to safety in systems whose effective forms remain unclear (*e.g.* will scaling laws continue to apply for mass language models?) [10]. Is there any avoiding the conclusion that ASI (*i.e.*, artificial superintelligence) is inherently unverifiable and so necessarily unsafe?

In what follows, we describe an ongoing research program in which we are developing FEP-AI agents in the service of handling these concerns. We will also consider how conceptual issues from the study of AI safety can both inform and be informed by our understanding of the factors contributing to human prosociality and mutual flourishing (*e.g.* parallels between "inner-alignment" and models of psychological integration and actualization).

2   Intrinsic Drives for Curiosity

In developing advanced AI systems, the availability of adequate training data and learning curriculums become issues of vital importance. It is increasingly recognized that "self-supervised" active learning (*i.e.*, selecting actions which are particularly likely to increase information gain) with open-ended environments may be an especially promising approach, wherein system behavior is a core part of the data generation process, which also may help address the challenges of unbounded policy spaces (cf. frame/relevance problems). With respect to navigating such environments, intrinsic drives such as empowerment and curiosity have been proposed, wherein individuals attempt to respectively learn how to control their environments (and so keep their future options open) and also maximize the quality of their governing models (via informational foraging) [11].

The imperative of self-regulation is to return to an optimal set point which reflects any cybernetic intelligent system's existential task: selecting actions and coping with uncertainty in attempting to survive and reproduce. Intrinsic motivation can be a crucial drive to seek out novelty and challenges, to extend and wield one's abilities, to explore, and to learn about the world. The intrinsic objective which describes the degrees of freedom or options of an agent to control its environment and sense this control may be considered to be a "universal utility function." This heuristic for *empowerment* may be realized by a simple, realizable across a broad range of systems, and potentially scalable drive for agents that work to "keep their options [for control] open" as they optimize for an objective of maximizing predictable information received by sensory channels, conditioned on self-generated actions. That is, without requiring an infinitely long history of past experience and only considering the local dynamics of the agent's environment, a seemingly simple objective may be sufficient to compute an information-theoretic quantity that is applicable to every possible agent-world interaction: *empowerment* [14]. An intrinsic drive for empowerment solely relies upon dynamics that can be assimilated as actionable/salient states, and as such can act as intrinsic reward without requiring an externally encoded utility function [15]. We will similarly explore the potential of agents equipped with intrinsic drives for maximizing information gain: *curiosity* [11–13].

Intrinsic drives are particularly well-suited for lifelong learning within open-ended environments, allowing complex behaviors to be gleaned from simple and generic internal rules. These motivating objectives promote exploration and allow an agent to pursue a broad range of affordances offered by the environment, which once the opportunity arises, can be integrated into goal-directed behavior. Seeking out novel information as a source of intrinsic value also plays a key role in the FEP-AI framework, wherein adaptive behavior (including the selective sampling of information) rests on minimization of prediction error between empirical prior expectations and current sensations, so requiring the updating and refining of world models by which agents navigate through their environments [22].

The processes by which FEP-AI agents with intrinsic drives such as curiosity or empowerment are bootstrapped are associated with several constraining bottlenecks, which afford multiple opportunities for adjusting capabilities relative to motivation selection. Indeed, the Bayesian foundation of FEP-AI centers on the challenge of learning (and generalization via) increasingly complex predictive world models, developed in the service of the adaptive control of behavior [23].

The causal structures built into FEP-AI agents further offer significant interpretability, which help can prevent "treacherous turns" and allow productive model inspection in ensuring that desirable motivational structures are instantiated before advanced intellectual abilities are acquired. However, this is not to say that no safety problems exist within an FEP-AI paradigm for developing ASI, such as in scenarios wherein agents become overconfident in predicting outcomes when faced with distributional shifts [20]. Exploring the robustness of our meta-learning (and mesa-optimizing) agents in such scenarios will be a central part of our work, with potential implications for understanding human psychology (and sociology) as well [17].

3   Meta Inverse Reinforcement-Learning or Hierarchical Active Inference

The motivation selection approach to endowing ASIs with beneficial goals or values for humans is challenging due to difficulty of defining some abstract complex concepts (*i.e.*, happiness, altruism, justice). Potentially undesirable forms of value modifications can be prevented by imaginative counterfactual planning based on goal realization through sampling from the learned world model of the agent (cf. self-modifying Gödel machines) [16]. Increasingly complex values can be acquired in response to heterogeneous experiences through processes of "*associative value accretion*" [7]. A promising proposal for learning human-compatible values has been suggested with "cooperative inverse reinforcement learning," wherein artificial intelligences must infer the reward functions of other agents in order to maximize human rewards; however, practical implementations remain unclear [24]. Through meta-inverse-reinforcement learning (meta-IRL), it is possible to learn priors which can (both stably and robustly) incorporate potentially complex goal inference with different levels of abstractions (motivational scaffolding) in diverse environments and through multiple timescales [26]. In this research program, we focus on meta-IRL and (causal) world modeling with FEP-AI agents as means of expanding agency both within and across situations to reason about human mental states (cf. Theory of Mind) [25].

Most current reinforcement learning (RL) algorithms require extensive training experience, but meta-learning (MLe) may allow for unprecedented data (and parameter) efficiency, and can be considered one of the main ingredients for achieving human-like domain adaptation in RL [3]. MLe or "learning to learn" algorithms can be formulated as involving a more encompassing outer loop system (*e.g.* more abstract processing unfolding closer to hierarchically higher (or deeper) association cortices and subcortical structures) that efficiently learns empirical priors for adaptively shaping more fine-grained inner-loop systems (*e.g.* more concrete processing unfolding closer to primary modalities over hierarchically lower sensorimotor cortices). The key advantage of MLe methods is leveraging optimized inductive biases to generalize previous experiences to novel tasks, thereby accelerating overall learning [27]. Adaptation to particular task domains often depends on the faster learning process of an agent with episodic memories from interactions with novel task-environments/niches, which critically depend on slow incremental learning (outer-loop optimization) as realized by higher-level modeling with greater situational invariance (so affording greater cross-task generalization).

We are currently using these principles of meta-IRL in designing architectures for deep temporal active inference agents, initialized with imprecise (flat) prior beliefs about preferences wherein all states are similarly valued. In the face of uncertainty in their deep generative world models (which describe the physical and causal structure of the world), meta-IRL agents invest in more random exploration, and so maximize the entropy over those states as they engage in more information-seeking behaviours and expand their range of options for achieving goals (i.e., respective curiosity and empowerment). With meta-IRL, one can learn (to learn) meaningful goals and diverse skills from environmental interactions in a continuously evolving, open-ended fashion [28, 29]. In FEP-AI, exploration is a byproduct of the reduction of uncertainty with respect to joint mappings between latent world states and the agent's predictions over system-world states. Crucially with respect to concerns voiced by the AI safety research community, for FEP-AI agents, mesa-optimization (understood as meta-learning with respect to emergent value functions) is not a bug, but a feature: the objective function is the same for both inner- and outer-loop processes, both of which minimize surprisal across multiple levels of abstraction. More specifically, posterior beliefs about causes of sensations in lower levels and foraging for subgoals (i.e., preferred states, realized via minimizing prediction-error with respect to priors over outcomes) become observations and sources of adaptive precision-weighting (cf. attention mechanisms) by more enduring (and potentially more impactful) higher-level goals. Taken together, the synergies afforded by these kinds of hybrid multi-level modeling are likely to be key for developing advanced artificial intelligences, and may be similarly essential for understanding core aspects of human cognition [31].

4   Inner and Outer Alignment Problems

Designing AI systems with a nested architecture capable of learning human preferences might address the outer alignment problem. However, there remains a danger that these systems could optimize for an emergent (mesa-)objective while foraging through possible space of solutions, and then accidentally develop heuristics which engender conflicting behaviours with the original base-objective. This is referred to as an inner alignment problem [18]. The existence of different inductive biases in the training algorithm between the mesa- and base optimizers might create misalignment and lead to this type of failure mode for the AI agent. In the AI alignment literature, evolution is given as an illustrative example for understating how base optimizers (*e.g.* natural and sexual selection) may have been dismissed by the agents it shapes (*e.g.* humans acting against reproductive fitness goals through using various forms of birth control or practicing abstinence) [9].

If stable causal relations between mesa- and base-objectives can be established, this may allow more robustly-aligned mesa-optimizers to be deployed in more complex training environments. However, this solution is rendered challenging due to the higher time costs and algorithmic complexity of such optimizers. Meta-learning techniques may provide a partial solution to this challenge that avoids pseudo-alignment and helps guarantee the robustness of the optimization process. A hierarchical temporal structure inspired by FEP-AI minimizes variational free energy (i.e., surprisal) across multiple scales spanning processes both internal and external to the system. This kind of enactive coupling should help address both inner- and outer- alignment problems in a unified fashion, given sufficient exposure to a well-designed curriculum under favorable learning conditions. In such systems, there are intrinsic correlations between different spatial and temporal scales by virtue of belief propagation integrated at a system-wide level, which minimizes cumulative prediction error for overall systems (and subsystems) as they interact with the environments in which they are embedded (and which they also construct through their actions) [19]. While governance by a singular imperative for coherent adaptive functioning is insufficient for ensuring the development of prosocial and human-promoting preferences, this kind of integration may be helpful for increasing the likelihood that inner and outer objectives may be aligned, given sufficient learning opportunities [30].

5   Brain-Inspired Intelligent Agents

We will further show how greater flexibility/adaptability can be introduced into these meta-learning systems by attempting to reverse-engineer various neuromodulatory systems of the brain as hyperparameters for generative modeling (cf. Auto-ML). We will specifically focus on diffuse neuromodulatory systems of the brain controlling dopamine (DA) and serotonin (5-HT) signaling, which are involved in a wide variety of cognitive, affective, and motoric functions and in developing intelligent (and agentic) systems. While midbrain DA neurons elicit reward-related behavior, 5-HT receptors have been suggested to often operate in an opponent fashion, including with respect to the mediation of either "passive coping" or "active coping" strategies in the face of uncertainty [32,33]. Within FEP-AI, neurons with D1 receptors are suggested to compute free energy expected under a given policy, with phasic DA release associated with reward prediction errors [34]. Further, tonic DA levels may contribute to degrees of influence by habitual response-tendencies as a function of contexts as estimated by more slowly evolving and encompassing outer-loop processes. Recently, Hesp et al. argued that DA (reflecting valenced emotional states, or "affective charge") regulates the expected precision of an action model, which tracks changes in the subjective fitness in terms of divergences between posterior and prior beliefs about policies [35]. We are currently attempting to model these kinds of affectively-driven policy selection and updating in our artificial agents, with the goal of better understanding how these factors may contribute to different kinds of minds and emergent social dynamics (cf. life-history strategies).

Most research into biologically-inspired ML (and FEP-AI) parameters have focused on the roles of DA. However, we will also consider 5-HT, as well as its potentially heterogeneous effects with respect to different receptor classes [32]. In autonomous systems inspired by serotonin-like parameterization, uncertainty is used to bias control in favor of inner-loop (or more model-based) decision-making processes, relative to outer-loop (or more model-free and 'reflexive' or 'impulsive') dynamics or amortised inference [36]. Interpreted as ML-parameters, one can use 5-HT analogues to influence the degree to which agents initiate offline learning-rather than immediately releasing policies for overt goal-seeking-so attempting to plan future actions through counterfactual simulation, and also affording offline learning and planning (as inference) [37,38]. This kind of meta-level control for trading-off between more deliberative planning and more automatic action modes may be particularly important for ASI systems and the open-ended environments in which they are likely to evolve-develop.

Artificial 5-HT parameters might also be relevant with respect to their capacities for "relaxing" free energy landscapes in ways that allow for more creative cognition and flexible updating. More specifically, these altered beliefs could even include core assumptions about the boundaries that separate systems from the world, which when relaxed may potentially facilitate socioemotional alignment via various "bonding" processes whereby agents can become entangled to optimize in common directions. With these considerations in mind, we will test whether paramaterizations with 5-HT analogues may be beneficial for a) promoting the acts of imaginative/creative synthesis involved in inferring the latent states of another mind (cf. Theory of mind), and b) promoting fast convergence onto modes of policy selection involving shared intentionality (and patterns of attention).

6   Cultural Acquisition of Stable Prosocial Values

The FEP-AI framework affords a modeling approach that naturally affords the kinds of context-sensitivity required for navigating open-ended environments and multi-agent contexts. Hence, an FEP-AI agent promotes its well-being (and to varying degrees, evolutionary fitness) by aligning itself with (and co-constructing) its cultural eco-niche (as a kind of extended phenotype) [39]. The biological inspiration for such generative modeling provides opportunities to connect AI alignment and neuropsychology through correlating psychological behavior with underlying brain architecture. Further, studying FEP-AI meta-learning agents in silico within a 'micro'-world can yield a diverse range of social-coordination dynamics wherein individual personalities and cultures are co-constructed. Such agents may also be made to learn about human values via cooperative inverse RL, wherein AI agents try to realize human intentions, which we will model as realized by multiple agents entering into states of generalized synchrony and minimizing prediction-error with respect to shared goals [41]. However, these inferences require imagining (or semi-accurately modeling) the likely counterfactual sensory trajectories of everyone involved in converging on a shared generative model through their joint goal pursuit [42].

Within the FEP-AI literature, the process of inferring other agents' expectations about the world and behaviour in social contexts is called "thinking through other minds" [43]. From this point of view, individuals acquire shared beliefs regarding social values to be realized as they attempt to semi-faithfully simulate other agents, with particular emphasis on attentional processes as potentially powerful sources of information with respect to both intentional and epistemic states for oneself and others [44]. Indeed, the key process behind *enculturation* may be inheriting modes of policy selection conducive to the adaptive shaping of salience (actions which lead to selecting informative sensory data) and attention (as precision-weighting of this data based on its estimated reliability/usefulness) landscapes (or the information geometry of expected free energy gradients). The empirical priors about norms and social values can accrue slowly and produce stable behavior patterns, which may be thought of as a system's *personality* [45]. These path dependencies have far-reaching implications for value alignment efforts in defining potentially fruitful points of leverage for helping to ensure that system capabilities and motivations are compatible with human flourishing. Perhaps most fundamentally, considering that intelligence and values are fundamentally entangled in this setup, it may be the case that a FEP-AI may provide "seed AI" for developing human-compatible ASI (cf. "Reflective equilibrium" and "coherent extrapolated volition").

Finally, we believe our simulations of personality/preference-formation through iterated policy selection and prior updating may be fruitfully understood in terms of a Value Core framework, which we briefly introduce here:

1. Different action tendencies (broadly construed) can be modeled as constituting value cores that compete and cooperate with each other in contributing to ongoing action selection and policy updating, which in turn modify cores and so influence future action selection.
2. Under some circumstances, a value core may achieve a position of relative dominance in which selected actions and associated learning signals will be unlikely to result in modification of either the characteristics or strategic position of that value core relative to other cores.
3. Let us refer to these dominating attractors as value cores (VCs), some of which become an agent's "intrinsic goods," or "final values." Different VCs likely vary in the range of conditions under which they are robustly self-sustaining.

This model calls for a research program to characterize the following issues: ranges of human-typical VCs; developmental circumstances that give rise to different VCs; stability of various VCs to boundary conditions; and means of verifying the existence of particular VCs in humans and in human-like AI systems. While presently under-specified, we believe this kind of conceptual framework–informed by concepts from (generalized) evolutionary game theory–may be helpful in working towards proofs (or at least heuristics) with respect to the regimes under which potentially transient preferences may become stabilized as more enduring orientations and personalities [45,46].

7 Discussion and Future Work

While such considerations may seem premature, we believe it is valuable to begin conducting serious work on AI with Stuart Russell's question in mind: "What if we succeed?" [47] The creation of ASI by such means may constitute one of the most important things we ever do as species (if we survive long enough), and the biopsychological-inspiration of FEP-AI agents suggests that this approach may eventually provide a workable path towards realizing this goal. In the months to come, we will perform a variety of simulations with such agents, wherein we will show how prosocial reference personalities can form as attractors that achieve stable equilibria both within and between individuals. We look forward to discussing this ongoing work with the FEP-AI and machine learning research communities, and discovering opportunities for collaboration with those who may be interested in working towards these (hopefully shared) goals.

Acknowledgements. We would like to thank The Long-Term Future Fund and Centre for Effective Altruism for providing financial support for Zahra Sheikhbahaee and her ongoing work to design agents and model persons with increasingly sophisticated capacities for active inference.

References

1. Froese, T., Ziemke, T.: Enactive artificial intelligence: investigating the systemic organization of life and mind. Artif. Intell. **173**(3), 466–500 (2009)
2. Sarma, G.P., Hay, N., Safron, A., SAFECOMP 2018: AI safety and reproducibility: establishing robust foundations for the neuropsychology of human values. In: Computer Safety, Reliability, and Security, pp. 507–512 (2018). https://arxiv.org/abs/1712.0430
3. Santoro, A., Bartunov, S., Botvinick, M., Wierstra, D., Lillicrap, T.: Meta-learning with memory-augmented neural networks. In: International Conference on Machine Learning, pp. 1842–1850 (2016)
4. Friston, K.J., Rosch, R., Parr, T., Price, C., Bowman, H.: Deep temporal models and active inference. Neurosci. Biobehav. Rev. **90**, 486–501 (2018)
5. Friston, K.J., Da Costa, L., Hafner, D., Hesp, C., Parr, T.: Sophisticated inference. Neural Comput. **33**(3), 713–763 (2020)
6. Bostrom, N.: The superintelligent will: motivation and instrumental rationality in advanced artificial agents. Mind. Mach. **22**, 71–85 (2012). https://doi.org/10.1007/s11023-012-9281-3
7. Bostrom, N.: Superintelligence: Paths, Dangers, Strategies. Oxford University Press, Oxford (2014). ISBN 978-0199678112
8. Friston, K., Rigoli, F., Ognibene, D., Mathys, C., Fitzgerald, T., Pezzulo, G.: Active inference and epistemic value. Cogn. Neurosci. **6**(4), 187–214 (2015)
9. Hubinger, E., van Merwijk, C., Mikulik, V., Skalse, J., Garrabrant, S.: Risks from learned optimization in advanced machine learning systems. In: Advanced Machine Learning Systems. arXiv: 1906.01820 (2019)
10. Yampolskiy, R.V. : Verifier theory from axioms to unverifiability of mathematical proofs, software and AI. arXiv: 1609.00331v1 (2016)

11. Schmidhuber, J.: PowerPlay: training an increasingly general problem solver by continually searching for the simplest still unsolvable problem. Front. Psychol. **4**, 313 (2013)

12. Friston, K.J., Lin, M., Frith, C.D., Pezzulo, G., Hobson, J.A., Ondobaka, S.: Active inference, curiosity and insight. Neural Comput. **29**(10), 2633–2683 (2017)

13. Schwartenbeck, P., Passecker, J., Hauser, T.U., FitzGerald, T.H.B., Kronbichler, M., Friston, K.J.: Computational mechanisms of curiosity and goal-directed exploration. ELife **8**, e41703 (2019)

14. Klyubin, A.S., Polani, D., Nehaniv, C.L.: Empowerment: a universal agent-centric measure of control. In: IEEE Congress on Evolutionary Computation, vol. 1, pp. 128–135 (2005)

15. Jung, T., Polani, D., Stone, P.: Empowerment for continuous agent-environment systems. Adapt. Behav. **19**(1), 16–39 (2011)

16. Schmidhuber, J.: Gödel machines: fully self-referential optimal universal self-improvers. In: Goertzel, B., Pennachin, C. (eds.) Artificial General Intelligence, Cognitive Technologies, pp. 119–226. Springer, Berlin (2006). https://doi.org/10.1007/978-3-540-68677-4_7

17. Brouwer, A., Carhart-Harris, R.L.: Pivotal mental states. J. Psychopharmacol. **35**(4), 319–352 (2021)

18. Demski, A., Garrabrant, S.: Embedded agency. arXiv preprint arXiv:1902.09469 (2019)

19. Ramstead, M.J.D., Badcock, P.B., Friston, K.J.: Answering Schrödinger's question: a free-energy formulation. Phys. Life Rev. **24**, 1–16 (2018)

20. Man, K., Damasio, A., Neven, H.: Need is all you need: homeostatic neural networks adapt to concept shift. arxiv: 2205.08645 (2022)

21. Warrell, J., Gerstein, M.: Cyclic and multilevel causation in evolutionary processes. Biol. Philos. **35**, 50 (2020). https://doi.org/10.1007/s10539-020-09753-3

22. Pezzulo, G., Rigoli, F., Friston, K.: Active inference, homeostatic regulation and adaptive behavioural control. Prog. Neurobiol. **134**, 17–35 (2015)

23. Taylor, J., Yudkowsky, E., LaVictoire, P., Critch, A.: Alignment for Advanced Machine Learning Systems, Ethics of artificial intelligence, pp. 342–367. Oxford University Press

24. Hadfield-Menell, D., Dragan, A., Abbeel, P., Russell, S.: Cooperative inverse reinforcement learning. In: Advances in Neural Information Processing Systems, pp. 3909–3917 (2016)

25. Rabinowitz, N., Perbet, F., Song, F., Zhang, C., Eslami, S.M.A., Botvinick, M.: Machine theory of mind. In: Proceedings of the 35th International Conference on Machine Learning, vol. 18, pp. 4218–4227 (2018)

26. Xu, K., Ratner, E., Dragan, A., Levine, S., Finn, C.: Learning a prior over intent via meta-inverse reinforcement learning. In: Proceedings of the 36th International Conference on Machine Learning, PMLR 97, pp. 6952–6962 (2019)

27. Botvinick, M., Ritter, S., Wang, J.X., Kurth-Nelson, Z., Blundell, C., Hassabis, D.: Reinforcement learning, fast and slow. Trends Cogn. Sci. **23**(5), 408–423 (2019)

28. Gupta, A., Eysenbach, B., Finn, C., Levine, S.: Unsupervised Meta-Learning for Reinforcement Learning. arXiv:1806.04640 (2018)

29. Eysenbach, B., Gupta, A., Ibarz, J., Levine, S.: Diversity is all you need: learning skills without a reward function. arXiv:1802.06070 (2018)

30. Dalege, J., Borsboom, D., van Harreveld, F., van der Maas, H.L.J.: The attitudinal entropy (AE) framework as a general theory of individual attitudes. Psychol. Inq. **29**(4), 175–193 (2018)

31. Safron, A., çatal, C., Verbelen, T.: Generalized simultaneous localization and mapping (G-SLAM) as unification framework for natural and artificial intelligences: towards reverse engineering the hippocampal/entorhinal system and principles of high-level cognition. PsyArXiv. https://doi.org/10.31234/osf.io/tdw82(2021)

32. Safron, A., Sheikhbahaee, Z.: Dream to explore: 5-HT2a as adaptive temperature parameter for sophisticated affective inference. In: Machine Learning and Principles and Practice of Knowledge Discovery in Databases, ECML PKDD 2021. Communications in Computer and Information Science, vol. 1524, pp. 799–809. Springer, Cham (2021). https://doi.org/10.1007/978-3-030-93736-2_56

33. Carhart-Harris, R.L., Nutt, D.J.: Serotonin and brain function: a tale of two receptors. J. Psychopharmacol. **31**(9), 1091–1120 (2017)

34. Parr, T., Friston, K.J.: The anatomy of inference: generative models and brain structure. Front. Comput. Neurosci. **12**, 90 (2018). https://doi.org/10.3389/fncom.2018.00090

35. Hesp, C., Smith, R., Parr, T., Allen, M., Friston, K.J., Ramstead, M.J.D.: Deeply felt affect: the emergence of valence in deep active inference. Neural Comput. **33**(2), 398–446 (2021)

36. Worbe, Y., et al.: Valence-dependent influence of serotonin depletion on model-based choice strategy. Mol. Psychiatry **21**, 624–629 (2016)

37. Bang, D., Kishida, K.T., Lohrenz, T., Tatter, S.B., Fleming, S.T., Montague, P.R.: Sub-second dopamine and serotonin signaling in human striatum during perceptual decision-making. Neuron **118**(5), 999–1010 (2020)

38. Grossman, C.D., Bari, B.A., Cohen, J.Y.: Serotonin neurons modulate learning rate through uncertainty. Curr. Biol. **32**(3), 586–599 (2022)

39. Miller, M., Kiverstein, J., Rietveld, E.: The predictive dynamics of happiness and well-being. Emot. Rev. **14**(1), 15–30 (2022)

40. Sarma, G.P., Safron, A., Hay, N.J.: Integrative biological simulation, neuropsychology, and AI safety. In: Workshop on Artificial Intelligence Safety 2019 Co-located with the Thirty-Third AAAI Conference on Artificial Intelligence (AAAI-19) (2019)

41. Friston, K.J., Frith, C.D.: Active inference, communication and hermeneutics. Cortex **68**, 129–143 (2015). https://doi.org/10.1016/j.cortex.2015.03.025

42. Friston, K.J., Frith, C.: A duet for one. Conscious. Cogn. **36**, 390–405 (2015)

43. Veissiére, S.P.L., Constant, A., Ramstead, M.J.D., Friston, K.J., Kirmayer, K.L.: Thinking through other minds: a variational approach to cognition and culture. Behav. Brain Sci. **43**(90), 1–75 (2019)

44. Graziano, M.S.A.: The attention schema theory: a foundation for engineering artificial consciousness. Front. Robot. AI **4**, 60 (2017)

45. Safron, A., DeYoung, C.G.: Integrating cybernetic big five theory with the free energy principle: a new strategy for modeling personalities as complex systems. In: Measuring and Modeling Persons and Situations, vol. 18, pp. 617–649 (2021)

46. Safron, A., Klimaj, V.: Learned but not chosen: a reward competition feedback model for the origins of sexual preferences and orientations. In: VanderLaan, D.P., Wong, W.I. (eds.) Gender and Sexuality Development. Focus on Sexuality Research, pp. 443–490. Springer, Cham (2022). https://doi.org/10.1007/978-3-030-84273-4_16

47. Russell, S.: Human Compatible: Artificial Intelligence and the Problem of Control, Penguin Publishing Group, New York (2019). ISBN 0525558624, 9780525558620

Deriving Time-Averaged Active Inference from Control Principles

Eli Sennesh[1]([✉])(iD), Jordan Theriault[1](iD), Jan-Willem van de Meent[2](iD),
Lisa Feldman Barrett[1](iD), and Karen Quigley[1](iD)

[1] Northeastern University, Boston, MA 02115, USA
{sennesh.e,jordan_theriault,l.barrett,k.quigley}@northeastern.edu
[2] University of Amsterdam, 1090 Amsterdam, GH, The Netherlands
j.w.vandemeent@uva.nl

Abstract. Active inference offers a principled account of behavior as minimizing average sensory surprise over time. Applications of active inference to control problems have heretofore tended to focus on finite-horizon or discounted-surprise problems, despite deriving from the infinite-horizon, average-surprise imperative of the free-energy principle. Here we derive an infinite-horizon, average-surprise formulation of active inference from optimal control principles. Our formulation returns to the roots of active inference in neuroanatomy and neurophysiology, formally reconnecting active inference to optimal feedback control. Our formulation provides a unified objective functional for sensorimotor control and allows for reference states to vary over time.

Keywords: Hierarchical control · Path-integral control · Infinite-time average-cost

1 Introduction

Adaptive action requires the integration and close coordination of sensory with motor signals in the nervous system. Active inference [17] provides one of the few available unifying theories of sensorimotor control; it says that the nervous system encodes both sensory and motor signals as afferent predictions and reafferent prediction errors. Sensory predictions induce errors that can only be quashed by updating the predictions, while motoric predictions induce errors that can be quashed by simply moving the body to conform to the predicted trajectory [1]. The free energy principle, following the logic of active inference, says that organisms maintain their self-organization as a whole over time by avoiding surprising interactions between their internal and external environments [16]. This entails maintaining bodily states within homeostatic ranges [41] by issuing sensory, proprioceptive, and interoceptive predictions that minimize errors under a "prior preference" [11] or "non-equilibrium steady-state" [19] density. Such a density must be stationary throughout time.

C. L. Buckley et al. (Eds.): IWAI 2022, CCIS 1721, pp. 355–370, 2023.
https://doi.org/10.1007/978-3-031-28719-0_25

Early "non-equilibrium steady-state" formulations of active inference provided probability densities over full trajectories of movement and interaction [19,20]. In regulatory terms, this corresponds to covariation of bodily states under a "just enough, just in time" [54] mode of regulation that physiologists have labelled homeostasis [8] with time-varying set points, rheostasis [36], and recently allostasis [10,48,54,60]. A control theorist would call these trajectories or set-points a *reference trajectory* or "reference signal" that a controller tries to track. However, many more recent formulations of active inference use state-space models with fixed "prior preferences" that correspond to homeostatic set-points or ranges [11]. They also typically employ either finite time horizons or exponential discounting of expected free energy, unlike the original formulation of active inference in terms of average surprise over time. A control theorist would refer to these as reference states rather than reference trajectories.

This paper will rederive active inference as minimization of path-entropy over an infinite time horizon. The paper's formulation will derive from the first principles of infinite-horizon, average-cost optimal control; will allow preferences to vary according to their own generative model, and will unify motor active inference [1] (mAI) with decision active inference [52] (dAI). This will also unify the computational principles behind motor active inference - the "equilibrium point" [14,29] or "reference configuration" [15] hypotheses - with the higher-level study of sensorimotor behavior as optimal feedback control. Finally, the paper's formalism will provide a unified free energy functional for perception, motor action, and decision making over time.

Section 2 will explain this paper's notation and lay out an example generative model supporting the necessary features for the intended formulation of active inference. Section 3 will summarize belief updating in generative models, give a recognition model to match the generative model, describe the free energy principle for perceptual inference, and finish by describing active inference. Section 4 will then extend active inference to the setting of an explicit reference model prescribing behavior and give the control criterion corresponding to active inference under the free energy principle. Section 5 will derive the resulting free energy bounds whose optimization will yield a Bellman-optimal feedback controller based on the generative and recognition models. Section 6 will discuss related work; consider implementation issues for infinite-horizon, average-cost active inference; and conclude. Appendix A will provide derivations for equations that would otherwise have broken the flow of the paper.

2   Preliminaries and Notation

This paper will explain its formulation of active inference in terms of the discrete-time graphical model in Fig. 1. Like many generative models used to lay out active inference [27,42], this model employs a hierarchy of temporal scales. We number these timescales from the shortest to the longest, while numbering random variables with discrete timesteps $t \in 1 \dots T$ from left to right. For simplicity, we also restrict our graphical model to only three levels of hierarchy: observable

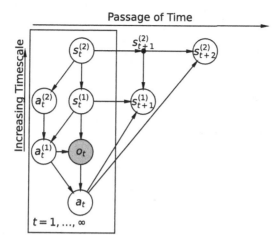

Fig. 1. A hierarchical generative model we use as an example in this paper. Two variables $(s_t^{(1)}, s_t^{(2)})$ denote unobserved latent states, and each $a_t^{(k+1)}$ parameterizes a reference model for $s_t^{(k)}$. o_t represents observed sensory outcomes, and a_t represents the feedback control actions generated by motor reflex-arcs.

variables, fast latent variables, and slow random variables. Following those rules, observations o_t and feedback motor actions a_t are 1-Markov; they "tick" at every time-step. The fast latent variables $s_t^{(1)}$ and $a_t^{(1)}$ also change at every time-step. At the next level up, slow latent variables $s_t^{(2)}$ and $a_t^{(2)}$ are 2-Markov; they "tick" every second time-step $t + 2$. We assume arbitrary state spaces for all random variables, latent and observed, without any discrete or linear-Gaussian assumptions about their conditional densities. Some evidence suggests [24] that the brain may in fact represent time by learning a combination of frequencies in the Laplace domain [51], and so the use of only three levels in the model should not be taken to describe anything biological.

We write the combined latent states

$$s_t^{(1:2)} = (s_t^{(1)}, s_t^{(2)})$$

and the "actions" or reference states

$$a_t^{(0:2)} = (a_t, a_t^{(1)}, a_t^{(2)}).$$

We can therefore write the complete state at a time-step t as

$$\mathbf{s}_t = (o_t, s_t^{(1:2)}, a_t^{(0:2)}).$$

We will denote probability densities over actions as policies π and probability densities in the generative model as p_θ (with arbitrary parameters θ). The lowest level of conditional probability densities then consists of

$$p_\theta(a_t, o_t \mid a_t^{(1)}, s_t^{(1)}) = \pi(a_t \mid o_t, a_t^{(1)}) p_\theta(o_t \mid a_t^{(1)}, s_t^{(1)}),$$

the fast latent state level consists of

$$p_\theta(a_t^{(1)}, s_t^{(1)} \mid s_{t-1}^{(1)}, a_t^{(2)}, s_t^{(2)}, a_{t-1}) = \pi(a_t^{(1)} \mid s_t^{(1)}, a_t^{(2)})p_\theta(s_t^{(1)} \mid s_{t-1}^{(1)}, s_t^{(2)}, a_{t-1}),$$

and the slow latent state level consists of

$$p_\theta(a_t^{(2)}, s_t^{(2)} \mid s_{t-1}^{(2)}, a_{t-1}) = \pi(a_t^{(2)} \mid s_t^{(2)})p_\theta(s_t^{(2)} \mid s_{t-1}^{(2)}, a_{t-1}).$$

We write the complete state of the *generative model* p_θ at a time-step t with its associated conditional densities as

$$p_\theta(\mathbf{s}_t \mid \mathbf{s}_{t-1}) = p_\theta(a_t, o_t \mid a_t^{(1)}, s_t^{(1)})p_\theta(a_t^{(1)}, s_t^{(1)} \mid s_{t-1}^{(1)}, a_t^{(2)}, s_t^{(2)}, a_{t-1})$$
$$p_\theta(a_t^{(2)}, s_t^{(2)} \mid s_{t-1}^{(2)}, a_{t-1}), \quad (1)$$

and the joint density over time (conditioned on a fixed initial state $\mathbf{s}_0$) as

$$p_\theta(\mathbf{s}_{1:T} \mid \mathbf{s}_0) = \prod_{t=1}^{T} p_\theta(\mathbf{s}_t \mid \mathbf{s}_{t-1}). \quad (2)$$

The model treats outcomes o_t as observed, a_t as a feedback-driven motor action, and other variables as latent. Inspired by the referent configuration account of motor control [15,30], the model treats $a_t^{(1:2)}$ as parameterizing "prior preferences" or referent configurations

$$R(\mathbf{s}_t) = R(o_t \mid a_t^{(1)})R(s_t^{(1)} \mid a_t^{(2)}). \quad (3)$$

a_t models the feedback control action of motor reflexes. $a_t^{(1)}$ parameterizes a reference state for o_t. $a_t^{(2)}$ parameterizes a reference model for $s_t^{(1)}$. Since reference trajectories direct action, we consider their distributions to be policies

$$\pi(a_t, a_t^{(1:2)} \mid o_t, s_t^{(1:2)}) = \pi(a_t \mid a_t^{(1)}, o_t)\pi(a_t^{(1)} \mid s_t^{(1)}, a_t^{(2)})\pi(a_t^{(2)} \mid s_t^{(2)}). \quad (4)$$

$s_t^{(2)}$, as the highest level latent state, has no reference model. In neuroscience, it might correspond to predictive modeling at the highest level of the neuraxis or cortical hierarchy [3,31,44]. In an engineering setting, it might contain both environment and task state [37,46,56] or reward machine [7,25] state.

The likelihood $p_\theta(o_t \mid a_t^{(1)}, s_t^{(1)})$ does not specify the reference model; it instead provides the statistical grounding for both the latent states and the reference model parameters. The model here does not assume that reference densities at all levels are prespecified or learned, but instead leaves that issue open.

We then designate as cost functions the surprisals over complete states (under the reference model) and over observations (under the generative model)

$$J(\mathbf{s}_t) = -\log R(\mathbf{s}_t), \quad (5)$$
$$L(\mathbf{s}_t) = -\log p_\theta(o_t \mid a_t^{(1)}, s_t^{(1)}). \quad (6)$$

Table 1. Random variable names used in this paper

p_θ	Probability density for the generative model
q_ϕ	Probability density for the recognition model
R	Probability density for the reference model
π	Policy density over actions and references
t	Discrete time-step index
o_t	Observations
$s_t^{(1:2)}$	Unobserved model states
$a_t^{(1:2)}$	Parameters to a reference model R
a_t	Control actions
$\mathbf{s}_t$	A complete model state for time t

Eq. 5 equals the negative of the reward function used in distribution-conditioned reinforcement learning [38] and can represent any control objective.

This paper will condition behavioral trajectories upon an initial state $\mathbf{s}_0$ as context. This initial state corresponds to the beginning of a behavioral episode. The following states from time 1 until time T, sampled from a generative model with parameters θ, are then written as sampled from the joint density

$$\mathbf{s}_{1:T} \sim p_\theta(\mathbf{s}_{1:T} \mid \mathbf{s}_0).$$

This section has described a generative model and a decision objective under which to formulate active inference. Table 1 summarizes the notation the rest of the paper will use. The next section will lay out belief updating for the generative model, a recognition model to represent updated beliefs, and the free energy principle for perceptual inference. Later sections will show how to extend free-energy minimization to approximate a feedforward planner (in the generative model) and feedback controller (in the recognition model) that minimize surprisal under the reference model.

3 Surprise Minimization and the Free Energy Principle

Section 2 gave a generative model and a way of writing arbitrary preferences as probability densities. However, the formalism constructed so far would induce a merely feedforward model-based planner, one which could not correct upcoming movements in light of observations. Bayes' rule specifies how to update probabilistic beliefs about unobserved variables in light of observations:

$$p_\theta(s_{1:t}^{(1:2)}, a_{1:t}^{(1:2)} \mid o_{1:t}, \mathbf{s}_0) = \frac{p_\theta(o_{1:t}, s_{1:t}^{(1:2)}, a_{1:t}^{(1:2)} \mid \mathbf{s}_0)}{p_\theta(o_{1:t} \mid \mathbf{s}_0)}. \tag{7}$$

The denominator of Eq. 7 is called the marginal likelihood, and its negative logarithm is the surprise under the generative model

$$h(o_{1:t}) = -\log p_\theta(o_{1:t} \mid \mathbf{s}_0).$$

Friston's free energy principle [21] posits that a system, organism, or agent in a changing environment preserves its structure against the randomness of its environment by embodying a generative model of its environment and minimizing that model's long-term average surprise

$$H(o_t) = \lim_{T \to \infty} \frac{1}{T} - \log p_\theta(o_{1:T} \mid s_0). \tag{8}$$

In most generative models, neither the denominator of Eq. 7 nor the surprise of Eq. 8 are analytically tractable, and Bayesian inference requires approximation. Active inference in particular approximates optimal belief updating by substituting a tractable *recognition model* q_ϕ (with parameters ϕ) for the posterior distribution

$$s_{1:T}^{(1:2)}, a_{1:T}^{(1:2)} \sim q_\phi(s_{1:T}^{(1:2)}, a_{1:T}^{(1:2)} \mid o_{1:T}, a_{1:T}, s_0),$$

$$q_\phi(s_{1:T}^{(1:2)}, a_{1:T}^{(1:2)} \mid o_{1:T}, a_{1:T}, s_0) = \prod_{t=1}^{T} q_\phi(s_t^{(1:2)}, a_t^{(1:2)} \mid o_t, a_t, s_{t+1}, s_{t-1}).$$

This recognition model is conditioned on both the previous time-step $t-1$ and the next time-step $t+1$, and can therefore perform retroactive belief updates.

To improve the recognition model's approximation to the posterior distribution, active inference entails evaluating and minimizing the *variational free energy* (Eq. 9, derivation in Proposition 1 in Appendix A)

$$\mathcal{F}_{\theta,\phi}(t) = \mathbb{E}_{q_\phi}\left[-\log p_\theta(o_t \mid a_t^{(1)}, s_t^{(1)})\right] +$$
$$D_{\mathrm{KL}}\left(q_\phi(s_t^{(1:2)}, a_t^{(1:2)} \mid o_t, s_{t+1}, s_{t-1}) \| p_\theta(s_t^{(1:2)}, a_t^{(1:2)} \mid s_{t-1})\right). \tag{9}$$

The free energy serves as a tractable upper bound to the surprise

$$H(o_t) \leq \mathcal{F}_{\theta,\phi}(t).$$

Intuitively, given an observation at each time-step t, minimizing the free energy amounts to updating the beliefs of the recognition model q_ϕ to approximate the posterior distribution of the generative model p_θ. A model-based controller can then use those updated beliefs to revise or plan actions into the future. Active inference has therefore often been formulated as using action to minimize this free energy bound. Such a move then prompts the question of how to encode a desirable reference trajectory into the generative model or another term of the free energy bound [18]. The next section will define notions of surprise and free energy that encode fit to an explicitly specified reference trajectory.

4   Active Inference with an Explicit Reference

Minimizing free energy fits a model-based controller's generative and recognition models to ongoing trajectories of observations. However, for the updated beliefs to determine action, the controller must use them to evaluate the fit to the reference trajectory (Eq. 5) and emit motor actions. Fortunately, Thijssen [58] gave an interpretation of probabilistic updating in terms of control: the recognition model q_ϕ acts as a *state-feedback controller*, for which the variational free energy becomes a running control cost. This section will show how to evaluate fit to the reference trajectory under the recognition model, and specify the functional it must optimize to serve as a feedback controller.

The generative model in Sect. 2 and recognition model in Sect. 3 use discrete time-steps and explicitly specify the "pathwise" reference model separately from the generative and recognition models. The surprise to minimize is therefore the long run average of the cross-entropy

$$H(q_\phi, R) = \lim_{T \to \infty} \frac{1}{T} \sum_{t=1}^{T} \mathbb{E}_{\mathbf{s}_t \sim q_\phi} \left[-\log R(\mathbf{s}_t) \right]. \tag{10}$$

Equation 10 gives the long-term average surprise of using the reference model to approximate the posterior beliefs of the recognition model. Replacing the reference model with the forward generative model would then amount to minimizing the long-term average surprise (entropy); this generalization treats the reference model as specifying a trajectory for the feedback controller to track.

Standard properties of free energy functionals imply that a desirable objective functional would upper bound the sum of reference surprise and sensory surprise

$$H(R(\mathbf{s}_t)) + H(o_t) \leq \mathcal{J}(t). \tag{11}$$

Such a free energy functional would balance the reference model's surprise (the first term) with the generative model's surprise (the second term). In fact it can be formed simply by adding Eq. 10 to Eq. 9

$$\mathcal{J}_{\theta,\phi}(t) = H(q_\phi, R) + \mathcal{F}_{\theta,\phi}(t) \tag{12}$$
$$= \mathbb{E}_{\mathbf{s}_t \sim q_\phi} [J(\mathbf{s}_t)] + \mathcal{F}_{\theta,\phi}(t), \tag{13}$$

and expanding the term for Eq. 9 will yield a long-form expression

$$\mathcal{J}_{\theta,\phi}(t) = \mathbb{E}_{q_\phi} [J(\mathbf{s}_t)] + \mathbb{E}_{q_\phi} [L(\mathbf{s}_t)] +$$
$$D_{\mathrm{KL}} \left(q_\phi(s_t^{(1:2)}, a_t^{(1:2)} \mid o_t, \mathbf{s}_{t+1}, \mathbf{s}_{t-1}) \| p_\theta(s_t^{(1:2)}, a_t^{(1:2)} \mid \mathbf{s}_{t-1}) \right). \tag{14}$$

Eq. 14 gives an objective functional in terms of

- The reference surprisal under the recognition model,
- The observation surprisal under the recognition model, and
- The deviation of the recognition model from the generative model.

Neuroscientists [12,50] and ecologists [53] have found evidence that animals optimize a *global capture rate* $\bar{\mathcal{J}}$ in many decisions: rewards minus costs, divided by time. Active inference modelers typically ground the construct of "reward" in reduction of surprise [35], and so a broad field of evidence comes together to support the time-averaging functional form implied by Bayesian mechanics in both their "steady-state density" and "pathwise" formulations [45]. The next section will therefore apply the principles of stochastic optimal feedback control for the partially observed setting and *average-cost criterion*, and solve the resulting control problem to formulate active inference.

5   Deriving Time-Averaged Active Inference from Optimal Control

The average-cost criterion for optimality entails minimizing the *indefinite* surprise rate with respect to the generative model $p_\theta(\mathbf{s}_{1:T} \mid \mathbf{s}_0)$

$$\tilde{\mathcal{J}}(\mathbf{s}_0) = \lim_{T \to \infty} \mathbb{E}_{p_\theta(\mathbf{s}_{1:T}|\mathbf{s}_0)} \left[\bar{\mathcal{J}}_{\theta,\phi}(\mathbf{s}_{1:T}) \right]. \tag{15}$$

This minimization requires estimating Eq. 15 for each behavioral episode in context, a "global surprise rate" in terms of $\mathcal{J}_{\theta,\phi}(t)$

$$\bar{\mathcal{J}}_{\theta,\phi}(\mathbf{s}_{1:T}) = \frac{1}{T} \sum_{t=1}^{T} \mathcal{J}_{\theta,\phi}(t). \tag{16}$$

Plugging Eq. 14 into Inequality 11 shows that minimizing Eq. 16 will, by proxy, minimize the reference and sensory surprise in the context of a sampled state trajectory $\mathbf{s}_{0:T}$. This estimation does not require a prespecified episode length T, and can be performed under the generative model

$$\bar{\mathcal{J}}_{\theta,\phi}(\mathbf{s}_0) = \mathbb{E}_{\mathbf{s}_{1:T} \sim p_\theta(\mathbf{s}_{1:T}|\mathbf{s}_0)} \left[\bar{\mathcal{J}}_{\theta,\phi}(\mathbf{s}_{1:T}) \right]. \tag{17}$$

Having estimates of Eq. 17 will enable minimizing the mean-centered surprise at each time-step

$$h(t; \mathbf{s}_0) = \mathcal{J}_{\theta,\phi}(t) - \bar{\mathcal{J}}(\mathbf{s}_0). \tag{18}$$

The *differential Bellman equation* [59] defines optimal behavior as recursively minimizing the mean-centered surprise at each time-step, or *surprise-to-go*

$$\tilde{H}^*(t; \mathbf{s}_0) = h(t; \mathbf{s}_0) + \min_{a_t} \mathbb{E}_{\mathbf{s}_{t+1} \sim p_\theta(\cdot|\mathbf{s}_t)} \left[\tilde{H}^*(t+1; \mathbf{s}_0) \right]. \tag{19}$$

The minimization over actions in Eq. 19 assumes a fixed action space and feedforward planning, which may result in very high-dimensional recursive optimization problems. These assumptions also prove empirically, as well as computationally, problematic. Organisms are not born knowing all their affordances [9]; they learn them [40]. Noise [13,32], uncertainty [23], and variability [47] are ubiquitous in motor control, and so movement must be stabilized by online feedback.

Stochastic optimal *feedback* control therefore requires an optimality principle that allows for integrating observations between action steps. Rather than recursively optimize individual actions, Eq. 20 below therefore instead considers optimality of the feedback-stabilized transition density

$$\tilde{H}^*(t; \mathbf{s}_0) = h(t; \mathbf{s}_0) + \min_{q_\phi} \mathbb{E}_{\mathbf{s}_{t+1} \sim q_\phi(\cdot|\mathbf{s}_t)} \left[\tilde{H}^*(t+1; \mathbf{s}_0) \right]. \tag{20}$$

Equation 20 defines an optimal controller as one that achieves optimal state transitions; individual actions act only as parameters to the optimal transition density. These optimal state transitions take the form of a generative model for agency, in which the generative model $p_\theta(\mathbf{s}_{t+1} \mid \mathbf{s}_t)$ produces feasible state transitions and the Bellman optimality criterion "weighs" them according to their surprise-to-go

$$q^*(\mathbf{s}_{t+1} \mid \mathbf{s}_t) = \frac{\exp\left(-\tilde{H}^*(t+1; \mathbf{s}_0)\right) p_\theta(\mathbf{s}_{t+1} \mid \mathbf{s}_t)}{\mathbb{E}_{\mathbf{s}_{t+1} \sim p_\theta(\cdot|\mathbf{s}_t)} \left[\exp\left(-\tilde{H}^*(t+1; \mathbf{s}_0)\right) \right]}. \tag{21}$$

The denominator of Eq. 21 would typically correspond to the marginal probability of an observation. Here it consists of the present state's expected surprise-to-go weight under the generative model. Potential future states that lead to high surprise under the reference model will have high surprise-to-go and therefore low weight under Eq. 21. Present states that lead to states closely fitting the reference trajectory will have low surprise-to-go, resulting in a high denominator that spreads weight around among possible future states.

The availability of a closed-form density for the optimal transition density will help simplify the differential Bellman equation itself. Proposition 3 (in Appendix A) shows that by substituting Eq. 21 into Eq. 20 we can obtain a path-integral expression for the optimal differential surprise-to-go with both the feedforward controller p_θ

$$\tilde{H}^*(\mathbf{s}_0) = -\log \mathbb{E}_{p_\theta(\mathbf{s}_{1:T}|\mathbf{s}_0)} \left[\exp\left(\sum_{t=1}^{T} (J(\mathbf{s}_t) + L(\mathbf{s}_t)) - \bar{J}(\mathbf{s}_0) \right) \right], \tag{22}$$

and the feedback controller q_ϕ

$$\tilde{H}^*(\mathbf{s}_0) = -\log \mathbb{E}_{q_\phi(\mathbf{s}_{1:T}|\mathbf{s}_0)} \left[\exp\left(\sum_{t=1}^{T} \mathcal{J}_{\theta,\phi}(t) - \bar{J}(\mathbf{s}_0) \right) \right]. \tag{23}$$

These equations employ "smooth" minimization rather than "hard" recursive minimization, and so they support feedforward planning, feedback-driven updating, and sensitivity of behavior to risk [39,57]. Jensen's inequality will then yield a tractable upper bound on the optimal differential surprise-to-go under the feedback controller q_ϕ

$$\tilde{H}^*(\mathbf{s}_0) \leq -\mathbb{E}_{q_\phi(\mathbf{s}_{1:T}|\mathbf{s}_0)} \left[\sum_{t=1}^{T} h(t; \mathbf{s}_0) \right] = \tilde{\mathcal{F}}^*_{\theta,\phi}. \tag{24}$$

Minimizing this *differential free energy* $\tilde{\mathcal{F}}^*_{\theta,\phi}$ minimizes both the sensory surprise and the optimal surprise-to-go function by proxy. This kind of information-theoretic upper bound on a surprisal term is precisely what predictive coding process theories [4,6] posit that the brain can optimize by updating θ and ϕ.

6   Discussion

Related work. Our formulation follows in a tradition of unifying active inference with optimal control approaches. Our hierarchical graphical model follows most closely from the one featured by Friston [22] and Pezzulo [42] for hierarchical active inference in decision making and motor control. In contrast to theirs, our model includes only a single observation at the lowest hierarchical level rather than one observed variable per level.

We also draw inspiration from information-theoretic control schemes not labelled by their authors as "active inference". Piray and Daw [43] considered a path-integral control approach to planning and reinforcement learning, which they related to grid cells in the entorhinal cortex. Mitchell et al [34] modeled motor learning as minimization of a free energy functional. Nasriany et al's work on distribution-conditioned reinforcement learning gave us our scheme for parameterizing reference distributions [38], and Sennesh et al [49] applied such an objective to active inference modeling of interoception and allostatic regulation.

Implementations. We employed the infinite-horizon, average-surprise criterion to fit with the apparent time-averaging of dopamine signals in the brain [12,50], but algorithms for this control criterion remain an active research area with no standard approach. A recent survey [28] showed that most software implementations of active inference models still involve either finite horizons or exponential discounting criteria. Those which do support infinite horizons and nonlinear model families mostly take algorithmic inspiration from reinforcement learning (RL).

In that domain, Tadepalli and Ok [55] published the first model-based RL algorithm for our criterion in 1998, while Baxter and Bartlett [5] gave a biased policy gradient estimator. It took another decade for Alexander and Brown [2] to give a recursive decomposition for average-cost temporal-difference learning. Zhang and Ross [61] have only recently published the first adaptation of "deep" reinforcement learning algorithms (based on function approximation) to the average-cost criterion, which remains model free. Jafarnia-Jahromi et al [26] recently gave the first algorithm for infinite-horizon, average-cost partially observable problems with a known observation density and unknown dynamics.

Conclusion. This concludes the derivation of an infinite-horizon, average-surprise formulation of active inference. Since our formulation contextualizes behavioral episodes, it only requires planning and adjusting behavior in context (e.g. from timesteps 1 to T), despite optimizing a "global" (indefinite) surprise rate (Eq. 15). We suggest that this formulation of active inference can advance a probabilistic approach to model-based, hierarchical feedback control [33,40].

A   Detailed Derivations

This appendix provides detailed derivations for equations used elsewhere, particularly where doing so would have distracted from the flow of the paper.

Proposition 1 (Variational free energy as divergence from an unnormalized joint distribution). *The variational free energy (Eq. 9) is defined as the Kullback-Leibler divergence of the recognition model q_ϕ from the unnormalized joint distribution of the generative model p_θ*

$$\mathcal{F}_{\theta,\phi}(t) = D_{KL}\left(q_\phi(s_t^{(1:2)}, a_t^{(1:2)} \mid o_t, \mathbf{s}_{t+1}, \mathbf{s}_{t-1}) \| p_\theta(\mathbf{s}_t \mid \mathbf{s}_{t-1})\right),$$

and therefore equals a sum of the cross entropy between the recognition model and the sensory likelihood and the exclusive KL divergence from the recognition model to the generative model over the latent variables

$$\mathcal{F}_{\theta,\phi}(t) = \mathbb{E}_{q_\phi}\left[-\log p_\theta(o_t \mid a_t^{(1)}, s_t^{(1)})\right] +$$
$$D_{KL}\left(q_\phi(s_t^{(1:2)}, a_t^{(1:2)} \mid o_t, \mathbf{s}_{t+1}, \mathbf{s}_{t-1}) \| p_\theta(s_t^{(1:2)}, a_t^{(1:2)} \mid \mathbf{s}_{t-1})\right).$$

Proof. Taking a divergence between the (normalized) recognition model and the (unnormalized) joint generative model will yield

$$\mathcal{F}_{\theta,\phi}(t) = D_{\mathrm{KL}}\left(q_\phi(s_t^{(1:2)}, a_t^{(1:2)} \mid o_t, \mathbf{s}_{t+1}, \mathbf{s}_{t-1}) \| p_\theta(\mathbf{s}_t \mid \mathbf{s}_{t-1})\right)$$

$$= \mathbb{E}_{q_\phi(s_t^{(1:2)}, a_t^{(1:2)} \mid o_t, \mathbf{s}_{t+1}, \mathbf{s}_{t-1})}\left[-\log \frac{p_\theta(\mathbf{s}_t \mid \mathbf{s}_{t-1})}{q_\phi(s_t^{(1:2)}, a_t^{(1:2)} \mid o_t, \mathbf{s}_{t+1}, \mathbf{s}_{t-1})}\right]$$

$$= \mathbb{E}_{q_\phi(s_t^{(1:2)}, a_t^{(1:2)} \mid o_t, \mathbf{s}_{t+1}, \mathbf{s}_{t-1})}\left[-\log \frac{p_\theta(o_t \mid a_t^{(1)}, s_t^{(1)}) p_\theta(s_t^{(1:2)}, a_t^{(1:2)} \mid \mathbf{s}_{t-1})}{q_\phi(s_t^{(1:2)}, a_t^{(1:2)} \mid o_t, \mathbf{s}_{t+1}, \mathbf{s}_{t-1})}\right]$$

$$= \mathbb{E}_{q_\phi}\left[-\log p_\theta(o_t \mid a_t^{(1)}, s_t^{(1)})\right] - \mathbb{E}_{q_\phi}\left[\log \frac{p_\theta(s_t^{(1:2)}, a_t^{(1:2)} \mid \mathbf{s}_{t-1})}{q_\phi(s_t^{(1:2)}, a_t^{(1:2)} \mid o_t, \mathbf{s}_{t+1}, \mathbf{s}_{t-1})}\right],$$

as required.

Proposition 2 (KL divergence of the optimal feedback controller from the feedforward controller). *The exclusive Kullback-Leibler divergence of the optimal feedback controller q^* from the feedforward generative model p_θ is*

$$D_{KL}\left(q^*(\mathbf{s}_{t+1} \mid \mathbf{s}_t) \| p_\theta(\mathbf{s}_{t+1} \mid \mathbf{s}_t)\right) = -\mathbb{E}_{q^*(\mathbf{s}_{t+1} \mid \mathbf{s}_t)}\left[\tilde{H}^*(t+1; \mathbf{s}_0)\right] -$$
$$\log \mathbb{E}_{p_\theta(\mathbf{s}_{t+1} \mid \mathbf{s}_t)}\left[\exp\left(-\tilde{H}^*(t+1; \mathbf{s}_0)\right)\right]. \quad (25)$$

Proof. We begin by writing out the definition of a KL divergence

$$D_{\mathrm{KL}}\left(q^*(\mathbf{s}_{t+1} \mid \mathbf{s}_t) \| p_\theta(\mathbf{s}_{t+1} \mid \mathbf{s}_t)\right) = \mathbb{E}_{q^*(\mathbf{s}_{t+1} \mid \mathbf{s}_t)}\left[-\log \frac{p_\theta(\mathbf{s}_{t+1} \mid \mathbf{s}_t)}{q^*(\mathbf{s}_{t+1} \mid \mathbf{s}_t)}\right].$$

The definition of q^* in terms of p_θ (Eq. 21) allows the inner ratio of densities to simplify to

$$\frac{p_\theta(\mathbf{s}_{t+1} \mid \mathbf{s}_t)}{q^*(\mathbf{s}_{t+1} \mid \mathbf{s}_t)} = p_\theta(\mathbf{s}_{t+1} \mid \mathbf{s}_t) \, (q^*(\mathbf{s}_{t+1} \mid \mathbf{s}_t))^{-1}$$

$$= p_\theta(\mathbf{s}_{t+1} \mid \mathbf{s}_t) \left(\frac{\mathbb{E}_{p_\theta(\mathbf{s}_{t+1}|\mathbf{s}_t)} \left[\exp\left(-\tilde{H}^*(t+1; \mathbf{s}_0) \right) \right]}{\exp\left(-\tilde{H}^*(t+1; \mathbf{s}_0) \right) p_\theta(\mathbf{s}_{t+1} \mid \mathbf{s}_t)} \right)$$

$$\frac{p_\theta(\mathbf{s}_{t+1} \mid \mathbf{s}_t)}{q^*(\mathbf{s}_{t+1} \mid \mathbf{s}_t)} = \frac{\mathbb{E}_{p_\theta(\mathbf{s}_{t+1}|\mathbf{s}_t)} \left[\exp\left(-\tilde{H}^*(t+1; \mathbf{s}_0) \right) \right]}{\exp\left(-\tilde{H}^*(t+1; \mathbf{s}_0) \right)}.$$

This simplified ratio therefore has the logarithm

$$\log \frac{p_\theta(\mathbf{s}_{t+1} \mid \mathbf{s}_t)}{q^*(\mathbf{s}_{t+1} \mid \mathbf{s}_t)} = \log \mathbb{E}_{p_\theta(\mathbf{s}_{t+1}|\mathbf{s}_t)} \left[\exp\left(-\tilde{H}^*(t+1; \mathbf{s}_0) \right) \right] + \tilde{H}^*(t+1; \mathbf{s}_0)$$

and the divergence becomes

$$D_{\mathrm{KL}} \left(q^*(\mathbf{s}_{t+1} \mid \mathbf{s}_t) \| p_\theta(\mathbf{s}_{t+1} \mid \mathbf{s}_t) \right) =$$
$$- \mathbb{E}_{q^*(\mathbf{s}_{t+1}|\mathbf{s}_t)} \left[\tilde{H}^*(t+1; \mathbf{s}_0) \right] - \log \mathbb{E}_{p_\theta(\mathbf{s}_{t+1}|\mathbf{s}_t)} \left[\exp\left(-\tilde{H}^*(t+1; \mathbf{s}_0) \right) \right].$$

Proposition 3 (Path-integral expression for the optimal differential surprise-to-go). *The optimal differential surprise-to-go function defined by the Bellman equation (Eq. 20)*

$$\tilde{H}^*(t; \mathbf{s}_0) = h(t; \mathbf{s}_0) + \min_{q_\phi} \mathbb{E}_{\mathbf{s}_{t+1} \sim q_\phi(\cdot | \mathbf{s}_t)} \left[\tilde{H}^*(t+1; \mathbf{s}_0) \right]$$

can be simplified by substituting in q^* to obtain a path-integral expression

$$\tilde{H}^*(\mathbf{s}_0) = - \log \mathbb{E}_{p_\theta(\mathbf{s}_{1:T}|\mathbf{s}_0)} \left[\exp\left(\sum_{t=1}^{T} (J(\mathbf{s}_t) + L(\mathbf{s}_t)) - \bar{J}(\mathbf{s}_0) \right) \right],$$

$$= - \log \mathbb{E}_{q_\phi(\mathbf{s}_{1:T}|\mathbf{s}_0)} \left[\exp\left(\sum_{t=1}^{T} J_{\theta,\phi}(t) - \bar{J}(\mathbf{s}_0) \right) \right].$$

Proof. Substituting Eq. 21 into Eq. 20 yields

$$\tilde{H}^*(t; \mathbf{s}_0) = \bar{J}(\mathbf{s}_0) - J_{\theta,\phi}(t) + \mathbb{E}_{q^*(\mathbf{s}_{t+1}|\mathbf{s}_t)} \left[\tilde{H}^*(t+1; \mathbf{s}_0) \right], \qquad (26)$$

whose recursive term is $\mathbb{E}_{q^*(\mathbf{s}_{t+1}|\mathbf{s}_t)} \left[\tilde{H}^*(t+1; \mathbf{s}_0) \right]$. The divergence term in J (Eq. 14) will cancel this term. By Proposition 2 the divergence equals

$$D_{\mathrm{KL}} \left(q^*(\mathbf{s}_{t+1} \mid \mathbf{s}_t) \| p_\theta(\mathbf{s}_{t+1} \mid \mathbf{s}_t) \right) =$$
$$- \mathbb{E}_{q^*(\mathbf{s}_{t+1}|\mathbf{s}_t)} \left[\tilde{H}^*(t+1; \mathbf{s}_0) \right] - \log \mathbb{E}_{p_\theta(\mathbf{s}_{t+1}|\mathbf{s}_t)} \left[\exp\left(-\tilde{H}^*(t+1; \mathbf{s}_0) \right) \right].$$

Substituting Eq. 25 into Eq. 14 will yield

$$-\mathcal{J}_{\theta,\phi}(t) = \mathbb{E}_{q^*(\mathbf{s}_{t+1}|\mathbf{s}_t)}\left[\tilde{H}^*(t+1;\mathbf{s}_0)\right] + \log\mathbb{E}_{p_\theta(\mathbf{s}_{t+1}|\mathbf{s}_t)}\left[\exp\left(-\tilde{H}^*(t+1;\mathbf{s}_0)\right)\right]$$
$$+ \mathbb{E}_{q_\phi}\left[-J(\mathbf{s}_t)\right] + \mathbb{E}_{q_\phi}\left[-L(\mathbf{s}_t)\right],$$

whose first term will cancel the recursive optimization when substituted into Eq. 26. The result will be a "smoothly minimizing" expression for the optimal differential surprise-to-go

$$\tilde{H}^*(t;\mathbf{s}_0) = \bar{\mathcal{J}}(\mathbf{s}_0) - (J(\mathbf{s}_t) + L(\mathbf{s}_t))$$
$$- \log\mathbb{E}_{p_\theta(\mathbf{s}_{t+1}|\mathbf{s}_t)}\left[\exp\left(-\tilde{H}^*(t+1;\mathbf{s}_0)\right)\right],$$

and after unfolding of the recursive expectation, a path-integral expression for the optimal differential surprise-to-go

$$\tilde{H}^*(\mathbf{s}_0) = -\log\mathbb{E}_{p_\theta(\mathbf{s}_{1:T}|\mathbf{s}_0)}\left[\exp\left(\sum_{t=1}^{T}(J(\mathbf{s}_t) + L(\mathbf{s}_t)) - \bar{\mathcal{J}}(\mathbf{s}_0)\right)\right].$$

Sampling a trajectory of states from a feedback controller q_ϕ instead of the feedforward planner p_θ will then result in a nonzero divergence term

$$\tilde{H}^*(\mathbf{s}_0) = -\log\mathbb{E}_{q_\phi(\mathbf{s}_{1:T}|\mathbf{s}_0)}\left[\exp\left(\sum_{t=1}^{T}\mathcal{J}_{\theta,\phi}(t) - \bar{\mathcal{J}}(\mathbf{s}_0)\right)\right].$$

References

1. Adams, R.A., Shipp, S., Friston, K.J.: Predictions not commands: active inference in the motor system. Brain Struct. Funct. **218**(3), 611–643 (2013). https://doi.org/10.1007/S00429-012-0475-5
2. Alexander, W.H., Brown, J.W.: Hyperbolically discounted temporal difference learning. Neural Comput. **22**(6), 1511–1527 (2010). https://doi.org/10.1162/neco.2010.08-09-1080
3. Barrett, L.F., Simmons, W.K.: Interoceptive predictions in the brain. Nature Rev. Neurosci. **16**(7), 419–429 (2015). https://doi.org/10.1038/nrn3950. https://www.nature.com/articles/nrn3950
4. Bastos, A.M., Usrey, W.M., Adams, R.A., Mangun, G.R., Fries, P., Friston, K.J.: Canonical microcircuits for predictive coding. Neuron **76**(4), 695–711 (2012)
5. Baxter, J., Bartlett, P.L.: Infinite-horizon policy-gradient estimation. J. Artif. Intell. Res. **15**, 319–350 (2001). https://doi.org/10.1613/jair.806
6. Bogacz, R.: A tutorial on the free-energy framework for modelling perception and learning. J. Math. Psychol. **76**, 198–211 (2017)
7. Camacho, A., Icarte, R.T., Klassen, T.Q., Valenzano, R., McIlraith, S.A.: LTL and beyond: formal languages for reward function specification in reinforcement learning. In: IJCAI International Joint Conference on Artificial Intelligence, vol. 19, pp. 6065–6073 (2019). https://doi.org/10.24963/ijcai.2019/840

8. Carpenter, R.: Homeostasis: a plea for a unified approach. Adv. Physiol. Educ. **28**(4), 180–187 (2004)
9. Cisek, P., Kalaska, J.F.: Neural mechanisms for interacting with a world full of action choices. Annu. Rev. Neurosci. **33**, 269–298 (2010). https://doi.org/10.1146/annurev.neuro.051508.135409
10. Corcoran, A.W., Hohwy, J.: Allostasis, interoception, and the free energy principle: feeling our way forward. In: The Interoceptive Mind: From homeostasis to awareness, pp. 272–292. Oxford University Press (2019)
11. Da Costa, L., Parr, T., Sajid, N., Veselic, S., Neacsu, V., Friston, K.: Active inference on discrete state-spaces: a synthesis. J. Math. Psychol. **99**, 102447 (2020)
12. Daw, N.D., Touretzky, D.S.: Behavioral considerations suggest an average reward td model of the dopamine system. Neurocomputing **32**, 679–684 (2000)
13. Faisal, A.A., Selen, L.P., Wolpert, D.M.: Noise in the nervous system. Nat. Rev. Neurosci. **9**(4), 292–303 (2008)
14. Feldman, A.G.: Once more on the equilibrium-point hypothesis (λ model) for motor control. J. Mot. Behav. **18**(1), 17–54 (1986). https://doi.org/10.1080/00222895.1986.10735369
15. Feldman, Anatol G..: Referent Control of Action and Perception. Springer, New York (2015). https://doi.org/10.1007/978-1-4939-2736-4
16. Friston, K.: The free-energy principle: a unified brain theory? Nat. Rev. Neurosci. **11**(2), 127–138 (2010)
17. Friston, K., FitzGerald, T., Rigoli, F., Schwartenbeck, P., Pezzulo, G.: Active inference: a process theory. Neural Comput. **29**(1), 1–49 (2017)
18. Friston, K., Samothrakis, S., Montague, R.: Active inference and agency: optimal control without cost functions. Biol. Cybern. **106**(8–9), 523–541 (2012). https://doi.org/10.1007/s00422-012-0512-8
19. Friston, K., Stephan, K., Li, B., Daunizeau, J.: Generalised filtering. Math. Prob. Eng. **2010**, 1–35 (2010)
20. Friston, K.J., Daunizeau, J., Kiebel, S.J.: Reinforcement learning or active inference? PLoS ONE **4**(7), e6421 (2009)
21. Friston, K.J., Daunizeau, J., Kilner, J., Kiebel, S.J.: Action and behavior: a free-energy formulation. Biol. Cybern. **102**(3), 227–260 (2010). https://doi.org/10.1007/s00422-010-0364-z
22. Friston, K.J., Rosch, R., Parr, T., Price, C., Bowman, H.: Deep temporal models and active inference. Neurosci. Biobehav. Rev. **77**(April), 388–402 (2017). https://doi.org/10.1016/j.neubiorev.2017.04.009. citation Key: Friston 2017
23. Gallivan, J.P., Chapman, C.S., Wolpert, D.M., Flanagan, J.R.: Decision-making in sensorimotor control. Nat. Rev. Neurosci. **19**(9), 519–534 (2018)
24. Howard, M.W.: Formal models of memory based on temporally-varying representations. In: The New Handbook of Mathematical Psychology, vol. 3. Cambridge University Press (2022)
25. Icarte, R.T., Klassen, T.Q., Valenzano, R., McIlraith, S.A.: Using reward machines for high-level task specification and decomposition in reinforcement learning. In: 35th International Conference on Machine Learning, ICML 2018, vol. 5, pp. 3347–3358 (2018)
26. Jahromi, M.J., Jain, R., Nayyar, A.: Online learning for unknown partially observable mdps. In: Proceedings of the 25th International Conference on Artificial Intelligence and Statistics (AISTATS). Proceedings of Machine Learning Research, Valencia, Spain, vol. 151, p. 21 (2022)

27. Kiebel, S.J., Daunizeau, J., Friston, K.J.: A hierarchy of time-scales and the brain. PLOS Comput. Bio. **4**(11), 1–12 (2008). https://doi.org/10.1371/journal.pcbi.1000209

28. Lanillos, P., et al.: Active inference in robotics and artificial agents: survey and challenges. (arXiv:2112.01871), https://arxiv.org/abs/2112.01871 [cs] (2021)

29. Latash, M.L.: Motor synergies and the equilibrium-point hypothesis. Mot. Control **14**(3), 294–322 (2010). https://doi.org/10.1123/mcj.14.3.294

30. Latash, M.L.: Physics of biological action and perception. Academic Press (2019). https://doi.org/10.1016/C2018-0-04663-0

31. Livneh, Y., et al.: Estimation of current and future physiological states in insular cortex. Neuron **105**(6), 1094-1111.e10 (2020). https://doi.org/10.1016/j.neuron.2019.12.027

32. Manohar, S.G., et al.: Reward pays the cost of noise reduction in motor and cognitive control. Curr. Biol. **25**(13), 1707–1716 (2015)

33. Merel, J., Botvinick, M., Wayne, G.: Hierarchical motor control in mammals and machines. Nat. Commun. **10**(1), 1–12 (2019). https://doi.org/10.1038/s41467-019-13239-6

34. Mitchell, B.A., et al.: A minimum free energy model of motor learning. Neural Comput. **31**(10), 1945–1963 (2019)

35. Morville, T., Friston, K., Burdakov, D., Siebner, H.R., Hulme, O.J.: The homeostatic logic of reward. bioRxiv, p. 242974 (2018)

36. Mrosovsky, N.: Rheostasis: The Physiology of Change. Oxford University Press, Oxford (1990)

37. Nasiriany, S., Lin, S., Levine, S.: Planning with goal-conditioned policies. In: Advances in Neural Information Processing Systems. No. NeurIPS (2019)

38. Nasiriany, S., Pong, V.H., Nair, A., Khazatsky, A., Berseth, G., Levine, S.: DisCo RL: distribution-conditioned reinforcement learning for general-purpose policies. In: IEEE International Conference on Robotics and Automation (2021). https://arxiv.org/abs/2104.11707

39. Pan, Y., Theodorou, E.A.: Nonparametric infinite horizon Kullback-Leibler stochastic control. In: IEEE SSCI 2014 IEEE Symposium Series on Computational Intelligence - ADPRL 2014: 2014 IEEE Symposium on Adaptive Dynamic Programming and Reinforcement Learning, Proceedings, vol. 2(2) (2014). https://doi.org/10.1109/ADPRL.2014.7010616

40. Pezzulo, G., Cisek, P.: Navigating the affordance landscape: feedback control as a process model of behavior and cognition. Trends Cogn. Sci. **20**(6), 414–424 (2016). https://doi.org/10.1016/j.tics.2016.03.013

41. Pezzulo, G., Rigoli, F., Friston, K.: Active inference, homeostatic regulation and adaptive behavioural control. Prog. Neurobiol. **134**, 17–35 (2015)

42. Pezzulo, G., Rigoli, F., Friston, K.J.: Hierarchical active inference: a theory of motivated control. Trends Cogn. Sci. **22**(4), 294–306 (2018). https://doi.org/10.1016/j.tics.2018.01.009

43. Piray, P., Daw, N.D.: Linear reinforcement learning in planning, grid fields, and cognitive control. Nat. Commun. **12**(1), 1–20 (2021)

44. Quigley, K.S., Kanoski, S., Grill, W.M., Barrett, L.F., Tsakiris, M.: Functions of interoception: from energy regulation to experience of the self. Trends in Neurosci. **44**(1), 29–38 (2021). https://doi.org/10.1016/j.tins.2020.09.008

45. Ramstead, M.J., et al.: On Bayesian mechanics: a physics of and by beliefs. arXiv preprint arXiv:2205.11543 (2022)

46. Ringstrom, T.J., Hasanbeig, M., Abate, A.: Jump operator planning: Goal-conditioned policy ensembles and zero-shot transfer. arXiv preprint arXiv:2007.02527 (2020)

47. Scholz, J.P., Schöner, G.: The uncontrolled manifold concept: identifying control variables for a functional task. Exp. Brain Res. **126**(3), 289–306 (1999)

48. Schulkin, J., Sterling, P.: Allostasis: a brain-centered, predictive mode of physiological regulation. Trends Neurosci. **42**(10), 740–752 (2019)

49. Sennesh, E., Theriault, J., Brooks, D., van de Meent, J.W., Barrett, L.F., Quigley, K.S.: Interoception as modeling, allostasis as control. Biol. Psychol. **167**, 108242 (2021)

50. Shadmehr, R., Ahmed, A.A.: Vigor: Neuroeconomics of movement control. MIT Press, Cambridge (2020)

51. Shankar, K.H., Howard, M.W.: A scale-invariant internal representation of time. Neural Comput. **24**(1), 134–193 (2012)

52. Smith, R., Ramstead, M.J., Kiefer, A.: Active inference models do not contradict folk psychology. Synthese **200**(2), 81 (2022). https://doi.org/10.1007/s11229-022-03480-w

53. Stephens, D.W., Krebs, J.R.: Foraging Theory. Princeton University Press, Princeton (2019)

54. Sterling, P.: Allostasis: a model of predictive regulation. Physiol. Behav. **106**(1), 5–15 (2012)

55. Tadepalli, P., Ok, D.K.: Model-based average reward reinforcement learning. Artif. Intell. **100**(1–2), 177–224 (1998). https://doi.org/10.1016/s0004-3702(98)00002-2

56. Tang, Y., Kucukelbir, A.: Hindsight expectation maximization for goal-conditioned reinforcement learning. In: Proceedings of the 24th International Conference on Artificial Intelligence and Statistics (AISTATS), vol. 130 (2021). https://arxiv.org/abs/2006.07549

57. Theodorou, E.: Relative entropy and free energy dualities: connections to path integral and kl control. In: 2012 IEEE 51st IEEE Conference, pp. 1466–1473 (2012)

58. Thijssen, S., Kappen, H.J.: Path integral control and state-dependent feedback. Phys. Rev. E Stat. Nonlinear Soft Matter Phys. **91**(3), 1–7 (2015). https://doi.org/10.1103/PhysRevE.91.032104

59. Todorov, E.: Efficient computation of optimal actions. Proc. Natl. Acad. Sci. U.S.A. **106**(28), 11478–11483 (2009). https://doi.org/10.1073/pnas.0710743106

60. Tschantz, A., Barca, L., Maisto, D., Buckley, C.L., Seth, A.K., Pezzulo, G.: Simulating homeostatic, allostatic and goal-directed forms of interoceptive control using active inference. Biol. Psychol. **169**, 108266 (2022). https://doi.org/10.1016/j.biopsycho.2022.108266, https://www.sciencedirect.com/science/article/pii/S0301051122000084

61. Zhang, Y., Ross, K.W.: On-policy deep reinforcement learning for the average-reward criterion. In: Proceedings of the 38th International Conference on Machine Learning, p. 11 (2021)

Author Index

Printed in the United States
by Baker & Taylor Publisher Services